The PCR Technique: DNA Sequencing II

AF737663

The PCR Technique: DNA Sequencing II

PERKIN ELMER
Applied Biosystems Division
LIBRARY
850 Lincoln Centre Drive, Foster City, CA 94404

Eaton Publishing Co.
154 E. Central Street
Natick, MA 01760 USA

Copyright© 1997 Eaton Publishing Co.

All rights reserved. No part of this publication may be reproduced, stored in a retrieval system, or transmitted, in any form or by any means, electronic, mechanical, photocopying, recording, or otherwise, without prior permission of the copyright owner.

Reproduction or translation of any part of this work beyond that permitted by Section 107 or 108 of the 1976 United States Copyright Act without the permission of the copyright owner is unlawful. Requests for permission or further information should be addressed to the Permissions Department, Eaton Publishing Co.

ISBN 1-881299-05-8

Printed in the United States of America

9 8 7 6 5 4 3 2 1

CONTENTS

CONTRIBUTORS

J. Albert
National Bacteriological Laboratory
Stockholm, Sweden

Vicki Baker
Division of Gynecologic Oncology
University of Michigan Medical Center
Ann Arbor, MI

Sylvia W. Bartlett
Department of Biochemistry
Memorial University of Newfoundland
St. Johns, Newfoundland, Canada

Einar S. Berg
National Institute of Public Health
Olso, Norway

M. Blackwell
Department of Botany
Louisiana State University
Baton Rouge, LA

B. Budowle
FBI Academy
Quantico, VA

J. Butler
University of Virginia
Charlottesville, VA

R.B. Chadwick
Perkin-Elmer/Applied Biosytems Div.
Foster City, CA

Thomas M. Chambers
University of Kentucky
Lexington, KY

Morgan P. Conrad
Perkin-Elmer/Applied Biosystems Div.
Foster City, CA

John D. Crissman
Department of Pathology
Wayne State University School of Medicine
Detroit, MI

Susan J. Dale
Amersham International plc
Amersham, Buckinghamshire, England

William S. Davidson
Department of Biochemistry
Memorial University of Newfoundland
St. John's, Newfoundland, Canada

Gianluca De Bellis
CNR-ITBA
Milano, Italy

Q. Deng
University of Washington School of Medicine
Seattle, WA

Kerry Dew-Jager
University of Utah
Salt Lake City, UT

J.A. DiZinno
FBI Academy
Quantico, VA

James J. Earley
Bockus Research Institute
Philadelphia, PA

Mari Espelund
Department of Biology
University of Oslo
Oslo, Norway

Qiong Fang
Department of Medicine
University of Pennsylvania
Philadelphia, PA

B.M. Fangan
Biological Institute
University of Oslo
Oslo, Norway

E.-M. Fenyo
Karolinska Institute
Stockholm, Sweden

Giora Z. Feuerstein
SmithKline Beecham Pharmaceuticals
King of Prussia, PA

Ricardo Fujita
W.K. Kellogg Eye Center
University of Michigan
Ann Arbor, MI

Raymond F. Gesteland
University of Utah
Salt Lake City, UT

Peter Gill
The Forensic Science Service
Birmingham, England

Patrick M. Gillevet
National Center for Human Genome Research and the National Institutes of Health
Bethesda, MD

E. Halapi
Karolinska Institute
Stockholm, Sweden

Romelle Hopgood (Piercy)
Metropolitan Police Services
London, England

Elvin Hsu
National Center for Human Genome Research and the National Institutes of Health
Bethesda, MD

C. Hu
University of Texas Health Sciences Center
Houston, TX

S.M. Hyder
University of Texas Health Sciences Center
Houston, TX

Kjetill S. Jakobsen
Department of Biology
University of Oslo
Oslo, Norway

E.S. Jensen
Dynal AS.
Oslo, Norway

Bernhard Kaltenboeck
College of Veterinary Medicine
Auburn University
Auburn, AL

Alexander Kamb
Myriad Genetics Inc.
Salt Lake City, UT

Motoya Katsuki
The Institute of Medical Science
The University of Tokyo
Minato-ku, Tokyo, Japan

Byung-Kook Kim
Department of Biology
Anjou University
Suwon, S. Korea

Han-Jip Kim
Department of Biology
Anjou University
Suwon, S. Korea

Minoru Kimura
Division of Molecular Life Science
School of Medicine
Tokai University
Bohseidai, Isehara, Japan

Gus Kousoulas
School of Veterinary Medicine
Louisiana State University
Baton Rouge, LA

Marie Krinks
Food and Drug Adminstration
Center for Biologics Evaluation and Research
Rockville, MD

Mel N. Kronick
Perkin-Elmer/Applied Biosystems Div.
Foster City, CA

Helena Kuivaniemi
Center for Molecular Medicine and Genetics
Wayne State University
Detroit, MI

P.-Y. Kwok
Washington University School of Medicine
St. Louis, MO

Alexander C.K. Lai
University of Kentucky
Lexington, KY

Youn-Ho Lee
California Institute of Technology
Pasadena, CA

T. Leitner
Royal Institute of Technology
Stockholm, Sweden

Yi-Wei Li
Department of Pathology
Wayne State University School of Medicine
Detroit, MI

Heather Lin
University of Utah
Salt Lake City, UT

Kwok-Wai Lo
Chinese University of Hong Kong
Hong Kong

David N. Louis
Massachusetts General Hospital and Harvard Medical School
Charlestown, MA

M. Luzzana
CNR-ITBA
Milano, Italy

M. Manoni
CNR-ITBA
Milano, Italy

Bernard Mariamé
Centre d'Immunopathologie et de Génėetique Humaine
Toulouse, France

Samuel Chi-Ho Mok
Brigham and Women's Hospital
Boston, MA

Malcolm Moos, Jr.
Food and Drug Adminstration
Center for Biologics Evaluation
and Research
Rockville, MD

D.S. Needleman
University of Texas Health Sciences Center
Houston, TX

D.A. Nickerson
University of Washington School of
Medicine
Seattle, WA

Daniel L. Nickrent
Southern Illinois University
Carbondale, IL

Bjørnar Olaisen
Institute of Forensic Medicine
University of Oslo
Oslo, Norway

Linda T. Parker
Monsanto/Searle
St. Louis, MO

R. Pergolizzi
CNR-ITBA
Milano, Italy

R.S. Phelps
Myriad Genetics Inc.
Salt Lake City, UT

D. Polanskey
FBI Academy
Quantico, VA

Darwin J. Prockop
Center for Gene Therapy
MCP-Hahnemann Medical School
Philadelphia, PA, USA

Hans Prydz
University of Oslo
Oslo, Norway

Rajinder S. Ranu
Department of Bioagricultural Sciences
and Pest Management
Colorado State University
Fort Collins, CO

M.E. Redolfi
CNR-ITBA
Milano, Italy

P. Rossi
Karolinska Institute
Stockholm, Sweden

Mati Reeben
Institute of Biotechnology
University of Helsinki
Helsinki, Finland

Steven A. Reeves
Massachusetts General Hospital
and Harvard Medical School
Charlestown, MA

J. Replogle
FBI Academy
Quantico, VA

Christiane M. Robbins
National Center for Human Genome
Research and the National Institutes of
Health
Bethesda, MD

Arndt Rolfs
Department of Neurology
University of Rostock
Rostock, Germany

Mari-Paz Rubio
Massachusetts General Hospital
and Harvard Medical School
Charlestown, MA

Fazlul H. Sarkar
Department of Pathology
Wayne State University School of
Medicine
Detroit, MI

G. Scarlatti
Karolinska Institute
Stockholm, Sweden

Robert Sinclair
University of Utah
Salt Lake City, UT

Y. Sonoda
University of Texas Health Sciences
Center
Houston, TX

J.W. Spatafora
Department of Botany
Louisiana State University
Baton Rouge, LA

S. Spurgeon
Perkin-Elmer/Applied Biosystems Div.
Foster City, CA

O.E. Stabbetorp
Botanical Garden and Museum
Oslo, Norway

Brita Stedje
Agricultural University of Norway
Oslo, Norway

J. Storz
School of Veterinary Medicine
Louisiana State University
Baton Rouge, LA

Kevin Sullivan
The Forensic Science Service
Birmingham, England

Anand Swaroop
W.K. Kellogg Eye Center
University of Michigan
Ann Arbor, MI

Jeff Swensen
University of Utah
Salt Lake City, UT

Harold Swerdlow
University of Utah
Salt Lake City, UT

Shigeru Takagi
NTT Basic Research Laboratories
Musashino-Shi, Tokyo, Japan

Gerard Tromp
Center for Molecular Medicine and Genetics
Wayne State University
Detroit, MI, USA

Sai-Wah Tsao
Brigham and Women's Hospital
Boston, MA

Gillian Tully
The Forensic Science Service
Birmingham, England

M. Uhlén
Royal Institute of Technology
Stockholm, Sweden

Victor D. Vacquier
Department of Marine Biology
University of California-San Diego
La Jolla, CA

Bin Wang
Sinogenesis Institute
Haidian, Beijing, China

Shouwen Wang
Food and Drug Administration
Center for Biologics Evaluation and Research
Rockville, MD

Xinkang Wang
SmithKline Beecham Pharmaceuticals
King of Prussia, PA

X.-Y. Wang
University of Texas Health Sciences Center
Houston, TX

Ines Weber
Free University of Berlin
Berlin, Germany

David B. Weiner
Department of Pathology and Laboratory Medicine
University of Pennsylvania
Philadelphia, PA

William V. Williams
Department of Medicine
University of Pennsylvania
Philadelphia, PA

Mark R. Wilson
FBI Headquarters
Washington, DC

H. Zakeri
Washington University School of Medicine
St. Louis, MO

X. Zhang
School of Veterinary Medicine
Louisiana State University
Baton Rouge, LA

Preface

BioTechniques is the one journal more likely to be found on the lab bench than the bookshelf. For the past fourteen years it has been a popular how-to-do-it publication for life science researchers. Compared to a lab manual, *BioTechniques* has the advantage of being updated with new articles every month. Although some methods papers may become classics, without much need for further improvement, most researchers read *BioTechniques* to learn about the latest, rapidly changing tools of the trade, whether the articles describe innovative breakthoughs or simply minor modifications of well-established protocols or helpful lab tips. Most investigators do not have access to all back issues of *BioTechniques*, and it would not be practical to search through stacks of past *BioTechniques* issues to find related articles on a particular subject.

Many of our readers have requested that we provide topic-specific compilations of *BioTechniques* papers. In 1992 we responded by publishing the book The PCR Technique: DNA Sequencing, a compilation of key articles on that topic published in *BioTechniques* between 1989 and 1991. An important feature of that book was the "update", an addendum to each paper usually written by the original authors that added important new information and corrected known errors. Five years have passed while hundreds of new methods papers have appeared in the pages of *BioTechniques*. Readers continue to ask for more "Update" books. Recently Eaton Publishing established its new book publishing division, BioTechniques Books. In addition to publishing original bioscience reference works and laboratory methods books, BioTechniques Books is continuing the "Update Series" with the publication of this volume, The PCR Technique: DNA Sequencing II, and still others to follow in 1997 and beyond.

This second book in the series starts where the last one left off. Professor Ulf Gyllensten of Uppsala University has again agreed to help select a variety of interesting papers on the topic of PCR and DNA sequencing. All papers have been culled from those published since 1992; most include "updates". I hope you will find this collection interesting and of practical value at the lab bench. Your feedback is always appreciated.

We, the editors and publisher, are grateful to the contributing authors and the many reviewers of the original manuscripts. Acknowledgment is also due to the editorial and production staffs of Eaton Publishing, especially to Sandra Lamont, Production Manager. We also wish to extend our special thanks to Christine McAndrews, Managing Editor, and Stephen Weaver, Editor-in-Chief, BioTechniques Books, whose creative hard work has been crucial to the publication of this book and the revival of the "BioTechniques Update Series".

James Ellingboe, PhD
Scientific Editor
BioTechniques

DNA Sequencing When Approaching The New Millennium

Ulf Gyllensten

Department of Medical Genetics, Biomedical Center, Uppsala, Sweden

Ever since techniques for determining the individual nucleotides in deoxyribonucleotide (DNA) molecules were developed, deciphering the hereditary material has become an increasingly important endeavor. The central role of DNA sequencing in the genome projects provides the final proof that DNA sequencing has passed from being but a laboratory tool to become the basis for the largest biological/medical research projects ever undertaken. It was forecast at the onset of the genome projects that the determination of the complete sequence of individual organisms would require as-yet-undeveloped techniques. For a number of years, technique development was performed along two parallel lines: one based on successive improvements of existing technology and the other relying on entirely novel approaches. Today the genome sequencing projects are relying on the standard technology invented some years ago—essentially fluorescent-based detection using the Sanger sequencing approach—albeit with some recent refinements that increase the throughput. Entirely novel approaches, such as microchip sequencing, have made very promising progress, but it remains to be seen whether they will make any substantial contribution to the completion of the first genome sequence of an organism. Clearly, technologies for rapid screening of sequencing variants in defined regions of the genome, like the microchip-based sequencing by hybridization, will be of enormous value in the next stage of genome sequencing, when the nature of sequence polymorphism in individual genes or the genes in a particular biochemical pathway will be assayed for variation, in order to understand the complex relationships between genotype and phenotype. Also, DNA sequencing is finally becoming established as an accepted technique in the clinical disciplines; methods for screening of putative disease-associated mutations in hereditary conditions, pathogen identification and genetic identification from biological evidence and remains have all been developed and are now being utilized in a rapidly growing field of DNA diagnostics.

Why this emerging acceptance for DNA sequencing as a major tool? Although the basic technology underlying the majority of sequencing projects was presented in the seminal paper by Sanger in 1977, now ten years ago, there have been a number of significant improvements over the years. Sequencing is usually conceived as a multistep process, initiated by template preparation, followed by one or several steps for the elongation/termination reactions and rounded off by separation and detection of the reaction products. The most important methodological improvements are undoubtedly the introduction of the PCR method for generation both of template and reaction products, solid-face technologies for template purification and the semiautomated fluorescent detection of sequencing reaction products. The PCR method immediately facilitated the generation of template from minuscule amounts of original genetic material and reduced the need for a pure template. The PCR method also provided a means to avoid template preparation entirely, using the cycle sequencing approach, where amplification of template is combined with the generation of sequencing reaction products. Today cycle sequencing using some of the novel sequencing enzymes has become the basis for several of the large-scale genome sequencing projects. Similarly, the automated fluorescent detection of products has streamlined the generation and handling of the large amounts of sequencing data. It is inconceivable that the data generated in many large to medium-sized laboratories could have been possible to handle, were it not for this automated data retrieval. Technical improvements that are likely to follow are the ability to analyze an increasing number of reactions per instrument and with a much shorter runtime, both of which are aimed at increasing the throughput of the sequencing operation.

This book is a compilation of papers that appeared in *BioTechniques* from 1992 to 1996. Clearly, a number of significant contributions in this field have appeared in other journals, but it is beyond the scope of this volume to cover the entire field of DNA sequencing. The unifying concepts in the contributions included here are that they describe a novel approach to solving a common problem in DNA sequencing or that they describe an application of DNA sequencing which is of general interest. Although the techniques in this field are constantly improving, scientists employing DNA sequencing for their particular project frequently find themselves left in the dark with their problem, and with no help provided by the user manual of the kit or instrument employed. The compilation of experiences printed in this volume is intended to help in those situations, as well as to broaden the experience of people with a general interest in DNA sequencing.

Hybrid PCR Sequencing: Sequencing of PCR Products Using a Universal Primer

Einar S. Berg and Bjørnar Olaisen[1]

National Institute of Public Health and [1]Institute of Forensic Medicine, University of Oslo, Oslo, Norway

SUMMARY

We describe a general method for making template DNA for sequencing of PCR products. The procedure may be particularly useful for PCR products where minimal sequence information is known or as an alternative to primer walking when sequencing long PCR products. A cassette containing the hybridization site for the M13 sequencing primer is ligated to a sample PCR product. Using one phosphorylated primer specific for the cassette together with one primer specific for the sample PCR product, subsequent PCR amplifies one hybrid construct directionally. This allows utilization of the universal M13 primer when sequencing of one strand after the removal of the complementary strand using λ-exonuclease.

INTRODUCTION

Direct sequencing of polymerase chain reaction (PCR) products has several advantages. The effects of polymerase-induced errors are minimized compared with M13 cloning where only one PCR product molecule, which may contain errors, is amplified in vivo before the sequencing. The template DNA in direct sequencing comes from a pool of DNA molecules. Random misincorporations during PCR will therefore be invisible among all the correct copies. A common approach for direct sequencing is solid-phase sequencing (2). This procedure requires a biotin group preattached to the 5′ end of one of the PCR primers and the use of a specially purified internal primer. Design of internal primers requires sequence information beyond the PCR primers, and the magnetic beads, biotin labeling and synthesis of purified primers are expensive. We, therefore, found it worthwhile to develop a sequencing procedure that may be regarded as a link between direct sequenc-

(Reprinted from BioTechniques 17:896-901)

ing and cloning of PCR products into a vector, e.g., M13. Our protocol has similarities to ligation-mediated PCR protocols. A PCR product (216 bp) from the *lacZ* region of M13/pUC is synthesized. This cassette contains both a multiple cloning site and the target sequence for the M13 forward sequencing primer. Ligation that forms a mixture of hybrids may be obtained either by *Sma*I-digested *lacZ*- and untreated sample PCR product or by restriction enzyme digests of the cassette and the sample PCR products creating complementary sticky ends. Each hybrid construct may be amplified selectively when using one cassette primer together with one sample primer in the subsequent PCR. The M13 universal primer site is thus a part of this hybrid PCR product. λ-Exonuclease digests a DNA strand with a 5′ phosphate group, as in LacL in our hybrid PCR product. A 5′ hydroxyl group, on the other hand, as in our sample PCR primers, gives protection against the enzyme. The resulting single-stranded DNA might be sequenced using routine methods. This is illustrated in Figure 1.

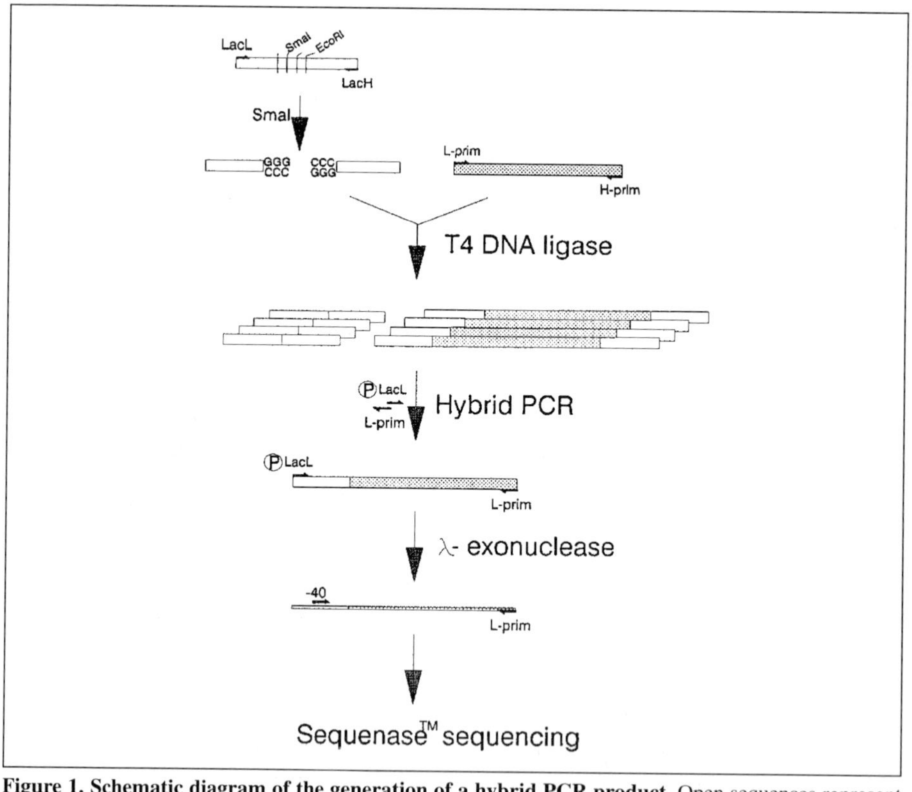

Figure 1. Schematic diagram of the generation of a hybrid PCR product. Open sequences represent cassette of the *lacZ* PCR product. Shaded sequences represent the sample PCR product. After *Sma*I digestion of the *lacZ* PCR product, this is ligated to the sample PCR product. PCR on the ligation product amplifies only one construct directionally. Treatment with λ-exonuclease gives single-stranded DNA containing a site for the M13 sequencing primer.

MATERIALS AND METHODS

Samples

The human astrovirus sample was kindly provided by Elisabeth Kjeldsberg (National Institute of Public Health, Oslo). The PCR product (1184 bp) from a human immunodeficiency virus (HIV-1) clone together with its PCR primers (positions 4155–4174 and positions 5353–5334 in Reference 3), was kindly provided by Enok Tjøtta (National Institute of Public Health, Oslo).

PCR

The PCR primers (0.2 µM of each, R&D Systems, Abingdon, England) used were designed by the aid of computer software (Oligo™, National Bioscience, Plymouth, MN, USA). LacL: 5′-GGCGAAAGGGGATGTGC-3′; LacH: 5′-CGGCTCGTATGTTGTGTGGAAT-3′; Astro_L: 5′-GCTTAGGA-

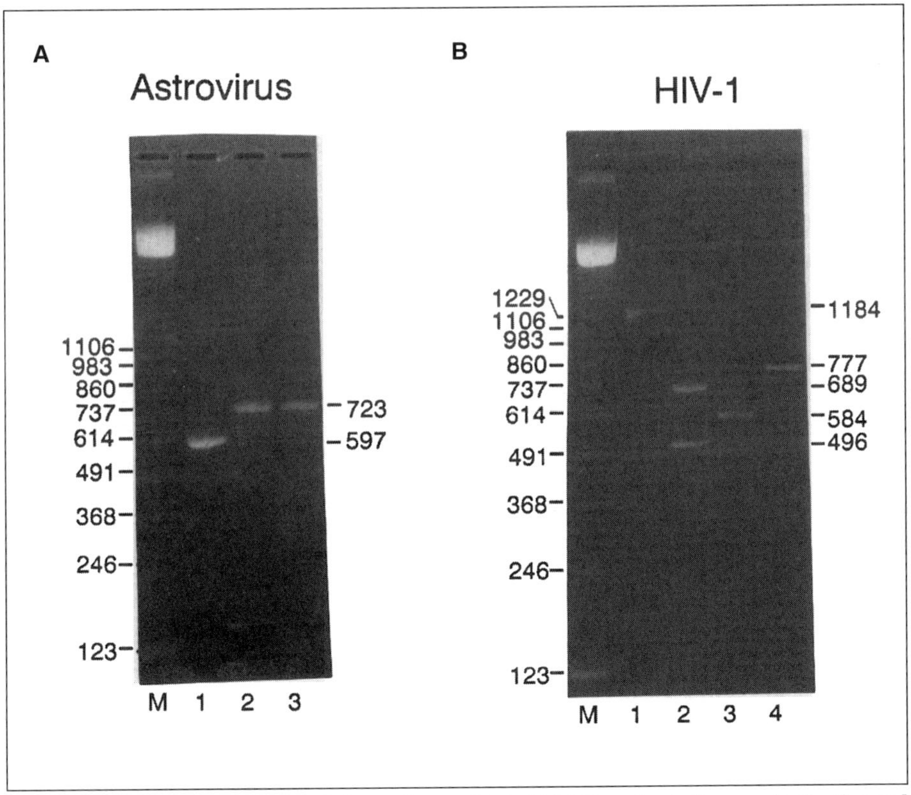

Figure 2. Agarose gel with original PCR product and hybrid PCR product after cassette ligation and amplification. Lane M. the molecular size standard (123-bp ladder; Life Technologies, Gaithersburg, MD, USA). (A) Lane 1, crude PCR product from human astrovirus; lanes 2 and 3, hybrid PCR products with different orientation of the sample PCR product relative to the cassette. (B) Lane 1, crude PCR products from an HIV-1 clone; lane 2, *Eco*RI digestion of the PCR product; lanes 3 and 4, hybrid PCR products with each fragment of the restricted sample PCR products ligated to the cassette.

CGATTATTGA-3′; Astro_H: 5′-TTCAGTATCAAAATAATTCAT-3′.

The PCR mixture used was made according Perkin-Elmer's (Norwalk, CT, USA) recommendation for their AmpliTaq™ DNA Polymerase. The *lacZ* cassette (216 bp) was amplified using standard 25 cycles PCRs with 3.5 mM Mg^{++} from M13mp18/mp19 RF I DNA (1 ng; Pharmacia Biotech, Freiburg, Germany) with annealing at 65°C. The astro PCR primers required 2.5 mM Mg^{++}, annealing at 55°C and gave a PCR product of 597 bp. The optimum magnesium concentrations and annealing temperatures were found after titration experiments (data not shown). To avoid "anemic" conditions in the PCR, the number of cycles were kept at a minimum (ordinarily 18–22). The sample PCR products were purified by gel filtration (Microspin™ S-300; Pharmacia Biotech). No further treatment of PCR products (e.g., T4 DNA polymerase) was performed before the ligation.

Restriction Enzyme Digestion

Restriction enzyme digestions were performed according to standard protocols. *Sma*I (New England Biolabs, Beverly, MA, USA) digestion of the

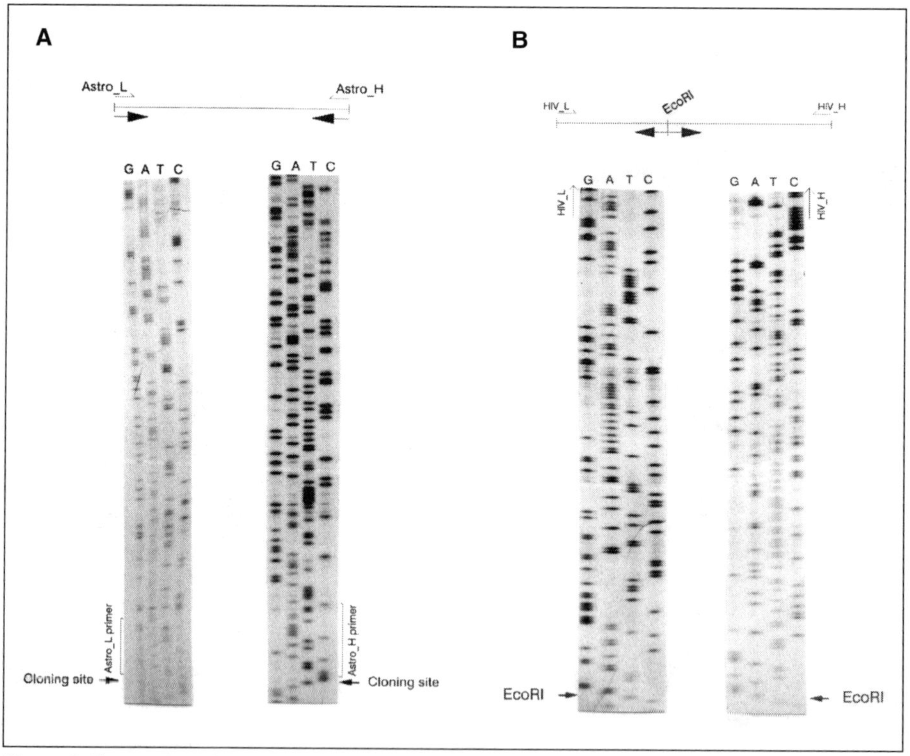

Figure 3. Sequencing of the hybrid PCR product. (A) Sequence ladder revealing both orientations of the hybrid PCR product from human astrovirus corresponding to lanes 2 and 3 in Figure 2A. (B) Sequencing of the hybrid PCR products from *Eco*RI-digested and *Eco*RI-ligated HIV-1 PCR product corresponding to lanes 3 and 4 in Figure 2B.

lacZ PCR product gave fragments of 90 and 126 bp. *Eco*RI (New England Biolabs) digestion of the *lacZ* PCR product gave fragments of 88 and 128 bp. The digestions were purified by gel filtration only (Microspin S-200; Pharmacia Biotech).

The Hybrid PCR Product

The ligation of *Sma*I-digested *lacZ* cassette (T4 DNA ligase; New England Biolabs) to the sample PCR product (molar ratio 1:1) was performed according to standard protocols (30 min incubation). *Sma*I was present in the blunt-end ligation mixture in order to reduce self-ligation of the cassette. When sticky end ligation was performed, both the cassette and the sample PCR product were predigested with *Eco*RI (cleaving the HIV-1 PCR product once).

A small aliquot (1 μL) of ligation reaction product was added to the hybrid PCR mixture (50 μL) containing kinased (5) LacL primer (T4 polynucleotide kinase; New England Biolabs). Only one of the hybrid PCR constructs was amplified when using LacL together with one of the sample primers. The other orientation of the sample PCR product relative to the *lacZ* cassette was obtained by using LacL in combination with the other sample PCR primer.

Conditions in the hybrid PCR were identical to that of the sample PCR.

DNA Sequencing

Generation of single-stranded template DNA was performed by λ-exonuclease (1) (PCR template prep kit; Pharmacia Biotech). The M13 primer (-40) (5′-GTTTTCCCAGTCACGAC-3′) and the reagents in the Sequenase™ Version 2.0 Kit (United States Biochemical, Cleveland, OH, USA) were used in the sequencing.

RESULTS AND DISCUSSION

We have developed a procedure that involves ligation-mediated PCR for sequencing a PCR product. Similar protocols have been published (4). These authors ligated synthetic linkers to fragments of genomic DNA and amplified one hybrid with a mixture of linker/sample PCR primers. The PCR product obtained was immobilized on paramagnetic beads and sequenced with a primer specific for the genomic DNA. Our aim was to develop an inexpensive procedure for sequencing any PCR product relying on routine methods. The *lacZ* PCR product containing both a multiple cloning site and the hybridization site for the universal M13 sequencing primer was chosen as linker. Single-stranded DNA was obtained by λ-exonuclease digestion of the hybrid PCR product.

HIV-1 and astrovirus PCR products appeared pure on the test gel (Figure 2A, lane 1, and 2B, lane 1). Therefore, purification by gel filtration

(Sephacryl® S-300) was chosen for removal of excess primers and invisible small size primer artifacts. Flushing and kinasing of the PCR product were not performed prior to ligation. Ligation of the fragments was expected to create a mixture of hybrids. Among them may be some constructs that amplify from the LacL primer only. Additional weak bands in Figure 2A, lanes 2 and 3, may be such constructs. Only one construct was amplified in the subsequent hybrid PCR after sticky end ligation (Figure 2B, lanes 3 and 4). Our hybrid PCR products carried a 5′ phosphate group in LacL allowing strand-specific digestion with λ-exonuclease to create single-stranded template DNA.

Alternatively, the sample PCR primers might have been biotinylated, which allowed the hybrid PCR product to be made single-stranded when bound on magnetic beads (1). Figure 3 is the sequencing autoradiogram of each of the hybrid PCR products after overnight exposure. The sequence pattern obtained using the robust universal "-40" primer has high quality.

The solid-phase sequencing protocol (2) is well suited for direct sequencing of one biotin-labeled PCR product using an internal primer. Our method, on the other hand, may be particularly suited for PCR products where minimal sequence information is known, e.g., virus diagnosis or sequencing of uncharacterized PCR products, as shown by the astrovirus sequencing. Among the first bases to be read in the sequence ladder is the cloning site followed by one of the PCR primers; a useful feature when using degenerate primers. In addition, as shown for the HIV-1 PCR product, the method may represent an alternative to primer walking when sequencing long PCR products. The only requirement is a site for a restriction enzyme compatible with one in the polylinker in the *lacZ* PCR product. The described method for making template DNA to sequence PCR products has been employed for other PCR products with success. In our experience, it is important to start with a PCR product of high quality, i.e., obtained after only a limited number of cycles in an optimized PCR. The procedure is rapid; the laboratory work may be performed within a few hours.

REFERENCES

1. **Higushi, R.G. and H. Ochman.** 1989. Production of single-stranded DNA templates by exonuclease digestion following the polymerase chain reaction. Nucleic Acid Res. *17*:5865.
2. **Hultman, T., S. Staahl, E. Hornes and M. Uhlén.** 1989. Solid phase in vitro sequencing of genomic and plasmid DNA using magnetic beads as solid support. Nucleic Acids Res. *17*:4937-4946.
3. **Hungnes, O., K. Jønsrud, E. Tjøtta and B. Grinde.** 1993. Sequence comparison and mutational analysis of elements that may be involved in the regulation of the DNA synthesis in HIV-1. J. Mol. Evol. *37*:198-203.
4. **Quivy, J-P. and P.B. Becker.** 1994. Direct dideoxy sequencing of genomic DNA by ligation-mediated PCR. BioTechniques *16*:238-241.
5. **Sambrook, J., E.F. Fritsch and T. Maniatis.** 1989. Molecular Cloning: A Laboratory Manual. Cold Spring Harbor Laboratory Press, Cold Spring Harbor, NY.

Update to:

Hybrid PCR Sequencing: Sequencing of PCR Products Using a Universal Primer

Einar S. Berg
National Institute of Public Health, Oslo, Norway

Design of PCR product-specific internal primers (that give the best sequencing results) requires sequence information beyond the PCR primers. Generation of single-stranded DNA using procedures involving, for example, magnetic bead capture and exonucleases requires further manipulations in addition to the use of expensive reagents. A substantial simplification is achieved by cycle sequencing with a thermostable DNA polymerase using the double-stranded PCR product as template. The hybrid PCR product is well suited for cycle sequencing.

The -21 M13 primer (forward) (5′-TGTAAAACGACGGCCAGT-3′) and the other reagents in the ABI PRISM™ Dye Terminator Cycle Sequencing Ready Reaction Kit from Perkin-Elmer (Foster City, CA, USA) were used in the sequencing of an astro hybrid PCR product. An ABI PRISM 310 Genetic Analyzer (Perkin-Elmer) revealed the electropherogram in Figure 1. Another approach might have been to perform the cycle sequencing with M13 universal dye primers before the electrophoresis. Overall, the quality of the sequence is fairly good, in defiance of a relatively dirty PCR product.

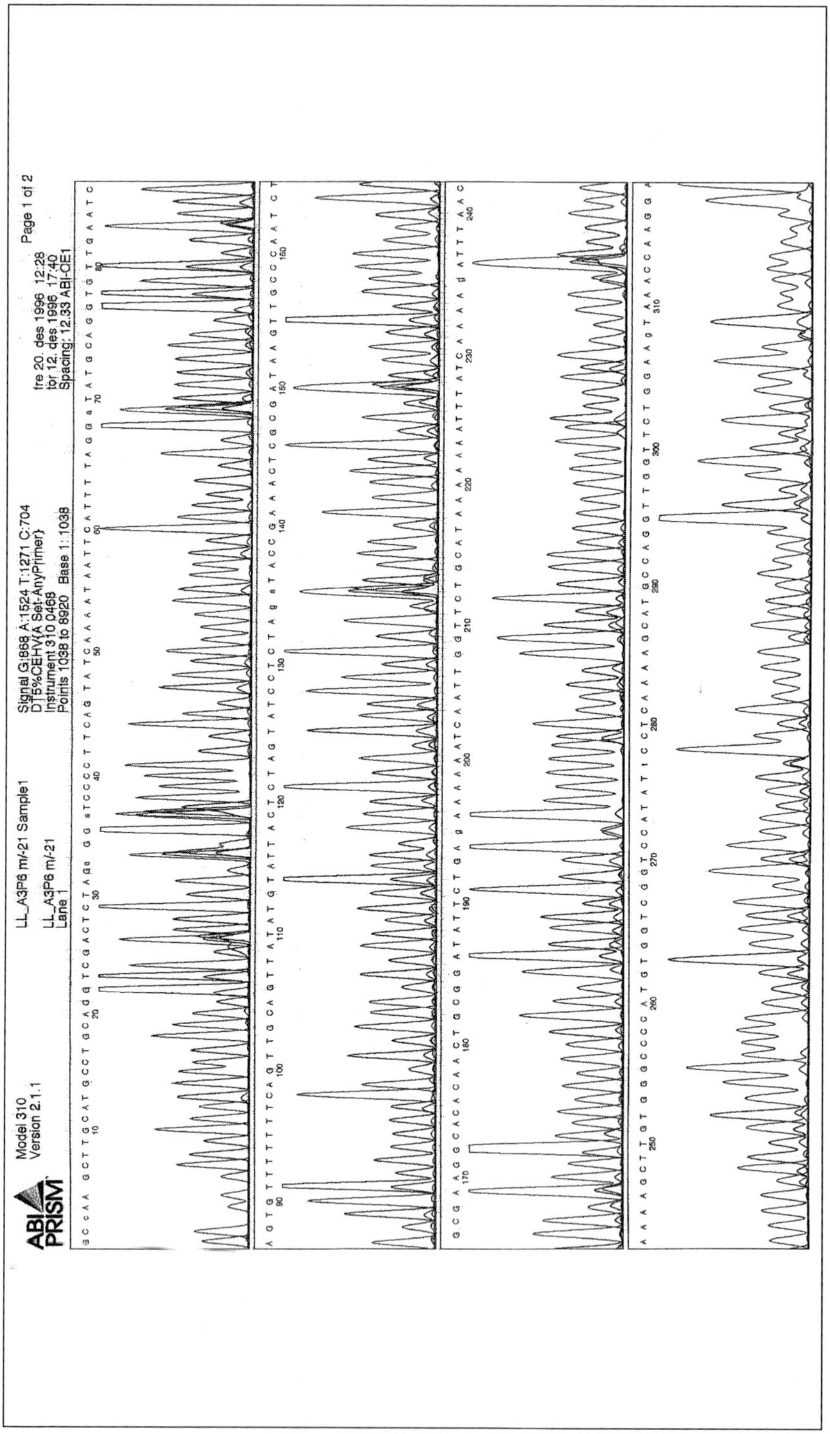

Figure 1A. The electropherogram after cycle sequencing of a double-stranded hybrid PCR product from astrovirus. The first series of bases to be read is the linker sequence with the cloning site followed by one of the astro PCR primers. The bases written in small letters, many of them involving weak G-bases, have been edited manually on the basis of earlier sequencing results from the isolate.

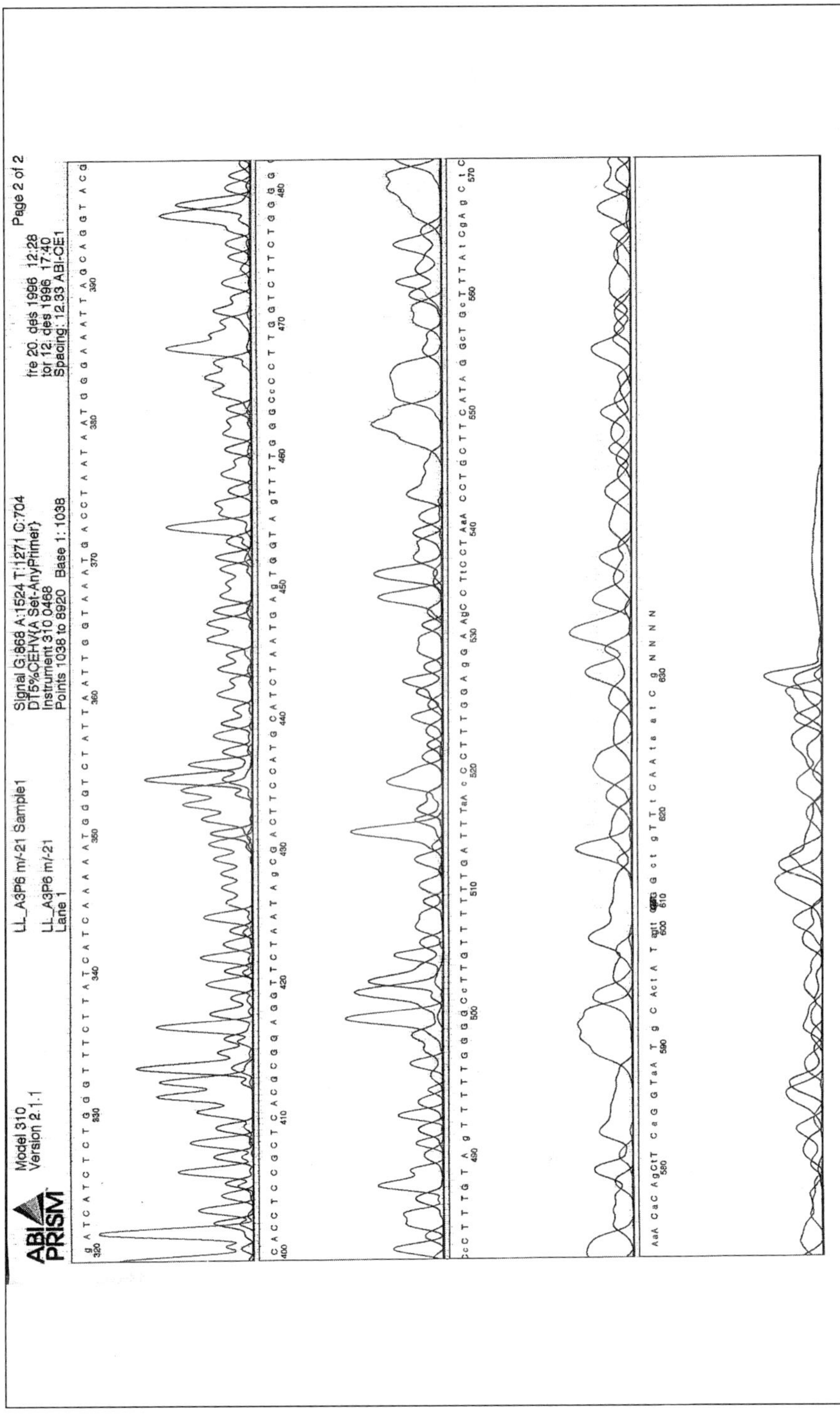

Figure 1B. Same as Figure 1A.

Improved Accuracy in Direct Automated DNA Sequencing of Small PCR Products by Optimizing the Template Concentration

S.M. Hyder, C. Hu, D.S. Needleman, Y. Sonoda, X.-Y. Wang and V.V. Baker

University of Texas Health Sciences Center, Houston, TX, USA

ABSTRACT

Data are presented illustrating the optimum concentration range of reverse transcription PCR-generated products under 500 bp for accurate base calling with direct automated DNA sequencing. A 357-bp fragment of the human estrogen receptor, which includes the DNA binding domain of the protein, was used as a representative example of a gene fragment that can be rapidly amplified and sequenced. Using the Taq *DNA polymerase dye terminator sequencing protocol and automated sequencing apparatus from Applied BioSystems, 0.1 to 1.0 pmol of PCR product in a 20-µL reaction volume provided >97% accurate base detection. Concentrations greater or lower than this range increased the number of ambiguous bases due to alterations in the signal-to-noise ratios. This procedure has been successfully utilized with 140–440-bp PCR products within the optimum concentration range. These results show that low amounts of PCR products are necessary and sufficient for direct sequence analysis.*

INTRODUCTION

Identification of gene mutations by polymerase chain reaction (PCR) amplification of the DNA, followed by sequence analysis is becoming increasingly popular (2,3). To avoid time-consuming vector construction for sequence analysis of the PCR product, direct sequencing of the PCR-amplified product is now being utilized by many investigators as the method of choice to detect mutations involving single base changes and abnormal splice sites that result in gross deletions of exons (17).

(Reprinted from BioTechniques 17:478-482, 1994)

Most techniques for DNA sequencing that use double-stranded purified PCR products are tedious, time-consuming and involve the use of hazardous radioactive or nonradioactive chemicals (5,6,10–12,14,15).

Recently, Applied Biosystems (ABI) (Foster City, CA, USA) has developed a technique for direct automated DNA sequencing of double-stranded DNA on an automated DNA sequencer using termination nucleotides tagged with fluorescent dyes (1). The reduction in time and cost associated with fluorescent-based sequencing of PCR products makes this an attractive alternative to the more traditional sequencing methods.

The sequencing protocols developed by ABI recommend using 1 μg of double-stranded DNA template per sequence reaction, regardless of the size of the template (1). However, their protocols have been developed for sequencing of plasmid DNA templates, which are generally greater than 3-kb in length. In our experience, the addition of this amount of template from small PCR products (<500 bp) is not optimal. ABI has not published recommendations for the molar ratios required for accurately sequencing small PCR products.

Review of the product literature also failed to provide any quantitative values of the PCR-amplified fragments that could be successfully analyzed by automated DNA sequencing using the fluorescent terminating nucleotides (1). Most studies describe the addition of a fixed volume of the PCR product to be analyzed for the DNA sequencing analysis (8,13,16). However, analysis based upon reaction volume does not permit quantitation of the optimum amount of DNA to achieve an accurate sequence data. Due to the nature of the signal generated from the automated DNA sequencing using the fluorescent dyes, suboptimal values of DNA could result in the detection of ambiguous bases.

The objective of this study was to define a concentration range of PCR-amplified templates under 500 bp that would provide the most accurate base calling in an automated DNA sequencing reaction system, thus reducing the analysis time for small PCR products. The template for this study was a 357-bp region of DNA encoding the human estrogen receptor DNA-binding domain. The sequence of this region has been previously published (7).

MATERIALS AND METHODS

Cell Culture and Preparation of RNA

MCF-7 cells (a human breast cancer cell line) were cultured in petri dishes with GIBCO BRL's Dulbecco's modified Eagle's medium (DMEM) (Life Technologies, Gaithersburg, MD, USA) to 50%–70% confluence. Cells were then lysed in 1 mL of denaturing solution per petri dish (4). Cell lysates from four petri dishes were pooled, and the total RNA was extracted as previously described (4,9).

PCR Primers

Primers were synthesized to analyze the DNA binding domain (the C region) of the estrogen receptor gene (7). The upstream primer (ERP-1^+ sequence is 5′ GGA AGT ATG GCT ATG GAA TCT GC 3′ (746–768 bp); it is located upstream of the C region and is complementary to exon 2. The downstream primer (ERP-4^-) sequence is 5′ GTT GGC AGC TCT CAT GTC TCC 3′ (1082–1102 bp); it is located downstream of the C region and is complementary to exon 4. Primers were synthesized by National Bioscience (NBI) (Plymouth, MN, USA). The expected PCR product was 357 bp in length.

cDNA Synthesis and PCR

Reverse transcription and PCR were carried out using the GeneAmp® PCR kit (Perkin-Elmer, Norwalk, CT, USA). The cDNA was synthesized in 20 µL of reaction mixture containing 10 mM Tris-HCl, pH 8.3; 50 mM KCl; 5 mM $MgCl_2$; 1 mM of each deoxyribonucleoside triphosphate (dNTP), 2.5 µM Oligo $d(T)_{16}$; 1 U/µL of RNase inhibitor; 2.5 U/µL of reverse transcriptase; and 1 µg of total RNA at 42°C for 1 h. The mixture was heated to 95°C for 5 min to inactivate the reverse transcriptase.

PCR was carried out in 100 µL of reaction mixture (10 mM Tris-HCl, pH 8.3; 50 mM KCl; 1.5 mM $MgCl_2$; 200 µm of each dNTP; 1 pmol/µL of each primer; and 2.5 U/100 µL of *Taq* DNA polymerase) for 35 cycles. Twenty microliters of cDNA were used in the PCR. The PCR parameters were denaturation at 95°C for 1 min, primer annealing at 65°C for 1 min and extension at 72°C for 30 s. The extension step was automatically increased by 2 s with each subsequent cycle. A negative control consisting of all reagents except the cDNA was always included during the thermal cycling.

Purification and Sequencing of PCR Product

PCR product was purified by the Wizard™ PCR Preps DNA purification system (Promega, Madison, WI, USA) according to the manufacturer's instructions, and the DNA concentration was determined using the OD_{260} value. Various concentrations of DNA ranging from 0.0075–4.5 pmol of purified DNA were used in the PCR sequencing reaction. The total reaction volume was 20 µL, as suggested by the ABI protocol (1).

The same primers used for generating the PCR product were also used in the sequencing reaction. Other reagents were supplied in the Taq DyeDeoxy™ Termination Sequencing Kit (ABI). The ABI protocol (1) was followed for the PCR sequencing reaction with the following exceptions: the 20× reaction mixture contained twice the amount of AmpliTaq® DNA polymerase (8 units; Perkin-Elmer). The use of twice the amount of AmpliTaq allows better sequencing results through any secondary structure and inverted repeats (ABI, personal communication). The sequencing reaction was performed in a PTC thermal cycler (MJ Research, Watertown, MA, USA) pre-

```
                                10         20         30         40         50         60
 1... 60  HER        AGCAGACCCC ACTTCACCCC TGCCCTCCCC ATCATCTCTC TGGCGCTTGT GTTTCAACAT
 1... 60  HER 12.5   AGCAGACCC- ACTTCACCCC TGCCCTCCCC ATCATCTCTC TGGCGCTTGT GTTTCAACAT
 1... 60             ---------D ---------- ---------- ---------- ---------- ----------
 1... 60  HER 0.5    -GCAGCCNC- ACTTCACCCC TGTCCTCCCC ATCATCTCTC TGGCGCTTGT GTTTCANCAT
 1... 60             D----*-*-D ---------- --*------- ---------- ---------- ------*---
 1... 60  HER 300    AGCAGACNC- ACTTCACCCC TGCCCTCCCC ANCATCTCTC TGGCGATTGT GTNTCAACAT
 1... 60             -------*-D ---------- ---------- -*-------- -----*---- --*-------

                                70         80         90        100        110        120
 61...120 HER        TCTCCCTCCT CTTCGGTCTT TTCGTATCCC ACC-TTTCAT CATTCCCACT TCGTAGCATT
 61...120 HER 12.5   TCTCCCTCCT CTTCGGTCTT TTCGTATCCC ACC-TTTCAT CATTCCCACT TCGTAGCATT
 61...120            ---------- ---------- ---------- ---------- ---------- ----------
 61...120 HER 0.5    TCTCCCTCCT CTTCGGTCTT TTCGTATCCC ACC-TATCAT CATTCCCACT TCGNAGCATT
 61...120            ---------- ---------- ---------- -----*---- ---------- ---*------
 61...120 HER 300    TCTCCCTCCT CTTCGGGCTT TTCGTGTCCC ACCNANTCAT CATTCTCAAT TCGTAGCATT
 61...120            ---------- ------*--- -----*---- ---I**---- -----*--*- ----------

                               130        140        150        160        170        180
121...180 HER        TGCGGAGCCG GCAGGCCTGG CAGCTCTTCC TCCTGTTTTT ATCAATGGTG CACTGGTTGG
121...180 HER 12.5   TGCGGAGNCG GCAGGCCTGG CAGCTCTTCC TCCTGTTTTT ATCAATGGTG CACTGGTTGG
121...180            -------*-- ---------- ---------- ---------- ---------- ----------
121...180 HER 0.5    TGGGGAGACG GCAGGACTGG NAGTTCTTCC TCCTGNTTTN ATCAATGGNG CACTGGTTGG
121...180            --*----*-- -----*---- *--*------ -----*---* --------*- ----------
121...180 HER 300    NGTGGNGACG GCAGGNCTGG AAGCTCTTCC NCCTGTTTTT ATCANTGGTG CACTGNTTGG
121...180            *-*--*-*-- -----*---- *--------- *--------- ----*----- -----*----

                               190        200        210        220        230        240
181...240 HER        TGGCTGGACA CATATAGTCG TTATGTCCTT GAATACTTCT CTTGAAGAAG GCCTTGCAGC
181...240 HER 12.5   TGGCTGGACA CATATAGTCG NTATGTCCTT GAATACTTCT CTTGAAGAAG GCCTTGCAGN
181...240            ---------- ---------- *--------- ---------- ---------- ---------*
181...240 HER 0.5    TGGTTGGACA CATATNGCCG TNATGTCCTT GAATACTTCT NTNGANGAAG GNCTTGCAGN
181...240            ---*------ -----*-*-- -*-------- ---------- *-*--*---- -*-------*
181...240 HER 300    NGGCTGGNCA CATATAGTCG GTATGTCCNT GNATACTNCT CTTGAAGAAG GNCTNGCAGC
181...240            *------*-- ---------- *-------*- -*-----*-- ---------- -*--*-----

                               250        260        270        280        290        300
241...300 HER        CCTCACAGGA CCAGACTCCA TAATGGTAGC CTGAAGCATA GTCATTGCAC ACTGCACAGT
241...300 HER 12.5   CCTCACAGGA CCAGACTCCA TAATGGTAGC CTGAAGCATA GTCATTGCAC ACTGCACAGT
241...300            ---------- ---------- ---------- ---------- ---------- ----------
241...300 HER 0.5    CCTCACAGGA NCAGACTCCA TAATGGNAGC CTGANGCATA GTCATTGCAC ACTGTACAGT
241...300            ---------- *--------- ------*--- ----*----- ---------- ----*-----
241...300 HER 300    CCTCACAGGA NCAGNCTCCA TNATGGAAGN CTGACGCATA GTCATTGGAN ANTGCACAGG
241...300            ---------- *---*----- -*----*--* ----*----- -------*-* -*-------*

                               310        320        330        340        350        360
301...338 HER        AGCGAGTCTC CTTGGCAGAT TCCATAGCCA TACTTCCC
301...339 HER 12.5   AGCGAGTCTC CTTGGCAGAT TCCATAGCAN CCCCCCCCC
301...339            ---------- ---------- --------** **-**---I
301...339 HER 0.5    AGCGAGTCTC CTTGGCAGNT NCCATAGCCA TATCCCCNC
301...339            ---------- --------*- *--------- --***--*I
301...337 HER 300    AGCGAGTCTC CTTGGNAGAT TCCATGANCC CCCNNNN
301...338            ---------- -----*---- -----***-* **-****D
```

Figure 1. Comparison of sequencing results of human estrogen receptor DNA binding domain using the downstream primer. The reverse complement sequence of the 357 bp was obtained using the four colored fluorescent Taq Dye Terminator Cycle sequencing protocol supplied by ABI, except that the amount of DNA template was varied to determine the optimum values for sequencing. Gel separation, scanning of the data and analysis was performed on ABI 373A DNA sequencer. PCR-generated DNA was sequenced. HER represents the published sequence data from the human estrogen receptor; HER 12.5 represents the sequence data obtained from 187.5 fmol of template; HER 0.5 is sequence obtained from a 7.5 fmol of template; HER 300 is sequence obtained from 4.5 pmol of template. Comparison utilized the SEQED 675 software (ABI).

Table 1. Determination of Percentage of Accurate Base Calling Following Sequencing with Different Amounts of PCR-Generated Estrogen Receptor Gene Fragment (357 bp)

DNA (fmol)	% Accurate Base Calling	
	Expt1	Expt2
7.5	92	87
30.0	91	98
75.0	98	98
187.5	99	98
375.0	99	99
750.0	98	99
1125.0	98	97
1500.0	97	96
3000.0	95	91
4500.0	89	83

Values from two different experiments are illustrated. Technical details for sequencing and generation of data are given in Materials and Methods.

heated to 96°C and held at this temperature for an additional 10 s. The thermal cycling was carried out as follows: 30 s at 98°C, 15 s at 50°C and 4 min at 60°C for 25 cycles.

Unincorporated nucleotides were removed from the extended products using Centri-Sep™ spin columns (Princeton Separations, Adelphia, NJ, USA). The sequencing products were then dried in a SpeedVac® Apparatus (Savant Instruments, Farmingdale, NY, USA) and resuspended in 4.0 µL deionized formamide: 50 mM EDTA, pH 8.0 (5:1). Immediately prior to loading on a 6% denaturing polyacrylamide gel, the samples were heated to 90°C for 2 min and transferred to ice. After loading the samples, the electrophoresis was carried out on a Model 373A DNA Sequencer (ABI) at a constant power of 30 W for 12 h, as suggested by the manufacturer. ABI 373A Sequence software, Version 1.1.1 for the Macintosh® (Apple Computer, Cupertino, CA, USA) was used to collect and analyze the data.

RESULTS AND DISCUSSION

Figure 1 illustrates the comparison of sequencing data obtained with three different amounts of the 357-bp PCR-amplified template. The sequence in all cases extends from the first base after the 3′-end of the 21-nucleotide (nt) primer to 357 bp. Overall, the quality of the signal was excellent and the peaks were well defined (original chromatograms, not shown) when 187.5 fmol of the PCR product (represented by HER 12.5) was used for sequencing. The only serious loss in resolution was confined to the last 10 bases. Since all DNA samples showed same loss of resolution, these nucleotides were excluded from any subsequent analysis.

Figure 1 further shows results from the sequencing reaction when either 7.5 fmol (represented by HER 0.5) or 4.5 pmol (represented by HER 300) were added to the sequencing mixture. In contrast to results obtained with 187.5 fmol of DNA, both 7.5 fmol or 4.5 pmol DNA showed a significant increase in the level of noise and a comparable decrease in base resolution, which is reflected in data analysis shown in Figure 1.

Most of the errors during the sequencing reaction occurred when either a very low (7.5 fmol) or very high amount of DNA (4.5 pmol) was used for sequencing. In most of the sequencing mistakes the base preceding the ambiguous N was almost always G, and this observation has been made before (16). Comparison of the sequence with the known sequence in the data bank indicated that N almost always represented a C or T. We were able to eliminate the N bases by sequencing with the opposite primer initially used in the PCR (not shown). It is imperative, therefore, that sequencing data of both strands be obtained in a range that we have described in this paper for complete accuracy, especially for the detection of any point mutations.

When the percentage accuracy of base calling was plotted against increasing amounts of template DNA as shown in Table 1, it is obvious that the best results were obtained when the DNA was maintained between 7.5 fmol to 1.1 pmol. Within this range the overall base-calling accuracy of >97% was achieved. In contrast, the accuracy of base calling decreased below 90% when 4.5 pmol of DNA template was added (1.1 µg DNA; approximate amount suggested by ABI) to the sequencing mixture. The same decrease in accuracy was observed when only 7.5 fmol of PCR-generated DNA template were added.

Using similar values to those described above for a 357-bp PCR product for optimal DNA sequencing, we have sequenced 140–440-bp PCR products of the estrogen receptor with approximately 98% base-calling accuracy, which illustrates the use of the technique with a wide range of PCR products (data not shown).

This report describes for the first time a quantitative study to sequence PCR-amplified products, and these guidelines will be useful in the analysis of amplified DNA fragments. The adaptation of the ABI method (1) described in this paper requires low amounts of purified DNA. This method does not require single-stranded DNA or radioactive primers. Apart from the rapid detection of mutations, this method can be used for rapid confirmatory analysis of DNA fragments or analysis of spliced variants.

ACKNOWLEDGMENTS

The work reported in this paper was supported in part by Grant HD-08615 from the NIH. S.M.H. wishes to acknowledge the generous support and encouragement of Dr. George Stancel during the course of this project.

REFERENCES

1. **Applied Biosystems, Inc.** 1993. Taq DyeDeoxy Terminator Cycle Kit Bulletin #91047. Foster City, CA.
2. **Boehm, C.D.** Use of polymerase chain reaction for diagnosis of inherited disorders. Clin. Chem. *35*:1843-1848.
3. **Carothers, A.M., G. Urlaub, J. Mucha, D. Grunberger and L.A. Chasin.** 1989. Point mutation analysis in a mammalian gene: rapid preparation of total RNA, PCR amplification of cDNA, and *Taq* sequencing by a novel method. BioTechniques *7*:494-499.
4. **Chomczynski, P. and N. Sacchi.** 1987. Single step method of RNA isolation by acid guanidinium thiocyanate-phenol-chloroform extraction. Anal. Biochem. *162*:156-159.
5. **Douglas, A.M., A.M. Georgalis and B.A. Atchison.** 1993. Direct sequencing of double-stranded PCR products incorporating a chemiluminescent detection procedure. BioTechniques *14*:824-828.
6. **Gold, P.** 1992. Use of a novel agarose gel-digesting enzyme for easy and rapid purification of PCR-amplified DNA for sequencing. BioTechniques *13*:132-134.
7. **Green, S., P. Walter, V. Kumar, A. Krust, J-M. Bornert, P. Argos and P. Chambon.** 1986. Human oestrogen receptor cDNA: sequence, expression and homology to v-erb-A. Nature *320*:134-139.
8. **Hunter, S.B., C. Bandea, D. Swan, K. Abbott and V.A. Varma.** 1993. Mutations in the p53 gene in human astrocytomas: detection by single-strand conformation polymorphism analysis and direct DNA sequencing. Mod. Pathol. *6*:442-445.
9. **Hyder, S.M., G.M. Stancel and D.S. Loose-Mitchell.** 1991. Presence of an estradiol response region in the mouse c-fos oncogene. Steroids *56*:498-504.
10. **Kalnoski, M.H., M.F. McCoy-Haman and G.F. Hollis.** 1991. A rapid method for purifying PCR products for direct sequence analysis. BioTechniques *11*:246-249.
11. **Kusukawa, N., T. Uemori, K. Asada and I. Kato.** 1990. Rapid and reliable protocol for direct sequencing of material amplified by the polymerase chain reaction. BioTechniques *9*:66-72.
12. **Medori, R., H.J. Tritschler and P. Gambetti.** 1992. Production of single-stranded DNA for sequencing: an alternative approach. BioTechniques *12*:346-350.
13. **Navone, N.M., P. Troncoso, L.L. Pisters, T.L. Goodrow, J.L. Palmer, W.W. Nichols, A.C. von Eschenbach and C.J. Conti.** 1993. p53 protein accumulation and gene mutation in the progression of human prostate carcinoma. J. Natl. Cancer Inst. *85*:1657-1669.
14. **Tahara, T., J.P. Kraus and L.E. Rosenberg.** 1990. Direct DNA sequencing of PCR amplified genomic DNA by the Maxam-Gilbert method. BioTechniques *8*:366-368.
15. **Takagi, S., M. Kimura and M. Katsuki.** 1993. Direct sequencing of PCR products using unlabeled primers. BioTechniques *14*:218-221.
16. **Tracy, T.E. and L.S. Mulcahy.** 1991. A simple method for direct automated sequencing of PCR fragments. BioTechniques *11*:68-75.
17. **Wilke, W.W., R.A. Robinson and C.D. Kennard.** 1993. H-ras-1 gene mutations in basal cell carcinoma: automated direct sequencing of clinical specimens. Mod. Pathol. *6*:15-19.

Update to:

Improved Accuracy in Direct Automated DNA Sequencing of Small PCR Products by Optimizing the Template Concentration

Vicki V. Baker

University of Michigan Medical Center, Ann Arbor, MI, USA

As reliance upon automated DNA sequencing increases, investigation of specific methodologic details that enhance accuracy assume increasing importance. Additional experience with the template concentration confirms its importance as well as the benefit of evaluating the sequence information from both strands as ways to improve sequence accuracy. Subsequent publications that cite these factors include Wilson, M.R., D. Polanskey, J. Butler, J.A. DiZinno, J. Replogle and B. Budowle. 1995. Extraction, PCR amplification and sequencing of mitochondrial DNA from human hair shafts. BioTechniques *18*:662-669, and Hyder, S.M., Z. Nanaz, C. Chiapetta, K. Yokoyama and G.M. Stancel. 1995. The proto-oncogene c-jun contains an unusual estrogen-inducible enhancer within the coding sequence. J. Biol. Chem. *270*:8506-8513.

Rapid Protocol for Sequencing RNA Virus Using Δ*Taq*® Version 2.0 DNA Polymerase

Alexander C.K. Lai and Thomas M. Chambers

University of Kentucky, Lexington, KY, USA

Due to the rapid evolution of influenza virus, large amounts of sequence information on the major antigenic molecule of the virus, hemagglutinin (HA), are collected for analysis in the design of vaccines and for the study of epidemiology (3,6). Traditional protocols require several days' work involving viral amplification, purification, reverse transcription and cloning before sequencing (4). We describe a rapid protocol using an RNase inhibitor, the polymerase chain reaction (PCR) and a direct noncycling sequencing of the PCR product using Δ*Taq*® Version 2.0 DNA Polymerase (United States Biochemical [USB], Cleveland, OH, USA). Sequence data can be obtained within a day from the isolation of viruses. This protocol has been successfully applied to virus amplified in eggs (allantoic fluid) or directly to original sample fluid (nasopharyngeal swab).

For extraction of viral RNA, 50 μL of 20% sodium dodecyl sulfate (SDS) and 2.5 μL of proteinase K (1.0 mg/mL) are added to 450 μL of virus sample (allantoic fluid or nasal swab fluid). The mixture is incubated at 37°C for 10 min; then 1/10 vol of 2.5 M sodium acetate, pH 4.5, is added, followed by phenol/chloroform extraction. The aqueous phase is then precipitated with 2× vol of absolute ethanol and chilled at -70°C for at least 15 min, followed by microcentrifugation at 14000× *g* for 10 min. The viral RNA pellet is washed at least once with 70% ethanol, then dried. Viral RNA is resuspended with either a minimum amount of DEPC-treated water or double-distilled water (ddH_2O) with 0.2 U RNasin® (Promega, Madison, WI, USA).

For reverse transcription, approximately 0.5 μg of viral RNA is mixed with 10 pmol of 3′ universal primer (AGCAAAAGCAGG) in a reaction volume of 20 μL (with 4 μL 5× reverse transcription buffer [USB], dNTP to 400 μM and 10 U of RNasin [Promega]). The mixture is incubated at 75°C for 3 min, then quickly cooled in ice for 5 min. Ten units of avian myeloblastosis virus (AMV) reverse transcriptase (USB) are added, followed by incubation at 42°C for 90 min. Volume is adjusted to 100 μL using ddH_2O, and 10 μL are then used for PCR amplification using HA-specific primers. The PCR product is purified using spin column chromatography (Chroma Spin™ +TE-400; CLONTECH Laboratories, Palo Alto, CA, USA).

(Reprinted from BioTechniques 19:704-706, 1995)

There are a number of published protocols for direct noncycling sequencing of PCR products (for review see References 2 and 5). Use of T7 DNA polymerase (Sequenase®; USB), however, often resulted in multi-tracking, "ghosting" and compression, especially at GC-rich regions or regions with secondary structures. The problem is probably due to the low reaction temperature for Sequenase. Recently, use of a genetically engineered thermostable DNA polymerase, Δ*Taq* DNA polymerase, for direct noncycling sequencing has been shown to generate sequence data comparable to Sequenase (7). The lack of 5′–3′ exonuclease activity in Δ*Taq* DNA poly-

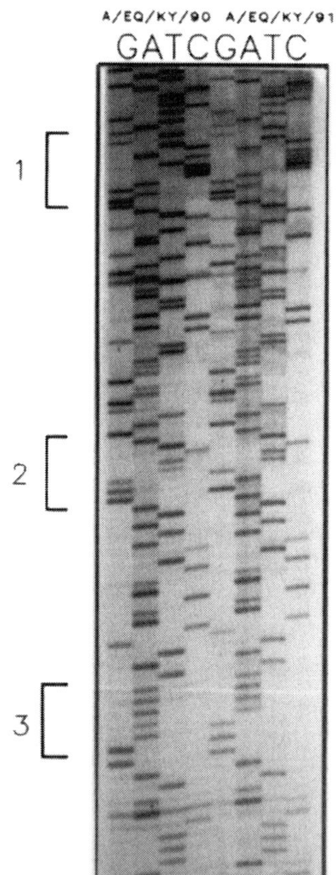

Figure 1. Sequence of equine influenza virus HA using the rapid protocol described. Both A/Equine/KY/90 and A/Equine/KY/91 vRNA were reverse-transcribed with primer UNI-12 (AGCAAAAGCAGG), then PCR-amplified (35 cycles at 95°C for 1 min, 40°C for 2 min and 72°C for 3 min) using primers EH3-1 (AGCAAAAGCAGGGGATATTTCTGTC) and EH3-832- (TTGAAATATCCCCGCGGTGC). The PCR product (832 bp) was purified by Chroma Spin +TE-400. Sequencing was performed using primer EH3-120+ (GGACACCATGCAGTAGCAAA) and [α-^{35}S]dATP. At least three nucleotide substitutions were detected in this autoradiogram for A/Equine/KY/91: region 1, T→C; region 2, G→A; region 3, A→G.

merase eliminates the "ghosting" effect on sequencing gels produced by native *Taq* DNA polymerase (8). Similar to sequencing with Sequenase, there are only three steps: annealing, extension and termination. Also, either [α-^{35}S]dATP or [α-^{32}P]dATP can be used for labeling during extension. The other advantage of using Δ*Taq* DNA polymerase for sequencing is that higher extension and termination temperatures (45° and 72°C, respectively) allow read-through of GC-rich regions, as discussed in Reference 7. Except for the primer annealing step, in which approximately 1.0 pmol of double-stranded DNA template (which can be 500 bp to 1.7 kb in size) is boiled with 0.5 pmol of sequencing primer for 3 min and followed by "quick-snap" in ice for 5 min, our sequencing protocol is essentially similar to that described by Barnes (1). Figure 1 is an autoradiograph of a sequencing gel (6%) electrophoresed at 65 W for 3 h. The quality of sequence data is comparable with that generated by Sequenase. The DNA bands are labeled evenly with very little background bands. At least three nucleotide changes were detected between A/ Equine/KY/90 and A/ Equine/KY/91 in this sequencing gel.

In summary, this protocol is simple, rapid and very reliable. It allows us to generate sequence data rapidly for newly isolated viruses in the course of an epidemiological study. It is expected that this protocol will also be applicable to other RNA viruses.

ACKNOWLEDGMENTS

This work is supported by the Grayson Jockey Club Research Foundation and by generous gifts from Mr. Paul Mellon and The Royal Hong Kong Jockey Club. The investigation reported in this paper (No. 95-14-036) is in connection with a project of the Kentucky Agricultural Experiment Station and is published with approval of the director.

REFERENCES

1. **Barnes, W.M.** 1993. TAQuence® version 2.0: *Taq* DNA polymerase sequencing. United States Biochemical, Cleveland, OH.
2. **Bevan, I.S., R. Rapley and M.R. Walker.** 1992. Sequencing of PCR-amplified DNA. PCR Methods Appl. *1*:222-228.
3. **Chambers, T.M., A.C.K. Lai, K.M. Franklin and D.G. Powell.** 1995. Recent evolution of the hemagglutinin of equine-2 influenza virus in the U.S.A, p. 175-180. *In* P.D. Rossdale and J.F. Wade (Eds.), Equine Infectious Diseases, Vol. 7. R & W Publications, Newmarket.
4. **Kawaoka, Y., W.J. Bean and R.G. Webster.** 1989. Evolution of the hemagglutinin of equine H3 influenza viruses. Virology *169*:283-292.
5. **Kusukawa, N., T. Uemori, K. Asada and I. Kato.** 1990. Rapid and reliable protocol for direct sequencing of material amplified by the polymerase chain reaction. BioTechniques *9*:66-72.
6. **Lai, A.C.K., Y.P. Lin, D.G. Powell, K.F. Shortridge, R.G. Webster, J. Daly and T.M. Chambers.** 1994. Genetic and antigenic analysis of influenza virus responsible for the 1992 Hong Kong equine influenza epizootic. Virology *204*:673-679.
7. **Ranu, R.S.** 1995. DNA Sequencing with Δ*Taq*® DNA polymerase. BioTechniques *18*:390-395.
8. **Samols, S.B. and C.W. Fuller.** 1995. Δ*Taq* DNA polymerase for DNA sequencing. United States Biochemical Editorial Comments *21*:57-78.

Update to:

Rapid Protocol for Sequencing RNA Virus Using Δ*Taq*® Version 2.0 DNA Polymerase

Alexander C.K. Lai
University of Kentucky, Lexington, KY, USA

The thermostable Δ*Taq*® Version 2.0 DNA Polymerase is available now only as a Δ*Taq* Cycle Sequencing Kit from Amersham (Arlington Heights, IL, USA). However, the enzyme and the reagents supplied can still be used for the rapid protocol originally described: annealing by "quick-snap", extension at 45°C and termination at 72°C.

DNA Sequencing with Δ*Taq* DNA Polymerase

Rajinder S. Ranu

Colorado State University, Fort Collins, CO, USA

DNA sequencing using the dideoxy chain termination method of Sanger et al. (12) has proven to be quite effective in obtaining sequence information from a single-stranded DNA template. A denatured double-stranded DNA template can also be used to obtain DNA sequencing data (2,4,6,11). Improvements in DNA sequencing techniques are being introduced constantly. These improvements include the introduction of T7 DNA polymerase by Tabor and Richardson (13), which, with its high processivity, allows generation of sequence information of extended length. More recently, *Taq* DNA polymerase has found increasing use in DNA sequencing, especially when a limited amount of DNA template is available (7). In this procedure, the primer is initially labeled with [^{32}P] or [^{33}P] using polynucleotide kinase and γ-labeled ATP. The labeled primer is then used in cycle sequencing in a thermal cycler using the principle of the polymerase chain reaction (PCR) (Reference 7; TAQuence® Cycle Sequencing System—United States Biochemical [USB], Cleveland, OH, USA); a very small amount of double-stranded or single-stranded DNA template is necessary. Because of the 5′→3′ exonuclease activity associated with standard *Taq* DNA polymerase, the standard [∝-^{35}S]dATP labeling reaction cannot be readily used without compromising the sequence information (7). Recently, however, a modified form of *Taq* DNA polymerase has been described (1); the enzyme is called Δ*Taq*® DNA polymerase (USB). The Δ*Taq* DNA polymerase (modified and cloned) lacks exonuclease activity and is stable at 90°C for at least one hour (1). The Δ*Taq* DNA polymerase offers several advantages over the standard *Taq* DNA polymerase. Δ*Taq* DNA polymerase can be used in a standard [∝-^{35}S]dATP-based sequencing protocol (1). Due to the absence of 5′→3′-exonuclease activity, there is little concern of extra DNA bands being introduced in the sequencing gel.

This paper describes the use of Δ*Taq* DNA polymerase in standard DNA sequencing from single-stranded (or double-stranded) template and also highlights some other advantages it may offer in sequencing over the standard Sequenase® system (modified T7 DNA polymerase; USB). We note that the Δ*Taq* DNA polymerase system is compatible with and complements the

(Reprinted from BioTechniques 18:390-396, 1995)

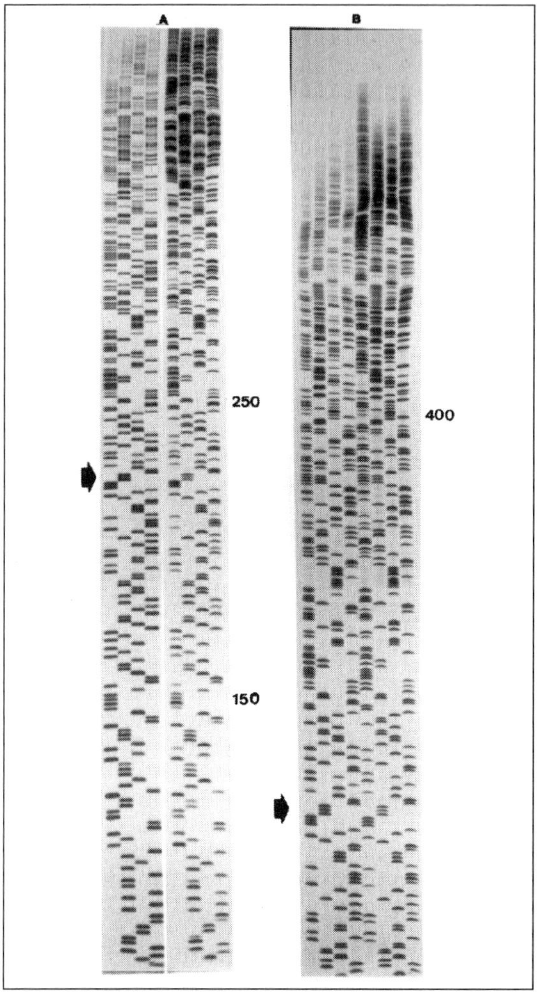

Figure 1. Comparison of sequencing results obtained with Sequenase Version 2.0 and TAQuence Version 2.0. DNA sequence ladder was generated using either Sequenase Version 2.0 (left sides of Panels A and B) or TAQuence Version 2.0 (right sides of Panels A and B) with M13mp18 DNA template and universal primer (17 nucleotides in length). Identical concentrations of template (0.5 pmol) and primer (0.5 pmol) were used. After annealing, the standard reaction conditions recommended for the Sequenase system were used. The protocol for Δ*Taq* DNA polymerase is described in the text. In each case, after termination of the sequencing reaction with stop solution, samples were heated at 95°C for 2 min and quickly cooled in an ice bath. Samples (3.2 µL from the Sequenase system and 4 µL from the Δ*Taq* system) were applied to a sequencing gel (6%) and electrophoresed at 2000 V for 5 h (A) or 7.5 h (B). The gel loading order, in each case, from left to right was GATC. Alignment of the sequences for 5 h and 7.5 h gel autoradiograms are indicated by arrows.

widely used Sequenase Version 2.0 system.

The Δ*Taq* DNA polymerase sequencing kit, known as TAQuence Version 2.0, and the Sequenase Version 2.0 sequencing kit were obtained from USB. [α-^{35}S] dATP (specific activity 1000 Ci/mmol) was from Amersham (Arlington Heights, IL, USA). The procedure for growth of MAC-1 bacteriophage and preparation of phage DNA has been described (10). The primers used in the present study for sequencing the MAC-1 DNA template were synthesized at the Colorado State University Macromolecular Resources facility. The protocol is similar to that described by Barnes (1) except that the labeling step was carried out at 55°C for 5–6 min (additional details below).

When the sequencing results obtained using Δ*Taq* DNA polymerase are compared with those obtained using the standard Sequenase Version 2.0 system under otherwise identical conditions (i.e., same levels of primer and template), the Δ*Taq* DNA polymerase sequencing system compares well with T7 DNA polymerase (Figure 1). Like the Sequenase system, background bands are absent (Figure 1). Labeling of DNA bands in the higher molecular weight range, 150

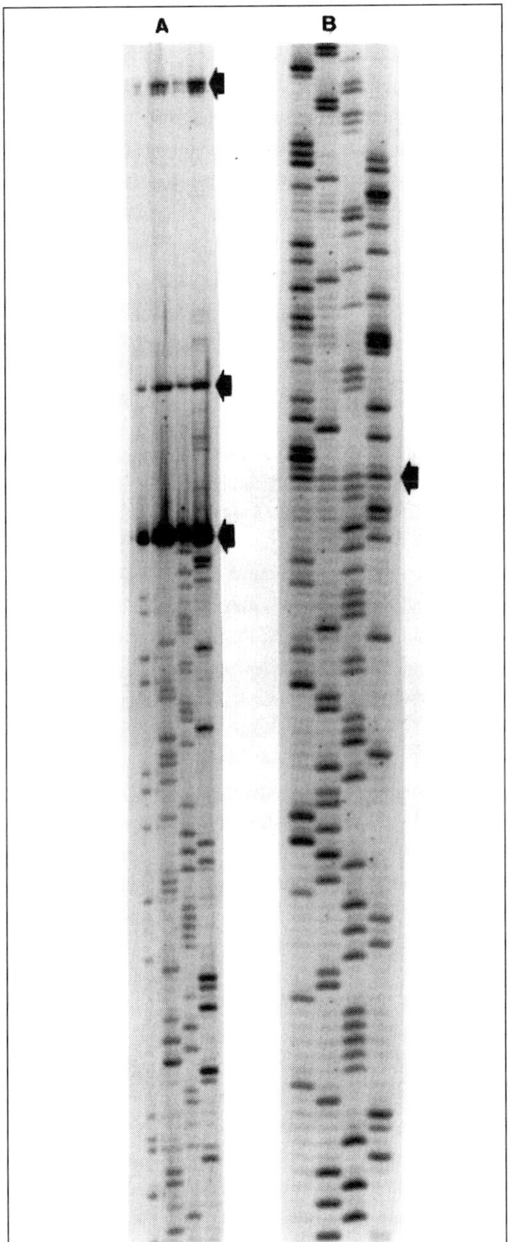

Figure 2. DNA sequence ladder generated by using either Sequenase Version 2.0 or TAQuence Version 2.0. DNA sequencing reactions were performed by using either standard Sequenase Version 2.0 protocol (A) or TAQuence Version 2.0 protocol (B) described in the text. Samples were electrophoresed at 2000 V. In each case, gel loading order from left to right was GATC.

to 200 bases and above, is superior, being more intense. This difference in intensity of labeling may be due to the fact that, with Δ*Taq* DNA polymerase, the primer and template are utilized more efficiently under incubation conditions. A certain degree of annealing may continue to take place under the labeling conditions, whereas, with the Sequenase system, only the primer already annealed to the template at the time of the labeling step is utilized by DNA polymerase, in spite of the processivity of the T7 DNA polymerase. We propose, therefore, that by modulating the initial labeling conditions according to the length of the primer (lower temperature for short primers [37°C] to higher temperature [55° to 60°C] for primer length of 17 plus nucleotides), sequencing data between 400 and 600 nucleotides can be readily obtained with Δ*Taq* DNA polymerase with routine equipment and reaction conditions. Staggered loading of the gel, at 2–3-h intervals, is necessary (9). Because of limited Δ*Taq* DNA polymerase activity at low temperature (37°C), short primers that are annealed to a template DNA are likely to be stabilized because of the extension of the primer by 5–10 nucleotides. Therefore, several minutes of initial incubation (2–5 min) at 37°C during the labeling step can be followed by incubation at 45°C and above. When the sequencing data (10–35 nucleotides) closer to the primer (from experiment in Figure 1) are compared between the two DNA polymerase systems, the Sequenase system appears to be superior (as seen from the intensity of bands on the autoradiograph) to the Δ*Taq* DNA polymerase system (results not shown).

We point out that Δ*Taq* DNA polymerase has an additional advantage, especially over the Sequenase system: It provides for relief of band compression that is introduced by DNA polymerase stop(s) because of the severity of the secondary structure of the DNA template during DNA sequencing. The standard band compression produces ambiguous and unreadable band patterns in polyacrylamide sequencing gels (3,5,8). This form of band compression is caused by the ability of G and C residues to form secondary structures in chain termination reaction products that are not fully denatured. These stable intrastrand structures can be readily prevented or eliminated by substituting dGTP with an analog, either 7-deaza-dGTP or dITP, that forms a weaker secondary structure (3,5,8). During sequencing of the single-stranded DNA bacteriophage MAC-1 from *Bdellovibrio bacteriovorus* (10), which involved walking along the single-stranded DNA by primer extension using the Sequenase Version 2.0 system, the product of the sequencing reaction upon autoradiography resulted in intensely thick DNA bands that were present in all four lanes (Figure 2A). Above these thick bands, there was little discernible sequence information. Initial interpretation of these bands as the result of standard band compression was not supported by the results of the sequencing reactions in which either 7-deaza-dGTP or dITP was substituted for dGTP (results not shown; they were identical to those shown in Figure 2A). These negative results raised the possibility that this problem may be due to the formation of stable structure(s) in the template. Sequenase is unable to continue beyond these structures. A limited number of T7 DNA polymerase molecules that manage to continue beyond the first full stop encounter, once again, similar structure(s), again resulting in bands across all four lanes (full stops, indicated by arrows in Figure 2A).

A sequencing method, using Δ*Taq* DNA polymerase, has been developed to relieve these strong stops, which are due to the severity of the template DNA secondary structure. The basic protocol of this method is similar to that described for DNA sequencing, except that the labeling step (step 2) was carried out between 45°C and 60°C. The best results were obtained at 60°C (Figure 2B). The sequence above the compression (full stop) can be read clearly. Some uncertainty of one or two residues remains at the point of compression (indicated by an arrow to the right of Figure 2B); i.e., bands across all four lanes. Any uncertainty at this point was resolved by sequencing from the complementary strand of a double-stranded DNA template (results not shown). We point out that even when the labeling step was carried out at 45°C, some relief from full-stop compression was observed, suggesting that a certain degree of "breathing" in the secondary structure allows some polymerase molecules to thread through. These structures melt more completely at higher temperatures (60°C). Based on these results, we suggest that these DNA polymerase full stop(s) may be lessened by varying the labeling temperature with the needs and severity of template secondary structure(s) or structures unique to a particular template.

Finally, it should be noted that the Δ*Taq* DNA polymerase and the T7 polymerase systems complement each other. Both provide sequence data that lack background bands. Basic steps in the sequencing reaction protocol are similar; in fact, the same dGTP-labeling deoxynucleoside triphosphate mixture (1:5 diluted) can be used with either Δ*Taq* DNA polymerase or Sequenase Version 2.0 from the Sequenase Version 2.0 kit. The Δ*Taq* DNA polymerase provides additional flexibility (or versatility) because the labeling temperature can be varied from 37°C to 60°C. Templates that pose special sequencing problems due to unique structural feature(s) or strong secondary structure(s) can be sequenced with Δ*Taq* DNA polymerase by increasing the incubation temperature during the labeling step (which is not possible with the Sequenase system). There is no reason why one cannot go to temperatures higher than even 60°C (such as 60°–70°C), especially with a longer primer. I have not tried it. In any case, the Δ*Taq* DNA polymerase sequencing system provides a useful alternative to the several sequencing systems available to investigators.

REFERENCES

1. **Barnes, W.M.** 1993. TAQuence Version 2.0: *Taq* DNA polymerase sequencing. United States Biochemical, Cleveland, OH.
2. **Barnes, W.M., M. Bevan and P.H. Son.** 1983. Kilo-sequencing: creation of an ordered nest of asymmetric deletions across a large target sequence carried by phage M13. Methods Enzymol. *101*:98-122.
3. **Barr, P.J., R.M. Thayer, P. Laybourn, R.C. Najarian, F. Seela and D.R. Tolan.** 1986. 7-Deaza-2′-deoxyguanosine-5′-triphosphate: enhanced resolution of M13 dideoxy sequencing. BioTechniques *4*:428-432.
4. **Chen, E.J. and P.H. Seeburg.** 1985. Supercoil sequencing: a fast and simple method for sequencing plasmid DNA. DNA *4*:165-170.
5. **Gough, J.A. and N.E. Murray.** 1983. Sequence diversity among related genes for recognition of specific targets in DNA molecules. J. Mol. Biol. *166*:1-19.
6. **Hattori, M. and Y. Sakaki.** 1986. Dideoxy-sequencing method using denatured plasmid templates. Anal. Biochem. *152*:232-238.
7. **Lee, J.-S.** 1991. Alternative dideoxy sequencing of double-stranded DNA by cyclic reactions using Δ*Taq* polymerase. DNA Cell Biol. *10*:67-73.
8. **Mizusawa, S., S. Nishimura and F. Seela.** 1986. Improvement of dideoxy chain termination method of DNA sequencing by use of deoxy-7-deazaguanosine triphosphate in place of dGTP. Nucleic Acids Res. *14*:1319-1324.
9. **Parkison, C. and S-Y. Cheng.** 1989. A convenient method to increase the number of readable bases in DNA sequencing. BioTechniques *7*:828-829.
10. **Roberts, R.C., M.A. Keefer and R.S. Ranu.** 1987. Characterization of *Bdellovibrio bacteriovorus* bacteriophage MAC-1. J. Gen. Microbiol. *133*:3065-3070.
11. **Sambrook, J., E.F. Fritsch and T. Maniatis.** 1989. Molecular Cloning: A Laboratory Manual. Cold Spring Harbor Laboratory Press, Cold Spring Harbor, NY.
12. **Sanger, F., S. Miklen and A.R. Coulson.** 1977. DNA sequencing with chain terminating inhibitors. Proc. Natl. Acad. Sci. USA *74*:5463-5467.
13. **Tabor, S. and C.C. Richardson.** 1989. Effect of manganese ions on the incorporation of dideoxynucleotides by bacteriophage T_7 DNA polymerase and *Escherichia coli* DNA polymerase. Proc. Natl. Acad. Sci. USA *86*:4076-4080.

Update to:

DNA Sequencing with Δ*Taq* DNA Polymerase

Rajinder S. Ranu

Department of Bioagricultural Sciences and Pest Management, Colorado State University, Fort Collins, CO, USA

DNA Sequencing and PCR Amplification with Δ*Taq* DNA Polymerase

In addition to the standard two-step DNA sequencing procedure described in our original paper (5), Δ*Taq*® DNA polymerase system has been further adopted for direct cycle sequencing (3). In this procedure, a prelabeling step is not required. The procedure can be used for sequencing of polymerase chain reaction (PCR)-amplified DNA, either from an asymmetrically amplified template (by PCR), double-stranded DNA template, PCR-amplified DNA cloned into a plasmid vector or from a single-stranded DNA template. Primer-to-template ratio and number of cycles necessary for best sequence data have been defined (3).

A model showing the secondary structures of a single-stranded DNA template that cause polymerase full stops has been developed (6). These secondary structures can be generated not only by a single-stranded DNA template but also from a denatured double-stranded DNA template (6). Such sequencing artifacts generated due to polymerase strong full stops (which result in intensely thick DNA bands across all four lanes in the DNA sequencing gel) are readily relieved with the Δ*Taq* DNA polymerase-based sequencing methods (3,5,6).

The utility of Δ*Taq* DNA in PCR (4) has also been investigated. Results show that, compared with *Taq* DNA polymerase, Δ*Taq* DNA Polymerase provides superior amplification of target DNA. Details are provided below.

REAGENTS

- 10× PCR buffer: 100 mM Tris-HCl (pH 8.3 at 20°C), 15 mM $MgCl_2$ and 500 mM KCl.
- Long-range (for 1000 bp and above) PCR amplification 10× buffer: 400 mM Tris-HCl (pH 9.1 at 25°C), 15 mM $MgCl_2$ and 500 mM KCl.
- dNTPs solution (pH 7.0): containing 2 mM each of dATP, dCTP, dGTP and dTTP.

- Primers: 100 μM each, prepared in TE buffer [Tris-HCl (pH 8.0) 10 mM and 1 mM EDTA].
- DNA molecular weight markers (Life Technologies, Gaithersburg, MD, USA).
- $MgCl_2$: 20 mM.
- Δ*Taq* DNA polymerase (version 2.0) 32 U/mL and polymerase dilution buffer; 10 mM Tris-HCl (pH 8.1), 1 mM β-mercaptoethanol, 0.5% Tween® 20 and 0.5% Nonidet® P-40 (Amersham Life Science, Arlington Heights, OH, USA).
- Mineral oil: (Sigma Chemical, St. Louis, MO, USA; M5904) autoclaved and then heated at 80°C for 2 h (to remove H_2O of condensation).

PROCEDURE

1. For each 50 μL PCR mixture, add the following components to a 0.5 mL thin-walled microcentrifuge tube:

*10× PCR buffer	5 μL
$MgCl_2$ 20 mM	3.75 μL
dNTPs 2 mM	5 μL
Primer #1 100 μM	0.5 μL
Primer #2 100 μM	0.5 μL
Target DNA	5-50 ng
Δ*Taq* DNA polymerase 32 U/μL	0.8 μL
H_2O to make the final volume	50 μL

   *For amplification of a target sequence larger than 1000 bp to about 6 kbp, the regular 10× PCR buffer should be substituted with the long-range 10× PCR buffer.
2. Samples are centrifuged briefly in a microcentrifuge (10 000 rpm for 15 s).
3. The sample is overlayed with a few drops of mineral oil (sufficient to cover the aqueous phase of meniscus). The tubes are placed in a thermocycler and the PCR program is initiated. Details are described in the legend for Figure 1.
4. At the end of the amplification program, an aliquot (5–10 μL) is removed from the reaction mixture and applied to an agarose gel (1%). Samples are subjected to electrophoresis and DNA in the gel visualized after staining with ethidium bromide (0.5 μg/mL).

NOTES

1. It is always advisable to include two control reaction mixtures; one that lacks target DNA and the other without the enzyme.

2. Cycling parameters, especially the annealing temperature, will vary with the length of primers and their melting temperature (Tm).
3. For long-range PCR it is desirable for the annealing temperature to be close to the Tm of the primer and the duration of annealing time be kept rather brief (10–15 s). It appears to improve the quality of the PCR product without generating artifacts.
4. It may be necessary to titrate the Mg^{++} concentration with some templates to determine best conditions for amplification (see results in Figure 1).

RESULTS AND DISCUSSION

Results of amplification of target DNA with primer #1 and #2, which borders a sequence of 336 nucleotides in plasmid ptACC2 (7) containing a 1.5-kbp insert, show that Δ*Taq* DNA polymerase was effective in amplification in the presence of 2–4.5 mM Mg^{++} (experimental details are described in

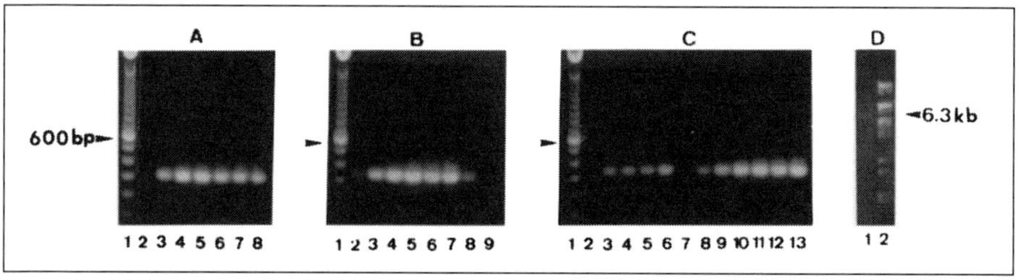

Figure 1. PCR amplification of DNA with Δ*Taq* DNA polymerase. Standard reaction mixture (100 µL) for PCR amplification contained 10 mM Tris-HCl (pH 8.3), 50 mM KCl, 3 mM $MgCl_2$, 1 µM of each of primers (#1: 5′-GGTGGAAAGCATACGATACGATCCTTTC-3′ and #2: 5′-CCAGGATCAGCCAAA-CAAAGTATGAGAGTCTC-3′), 50 ng of template DNA (ptACC2), 200 µM each of the four dNTPs and 5 units of *Taq* DNA polymerase or 50 units of Δ*Taq* DNA polymerase. In Panel A, the Mg^{++} concentration was varied; lane 2, 1.5 mM Mg^{++}; lane 3, 2 mM Mg^{++}; lane 4, 2.5 mM Mg^{++}; lane 5, 3 mM Mg^{++}; lane 6, 3.5 mM Mg^{++}; lane 7, 4 mM Mg^{++}; lane 8, 4.5 mM Mg^{++}. In Panel B, the KCl concentration in reaction mixture was changed but the Mg^{++} concentration was maintained at 3 mM; lane 2, no KCl; lane 3, 10 mM KCl; lane 4, 20 mM KCl; lane 5, 30 mM KCl; lane 6, 40 mM KCl; lane 7, 50 mM KCl; lane 8, 60 mM KCl; lane 9, 70 mM KCl. In Panel C, the enzyme concentration in reaction mixture was varied: Lanes 2–6, samples contained *Taq* DNA polymerase; lane 2, 1 unit; lane 3, 2 units; lane 4, 3 units; lane 5, 4 units; lane 6, 5 units. The samples in lanes 7–13 contained Δ*Taq* DNA polymerase; lane 7, 5 units; lane 8, 10 units; lane 9, 20 units; lane 10, 30 units; lane 11, 40 units; lane 12, 50 units; lane 13, 60 units. The reaction samples were overlaid with mineral oil. After initial denaturation step at 94°C for 5 min, samples were subjected to 2 cycles in which the conditions were: 1 min at 94°C for denaturation, 2 min at 60°C for annealing, and 1 min at 72°C for extension, followed by 30 cycles at 94°C for 30 s, 60°C for 1 min and 72°C for 45 s. The final cycle was at 72°C for 5 min. Reaction conditions for experiment in Panel D were: Tris-HCl (pH 9.1), 40 mM; $MgCl_2$, 3.5 mM; 200 µM each of four dNTPs; 1 µM of each of the primers (#1: 5′-CGAAATTAACCCTCACTAAAGGGAAC-3′ and #2: 5′-CTTGGCGTAATCAAGGTCATAGCTC), 50 units of Δ*Taq* DNA polymerase and 10 ng of template DNA (pcPH 41; size 6.2 kb). Samples were subjected to an initial denaturation at 94°C for 1 min, followed by 35 cycles at 94°C for 10 s and annealing/extension at 72°C for 3 min. The final cycle was at 72°C for 10 min. In Panel A to C, lane 1 contained 0.8 µg of 100-bp ladder (Life Technologies) and Panel D, lane 2 contained λ DNA digested with *Bst*EII as DNA standards. In each case, 10 µL of PCR mixture was subjected to electrophoresis in a 1.5% (1% in Panel D) agarose gel. Ethidium bromide-stained amplified DNA (336 nucleotides in length) in gel was located under UV light.

Figure 1 legend). Best results are observed at 3 mM Mg^{++} (Figure 1; panel A, lane 5), and KCl from 20–50 mM had no detectable effect on amplification (Figure 1, panel B). In the presence of low levels of Δ*Taq* DNA polymerase (5 units), there was little amplification of target DNA (Figure 1; panel 3, lane 7); as the amount of enzyme in the reaction mixture was increased, there was a dramatic increase in the target DNA amplification (Figure 1; panel C, lane 7–13). When the degree of amplification (based on intensity of ethidium bromide staining) is compared with standard *Taq* DNA polymerase, Δ*Taq* DNA polymerase provided three to five times better amplification of target DNA (Figure 1; panel C, lanes 2–6 vs. lanes 7–13). With a different set of primers, we have also amplified the full length of the plasmid along with the insert, which makes it about 6.2 kbp (Figure 1; panel D). A different set of buffer conditions (long-range buffer) are required for this amplification.

Δ*Taq* DNA polymerase provides superior PCR amplification to *Taq* DNA polymerase, albeit at somewhat higher enzyme levels. It is important to emphasize, however, that in our studies, in instances where standard *Taq* DNA polymerase failed to amplify target DNA in spite of the near-optimal annealing conditions (of the primer to the template), Δ*Taq* DNA polymerase was remarkably successful. This difference may be directly related to the lack of 5′→3′ exonuclease activity in Δ*Taq* DNA polymerase. With standard *Taq* DNA polymerase 5′→3′ exonuclease activity may shorten the annealed primers, thus reducing their stability, and may be responsible for the failed amplification. Since in most PCR-based studies, the important consideration is the amplification of target DNA rather than the amount of amplified DNA, Δ*Taq* DNA polymerase offers a most desirable alternative. It has the potential to save considerable investment in time otherwise necessary in initial trials.

We note that the Δ*Taq* DNA polymerase, in spite of the deletion of 235 amino acids from the amino terminus (1), still retains the template independent extendase activity (2) that results in a 3′-end with an A terminus (results not shown). This finding is particularly useful and can be exploited for subsequent cloning of the PCR product in a TA cloning vector such as pCRII (Invitrogen, Carlsbad, CA, USA).

ACKNOWLEDGMENTS

This work is supported by a grant from the Colorado Agriculture Experiment Station. We thank Dr. A. Theologis, ARS, Albany, CA, USA for providing the ptACC2 plasmid, and Peggy Godin for secretarial help.

REFERENCES

1. **Barnes, W.M.** 1992. The fidelity of *Taq* DNA polymerase catalyzing PCR is improved by an N-termi-

nal deletion. Gene *112*:29-35.
2.**Clark, J.M.** 1988. Novel non-templated nucleotide addition catalyzed by prokaryotic and eukaryotic DNA polymerase. Nucleic Acids Res. *16*:9677-9686.
3.**Fan, J. and R.S. Ranu.** DNA cycle sequencing with Δ*Taq* DNA polymerase. DNA Seq. (In press).
4.**Mullis, K.B. and F.A. Faloona.** 1987. Specific synthesis of DNA *in vitro* via a polymerase-catalyzed chain reaction. Methods Enzymol. *155*:335-350.
5.**Ranu, R.S.** 1995. DNA sequencing with Δ*Ta*q DNA polymerase. BioTechniques *18*:390-393.
6.**Ranu, R.S.** 1996. Δ*Taq* DNA polymerase: a new and versatile enzyme for DNA sequencing, p. 117-121. *In* J.M. Burke (Ed.), PCR: Essential Techniques. John Wiley & Sons, New York.
7.**Rottman, W.H., G.F. Peter, P.W. Oeller, J.A. Keller, F.F. Shen, B.P. Nagy, L.P. Taylor, A.D. Campbell and A. Theologis.** 1991. 1-Aminocyclopropane-1-carboxylate synthase in tomato is encoded by a multigene family whose transcription is induced during fruit and floral senescence. J. Mol. Biol. *222*:937-971.

Efficient Production of Single-Stranded DNA as Long as 2 kb for Sequencing of PCR-Amplified DNA

B. Kaltenboeck, J.W. Spatafora[1], X. Zhang, K.G. Kousoulas, M. Blackwell[1] and J. Storz

Department of Veterinary Microbiology and Parasitology, School of Veterinary Medicine, and [1]Department of Botany, Louisiana State University, Baton Rouge, LA, USA

ABSTRACT

A modification of the asymmetric PCR method is described, which reliably facilitates sequencing of PCR-amplified DNA. This procedure produces single-stranded DNA fragments as long as two kilobases that are suitable for dideoxy DNA sequencing. First, a PCR for double-stranded DNA is performed under optimal conditions (double-stranded PCR). Then, a 5–10-µL fraction of the double-stranded PCR and a single primer are used to generate single-stranded DNA in a separate PCR (single-stranded PCR). The concentration of the single primer is approximately 0.4 µM. Usually 15 to 25 cycles of single-stranded PCR are optimal to produce single-stranded DNA for four to eight sequencing reactions. The single-stranded DNA is purified by centrifugal ultrafiltration and used directly in dideoxy sequencing. This method was employed to produce high-quality single-stranded DNA templates from a variety of organisms for efficient DNA sequencing of PCR-amplified DNA.

INTRODUCTION

The technique of the PCR has had a profound impact on molecular biology. Sequencing PCR-amplified DNA fragments without cloning them into suitable vectors has proven more difficult than anticipated. Several approaches for direct sequencing of PCR products are feasible (4). Rapid reassociation of the linear DNA template strands inherently complicates direct sequencing of double-stranded (dsDNA) fragments (2,3,9,15). A linear PCR has been employed to generate directly labeled DNA fragments suitable for resolution in polyacrylamide-urea sequencing gels (12,14).

(Reprinted from BioTechniques 12:164-171, 1992)

Numerous DNA sequencing protocols are well established that use single-stranded DNA (ssDNA) as template and dideoxynucleoside triphosphates (ddNTPs) for chain termination. Several methods to generate ssDNA sequencing templates from dsPCR products are described. Affinity purification (7,11) and exonuclease digestion (6) employ modified primers to remove one DNA strand from dsDNA. Other methods modify the PCR such that ssDNA of a chosen strand is produced (1,5,10,13). While these asymmetric PCRs are simple in principle, adaptation to new primers and templates is often difficult and time-consuming, and the results are inconsistent.

We describe here a two-step PCR method that reliably produces ssDNA as long as 2 kb. First, a PCR for double-stranded DNA (dsPCR) is performed under optimal conditions. The ssDNA is produced by a second PCR, termed ssPCR, using a single primer and a small fraction of the previous dsPCR. The number of ssPCR cycles is the most critical parameter for production of ss-DNA. The ssPCR products are separated from primers and deoxynucleoside triphosphates (dNTPs) by spin dialysis and directly used in high-quality dideoxy DNA sequencing. This method is easily adaptable to new primers and DNA templates.

MATERIALS AND METHODS

DNA fragments from the major outer membrane protein (MOMP) gene of strains of *Chlamydia*, the fungal nuclear-encoded, small-subunit ribosomal (r) DNA (ssu rDNA) and the reversely transcribed S-glycoprotein gene of strains of bovine coronavirus were amplified by PCR. Primers for fungal ssu rDNA were described (16) and custom synthesized; all other primers were designed and synthesized in our laboratory on a Gene Assemblers® Plus DNA Synthesizer (Pharmacia LKB Biotechnology, Piscataway, NJ, USA). The buffer used in ds and ssPCRs of the chlamydial MOMP gene and the coronavirus S-glycoprotein gene was 20 mM Tris-HCl, pH 8.3, 25 mM KCl, 2 mM $MgCl_2$, 0.01% bovine serum albumin and 0.05% Tween 20. The buffer for the fungal ssu rDNA was 10 mM Tris-HCl, pH 8.3, 50 mM KCl, 2 mM $MgCl_2$, 0.01% gelatin, 0.01% Tween 20 and 0.01% NP-40®. Further data for optimal (8) dsPCRs are given in Table 1.

Volume amounts of dsPCRs were varied as input DNA for the production of ssDNA. Reaction parameters of ssPCRs that differed from dsPCRs are given in Table 2. Optimal number of cycles in ssPCR was determined by removing 10-µL aliquots from the reaction after various numbers of cycles. These aliquots were electrophoresed in 1.5% agarose gels (1:1 ratio of DNA grade agarose to NuSieve® agarose; FMC BioProducts, Rockland, ME, USA). PCR products were visualized by staining with ethidium bromide. Subsequent ssPCRs were performed using the derived optimum conditions.

After agarose gel electrophoresis, the remaining 90-µL volume of ssPCRs was diluted in 1.8 mL distilled water and subjected to three rounds of cen-

Table 1. PCR Parameters for dsDNA Amplification

	Chlamydial MOMP gene		Fungal 16S-like rRNA gene		Bovine coronavirus S-glycoprotein gene	
DNA source	**Infected Tissue**		**Genomic**		**Reversely transcribed purified viral RNA**	
Approx. fragment size (varies among strains)	1200 bp		1780 bp		2000 bp	
Primers	-20CHOMP	CHOMP371	NS1	NS8	5′2187 3′ *Eco*RI	3′ *Bam*HI
Length	20-mer	33-mer	19-mer	20-mer	22-mer[b]	22-mer[b]
# of inosines	—	5	—	—	—	—
Degeneracy (without inosines)	8-fold	8-fold	—	—	—	—
Calculated T_m[a]	50°–54°C	80°–86°C	54°C	64°C	66°C[c]	60°C[c]
Concentration	0.2 µM	0.2 µM	0.5 µM	0.5 µM	0.2 µM	0.2 µM
Thermocycler	Eppendorf North America[d]		Perkin-Elmer Cetus[e]		Eppendorf North America[d]	
Reaction Volume	100 µL		100 µL		100 µL	
[each dNTP]	200 µM		100 µM		200 µM	
Taq polymerase	2.5 units[f]		2.5 units[g]		2.5 units[f]	
Oil overlay	yes		no		yes	
Melting	96°C, 1 s		93°C, 1 min		96°C, 1 s	
Annealing	42°C, 1 min		53°C, 1 min		59°C, 30 s	
Slow ramping	yes		no		no	
Extension	72°C, 1 min		72°C, 2 min		72°C, 2 min	
Number of cycles	30		35		30	
Final extension	72°C, 7 min		72°C, 7 min		72°C, 7 min	

[a]$Tm = 4 \times \#(C+G) + 2 \times \#(A+T) + 0 \times \#I$, actual values for oligonucleotides >20-mers are lower than calculated.
[b]Restriction sites included in the 5′ end of the primer are not considered.
[c]T_m is calculated excluding the restriction sites.
[d]Eppendorf North America, Madison, WI, USA
[e]Perkin-Elmer Cetus, Norwalk, CT, USA
[f]GIBCO BRL/Life Technologies, Gaithersburg, MD, USA
[g]International Biotechnologies, New Haven, CT, USA

trifugal ultrafiltration (Centricon-100; Amicon, Danvers, MA, USA). Two milliliters of distilled water were added after the first and second round of filtration. If necessary, the retentate was concentrated to contain 12% to 30% of the ssDNA per 7 µL. These 7-µL aliquots served as the template for DNA sequencing. T7 DNA polymerase was used in the dideoxy chain termination method according to manufacturer's instructions (United States Biochemical,

Cleveland, OH, USA). Best results were obtained using 5–10 pmol of sequencing primer (limiting PCR primer or other primer complementary to internal sequence of the ssDNA template) per reaction, and dGTP, c^7dGTP or dITP labeling solutions were diluted 5-fold to 50-fold. ^{35}S-dATP-labeled reaction products were separated by 6% polyacrylamide-urea gel electrophoresis and visualized by autoradiography.

RESULTS AND DISCUSSION

Inconsistent and often unsatisfactory results produced in asymmetric PCRs (5,10,13) led us to consider possible improvements. We reasoned that the major problem in the asymmetric reaction is the high number of cycles required to generate sufficient dsDNA template necessary for an efficient linear increase of ssDNA fragments during the last cycles. Skewed primer concentrations adversely affect previously optimized parameters of dsPCRs (8) in an unpredictable fashion. Frequently, low yields of ssDNA and aberrant products are the result. Our hypothesis was that it should be possible to generate sufficient amounts of ssDNA by PCR using a single primer and a fraction of a dsPCR sample. In the ssPCR the dsPCR product functions directly as template for the production of the ssDNA. The ssPCR requires relatively few cycles. This approach uncouples the exponential accumulation of dsPCR products from the linear increase of ssPCR products resulting in an enhanced signal (ssDNA)-to-noise (aberrant products) ratio.

Initial results achieved by using 5–10 µL of dsPCR sample as input DNA in the ssPCR appeared to validate this hypothesis. We systematically varied optimal dsPCR parameters in an attempt to define general guidelines for op-

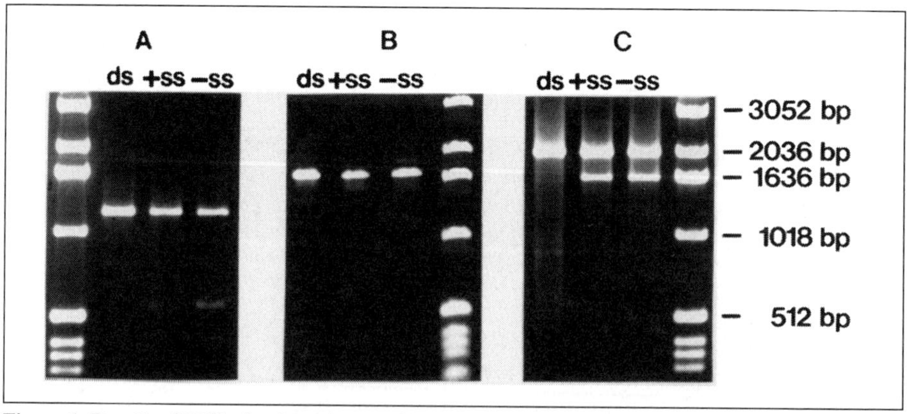

Figure 1. Results of PCRs for ds and sense (+) and antisense (-) ssDNA. Panel A shows the amplified MOMP gene fragment of C. psittaci strain L-14, Panel B the fragment of the 16-S-like rRNA gene of Ambroziozyma platypodis,and Panel C the fragment of the S-glycoprotein gene of bovine coronavirus strain L9. Ten-microliter aliquots of PCRs were electrophoresed (1.5% agarose) in 0.5 × Tris-Phosphate-EDTA BUFFER (Panels A and C) or 1 × Tris-Borate-EDTA buffer (Panel B), and DNA fragments were visualized by ethidium bromide staining. DNA size marker is the 1-kb DNA Ladder (GIBCO BRL).

timization of the ssPCR. The parameters investigated were input of dsPCR (µL), number of cycles in ssPCR, single primer and *Taq* DNA polymerase concentrations, annealing temperature and time, extension time, and ramping time between annealing and extension. We found that the volume-amount of input dsPCR sample and the number of ssPCR cycles were the most important factors. We produced ssDNA with all primers used to amplify three target DNA fragments and successfully employed it for sequencing. Results obtained with optimized ssPCRs (Table 2) are given in Figure 1.

Primer -20CHOMP, used for +ssDNA synthesis of the chlamydial MOMP gene fragment, was less effective than CHOMP371, which is used for -ssDNA synthesis (Panel A of Figure 1). This low efficiency, presumably due to the low T_m, was partially compensated for by using 30 cycles for production of +ssDNA as compared with 18 cycles for -ssDNA. Although this reaction produced sufficient +ssDNA for sequencing, more +ssDNA could be produced by increasing the number of cycles and lowering the annealing temperature. The fungal ssu rDNA fragments demonstrated differential mobility of +ss and -ssDNAs presumably due to their secondary structure (Panel B, Figure 1). Reduction of the dsPCR annealing temperature of 53°C to 50°C in the +ssPCR using primer NS1 (T_m = 54°C) compensated for the low efficiency. Both +ss and -ssDNAs yielded excellent results in DNA sequencing despite the faint staining of the ssDNA bands with ethidium bromide. The +ss and -ssDNA of the bovine coronavirus S-glycoprotein gene fragment migrated similarly (Panel C, Figure 1), and aberrant lower molecular weight fragments did not visibly influence sequencing results.

The influence of the critical parameters in ssPCR, volume-amount (µL) of dsPCR input and number of cycles, is emphasized in Figure 2. Ten microliters

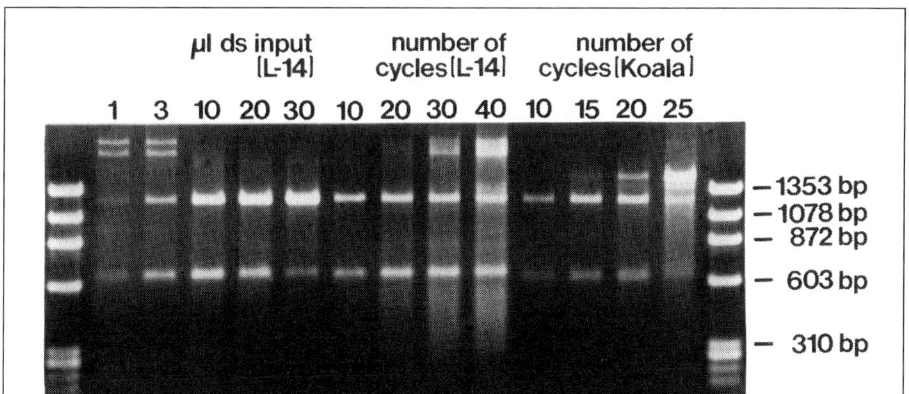

Figure 2. Determination of optimal dsPCR input (µL) and number of cycles in -ssPCRs. In Panels "µL ds input (L-14)" (1–30) and "number of cycles (L-14)" (10–40) the MOMP gene fragment of *C. psittaci* strain L-14, in Panel "Number of cycles (Koala)" the fragment of a Koala isolate of *C. psittaci* was amplified. DNA size markers is *Hae*III-digested ϕX174 RF DNA (GIBCO BRL).

Table 2. PCR Parameters for ssDNA Production (Only Parameters Different from dsDNA Amplification are Listed)

	Chlamydial MOMP Gene		Fungal ssu rDNA		Bovine Coronavirus S-Glycoprotein Gene	
Sense of Single-Stranded DNA	+ss (coding)	-ss (noncoding)	+ss	-ss	+ss	-ss
Primers	-20CHOMP	CHOMP371	NS1	NS8	5′ 2187 EcoRI	3′ BamHI
Concentration	0.4 μM	0.4 μM	0.4 μM	0.4 μM	0.4 μM	0.4 μM
[Each dNTP]	200 μM	200 μM	200 μM	200 μM	200 μM	200 μM
Annealing	42°C, 1 min	42°C, 1 min	50°C, 1 min	53°C, 1 min	50°C, 30 s	50°C, 30 s
Extension	72°C, 2 min	72°C, 2 min	72°C, 6 min	72°C, 6 min	72°C, 2 min	72°C, 2 min
Number of Cycles	30	18	20	20	25	25

of the dsPCR produced the best results [Figure 2, Panel "μL ds input (L-14)"]. In these reactions -ssDNA of the MOMP gene fragment of *C. psitttaci* strain L-14 was synthesized for 20 cycles using 0.4 μM primer concentration and 2 min extension at 72°C. The quantity of ds-DNA template fragment of approximately 1200 bp was proportional to input of dsPCR (μL). However, the -ssDNA fragment with an apparent mobility of approximately 650 bp peaked at 10 μL dsPCR sample and displayed reduced intensity when low and high amounts of input were used. This result suggests that the small amount of limiting primer (-20CHOMP) carried over from the dsPCR resulted in the production of some dsDNA during the ssPCR, thereby increasing the pool of dsDNA that functioned as template in ssPCR. A very small amount of dsPCR was not sufficient to allow efficient dsDNA and consecutive ssDNA synthesis. When a high amount of dsPCR was used, overproduction of dsDNA inhibited efficient ssDNA production.

A general conclusion can be drawn from these observations: the volume-amount (μL) of dsPCR required in an ssPCR is inversely proportional to the primer concentration in the dsPCR. The fungal and chlamydial examples illustrate this point. Fungal dsPCR employed primers at 0.5 μM and required

5 μL of dsPCR sample as input in the ssPCR for optimum results. The chlamydial dsPCR employed primers at 0.2 μM requiring 10 μL input for optimum ssPCR. Using primers at 0.2 μM in the dsPCR, at 10 μL input of dsPCR a balance was achieved between quantity of input dsDNA fragment and quantity of limiting primer. Sufficient limiting primer was provided to produce dsDNA, yet ssDNA dominated over dsDNA production. This effect was observed reproducibly in the ssPCRs reported here as well as in others using different primers.

The influence of cycle number in ssPCR on the amount and quality of the ssDNA is demonstrated in Panel "number of cycles (L-14)" in Figure 2. Although the amount of ssDNA increased up to 30 cycles, aberrant low and high molecular weight products accumulated after 20 cycles. Furthermore, the intensity of the ethidium bromide-stained ssDNA fragment was reduced at 40 cycles as compared with 30 cycles. It was important for optimal sequencing results to use high-quality ssDNA templates and to avoid overproduction. Empirically, slightly less than 20 cycles in this titration were the best compromise between quantity and quality of ssDNA.

Another possible outcome of cycle titration in ssPCR is demonstrated in Panel "number of cycles (Koala)" in Figure 2. The dsPCR of isolates of *C. psittaci* from Koala consistently yielded an approximately 1550-bp aberrant fragment in addition to the specific MOMP gene fragment. Gel purified, dsPCR MOMP gene fragment (ca. 150 ng) was initially used as input for an ssPCR to alleviate this problem. Surprisingly, only the aberrant fragment of 1550 bp was produced and not the ssDNA. We hypothesized that internal complementary sequences within the Koala MOMP gene fragment might give rise to a partial head-to-tail dimer through in vitro recombination. This dimer would be exponentially amplified by a single primer, and the MOMP gene input fragment would be incorporated into the aberrant product. This hypothesis was supported by using other primers and by DNA sequence analysis of the Koala MOMP gene fragment, which identified two regions with a high degree of complementarity when aligned in opposite directions. To solve this problem, the number of cycles in a 100-μL ssPCR was optimized using approximately 40 ng dsDNA input Koala MOMP gene fragment, an additional 2 pmol of -20CHOMP to increase the pool of dsDNA and other parameters as described for the chlamydial MOMP gene fragment. Sufficient -ssDNA was produced using 15 and 20 cycles as shown in Panel "number of cycles (Koala)" in Figure 2. More PCR cycles led to the exponential accumulation of the aberrant 1550-bp fragment that dominated the PCR, and the ssDNA was lost.

The autoradiogram of a typical sequencing gel is shown in Figure 3. Typically, sequences can be determined between 30 and 400 nucleotides from the 3′ end of the primers with field gradient (wedge) 6% polyacrylamide-urea gel electrophoresis using a Bio-Rad sequencing electrophoresis apparatus (Bio-Rad, Richmond, CA, USA). As much as 30% of a 100-μL ssPCR is required

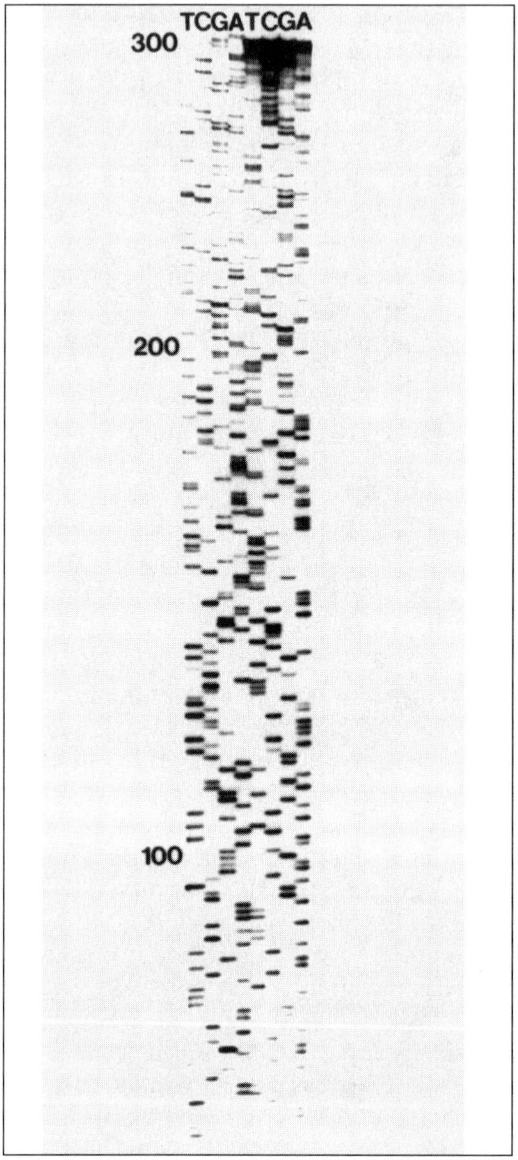

Figure 3. Partial nucleotide sequence of the MOMP gene fragment of *C. psittaci* strain L-14. Ten pmol of degenerate, inosine-containing primers, which are chlamydia genus-specific, were used per sequencing reaction. The primers in the right and left set of sequencing reactions are complementary to regions located about 300 and 570 nucleotides, respectively, downstream of the 5′ end of the ssDNA template. ^{35}S-dATP-labeled, dideoxy-terminated, extension products were resolved on a 6% polyacrylamide-urea field gradient sequencing gel. Autoradiographic images were obtained by exposing sequencing gels to Kodak X-OMAT AR film for 16 h.

per sequencing reaction for optimal reading of sequences close to the primer. This is especially true with large fragments.

These results prove that the precise volume-amount (µL) of dsPCR input and the determination of optimal number of cycles in ssPCR are critical in establishing ssDNA production for a given dsPCR. It is important to optimize the dsPCR before proceeding to ssPCR. More ssDNA was produced with increased single primer concentration. Prolonged extension time, high *Taq* DNA polymerase concentration, reduced annealing temperature and slow ramping between annealing and extension also increased ssDNA yield (data not shown). However, these relatively minor effects on the quantity of ssPCR product could be compensated for by adjusting the number of ssPCR cycles. In practice, reproducible results were obtained when the single primer was used at a concentration of about 0.4 µM and extension time was approximately twice the time used in dsPCR. Other reaction parameters probably need not be modified once they have been optimized for dsPCR.

The fact that ssDNA as long as 2 kb could be produced in surprisingly large quantity from dsDNA templates underscores the robustness and efficiency of the described method. Also, this method has been applied successfully for ssDNA production from dsPCR fragments amplified from genomic and organellar DNA of birds, gophers, green algae, diatoms and angiosperms (data not shown). This suggests a general applicability of the described method for ssDNA template production from dsPCR products.

ACKNOWLEDGMENT

This work was supported by Grants 86-CRSR-2-2871 from the United States Department of Agriculture and 89-34116-4675 from the Louisiana Education Quality Support Fund of the Board of Regents to J.S. and K.G.K., by Grants BSR-8918167 from the National Science Foundation and RD-A 137-99-5120 from the Louisiana Education Quality Support Fund of the Board of Regents to M.B. and by Grant AI 27886 from the National Institutes of Health to K.G.K.

REFERENCES

1. **Allard, M.W., D.L. Ellsworth and R.L. Honeycutt.** 1991. The production of single-stranded DNA suitable for sequencing using the polymerase chain reaction. BioTechniques *10*:24-26.
2. **Bachmann, B., W. Lueke and G. Hunsmann.** 1990. Improvement of PCR amplified DNA sequencing with the aid of detergents. Nucleic Acids Res. *18*:1309.
3. **Casanova, J.-L., C. Pannetier, C. Javlin and P. Kourilski.** 1990. Optimal conditions for directly sequencing double-stranded PCR products with Sequenase®. Nucleic Acids Res. *18*:4028.
4. **Gyllensten, U.B.** 1989. PCR and DNA sequencing. BioTechniques *7*:700-708.
5. **Gyllensten, U.B. and H.A. Erlich.** 1988. Generation of single-stranded DNA by the polymerase chain reaction and its application to direct sequencing of the HLA-DQA locus. Proc. Natl. Acad. Sci. USA *85*:7652-7656.
6. **Higuchi, R.G. and H. Ochman.** 1989. Production of single-stranded DNA templates by exonuclease digestion following the polymerase chain reaction. Nucleic Acids Res. *17*:5865.
7. **Hultman, T., S. Stahl, E. Hornes and M. Uhlen.** 1989. Direct solid phase sequencing of genomic and plasmid DNA using magnetic beads as solid support. Nucleic Acids Res. *17*:4937-4946.
8. **Innis, M.A. and D.H. Gelfand.** 1990. Optimization of PCRs, p. 3-12. *In* M.A. Innis, D.H. Gelfand, J.J. Sninsky and T.J. White (Eds.), PCR Protocols. A Guide to Methods and Applications. Academic Press, New York.
9. **Kusukawa, N., T. Vemori, K. Asada and I. Kato.** 1990. Rapid and reliable protocol for direct sequencing of material amplified by the polymerase chain reaction. BioTechniques *9*:66-72.
10. **McCabe, P.C.** 1990. Production of single-stranded DNA by asymmetric PCR, p. 76-83. *In* M.A. Innis, D.H. Gelfand, J.J. Sninsky and T.J. White (Eds.), PCR Protocols. A Guide to Methods and Applications. Academic Press, New York.
11. **Mitchell, L.G. and C.R. Merril.** 1989. Affinity generation of single-stranded DNA for dideoxy sequencing following the polymerase chain reaction. Anal. Biochem. *178*:239-242.
12. **Murray, V.** 1989. Improved double-stranded DNA sequencing using the linear polymerase chain reaction. Nucleic Acids Res. *17*:8889.
13. **Shyamala, V. and G. Ferro-Luzzi Ames.** 1989. Amplification of bacterial genomic DNA by the polymerase chain reaction and direct sequencing after asymmetric amplification: application to the study of periplasmic permeases. J. Bacteriol. *171*:1602-1608.
14. **Smith, D.P., E.M. Johnstone, S.P. Little and H.M. Hsiung.** 1990. Direct DNA sequencing of cDNA inserts from plaques using the linear polymerase chain reaction. BioTechniques *9*:48-54.
15. **Thein, S.L.** 1989. A simplified method of direct sequencing of PCR amplified DNA with Sequenase® T7 DNA polymerase. Comments, USB Corp., Cleveland, OH *16*:8.
16. **White, T.J., T. Bruns, S. Lee and J. Taylor.** 1990. Amplification and direct sequencing of fungal ribosomal RNA genes for phylogenetics, p. 315-324. *In* M.A. Innis, D.H. Gelfand, J.J. Sninsky and T.J. White (Eds.), PCR Protocols. A Guide to Methods and Applications. Academic Press, New York.

Update to:

Efficient Production of Single-Stranded DNA as Long as 2 kb for Sequencing of PCR-Amplified DNA

Bernhard Kaltenboeck
College of Veterinary Medicine, Auburn University, Auburn, AL, USA

Additional references, published after the original article, with relevant information are:

1. **Kaltenboeck, B. and K.G. Kousoulas.** 1996. Efficient PCR production of single-stranded DNA sequencing templates. Methods Mol. Biol. *65*:149-154.
This book chapter gives a step-by-step protocol for production of ssDNA by the approach described in the original article.

2. **Landweber, L.F. and M. Kreitman.** 1993. Producing single-stranded DNA in polymerase chain reaction for direct genomic sequencing. Methods Enzymol. *218*:17-26.
This book chapter describes the purification of the dsPCR product and subsequent production of ssDNA in a linear PCR using one primer and the purified dsPCR product. Hints are also given on the synthesis of ssDNA by the limiting primer method.

3. **Medori, R., H.-J. Tritschler and P. Gambetti.** 1992. Production of single-stranded DNA for sequencing: an alternative approach. BioTechniques *12*:347-349.
This paper describes PCR production and sequencing of ssDNA after purification of the dsDNA template.

4. **Nichols, R.C. and N. Raben.** 1994. Hints for direct sequencing of PCR-generated single-stranded DNA. BioTechniques *17*:412-414.
This paper uses a small aliquot of dsDNA PCR to generate ssDNA in a second PCR with a single primer. The method is similar to the one used in the original paper, but an excess of primers is used in the dsPCR, and consequently a smaller aliquot of the dsPCR is transferred to the ssPCR.

5. **Zhang, X. and K.G. Kousoulas.** 1993. Direct sequencing of PCR-amplified high GC DNA. BioTechniques *14*:376-377.
This paper describes modifications to the method used in the original paper, which improve the PCR production and sequencing of ssDNA with high GC content.

Direct Sequencing of PCR Products Using Unlabeled Primers

Shigeru Takagi, Minoru Kimura[1] and Motoya Katsuki[2]

NTT Basic Research Laboratories, Musashino-Shi, Tokyo; [1]Division of Molecular Life Science, School of Medicine, Tokai University, Bohseidai, Isehara; [2]The Institute of Medical Science, The University of Tokyo, Minato-ku, Tokyo, Japan

ABSTRACT

An improved protocol is described for using lambda exonuclease to directly sequence PCR products. It is important not to execute PCR cycles beyond the plateau of amplification. The asymmetric PCR and double-stranded DNA sequencing by a snap-cooling procedure were also performed using the same DNA samples and primers. The improved method was the most reliable and produced the best results.

INTROsDUCTION

Directly sequencing polymerase chain reaction (PCR) products without any additional cloning steps shortens processing time and minimizes potential errors due to imperfect PCR fidelity. Although many direct methods have been described, none are widely accepted because they do not always produce a clean sequence pattern.

We improved the method for producing single-stranded DNA from double-stranded PCR products using lambda exonuclease (2) in four ways. First, double-stranded DNAs with a 5′-phosphoryl terminus on only one strand are produced not only by PCR with the 5′-phosphoryl primer and the 5′-hydroxyl primer, but also by restriction endonuclease digestion of PCR products synthesized with a pair of 5′-hydroxyl primers attached by a restriction sequence. This is useful if amplified samples are to be cloned into other DNAs, and primers with restriction sites have already been synthesized. Second, PCR amplification conditions are optimized to produce clean sequence ladders. Third, the procedure excludes phenol extraction steps, which are the main causes of DNA loss. This greatly increases reliability and reduces the processing time. Fourth, the concentration of 5′-phosphoryl

A1	5'	TCCTGAGTTGCTCACTGTAC	3'
A2	5'	CACGGATCCTCCTGAGTTGCTCACTGTAC	3'
A3	5'	ACAGCAAGTACCATGGACCA	3'
B1	5'	TGCTTAGCACGCATCCTACT	3'
B2	5'	TTACAGAGGAGCCAGATCCA	3'

Figure 1. Primers used.

(Reprinted from BioTechniques 14:218-221, 1993)

primer is lowered to reduce the amount of T4 polynucleotide kinase used and eliminate the need for concentrating the primer by ethanol precipitation.

We also performed direct sequencing of double-stranded DNA by the typical snap-cooling procedure (1,3) and the typical asymmetric PCR procedure (4) using the same DNA samples and the same unlabeled primers. The improved lambda exonuclease method was the most reliable and produced the best results.

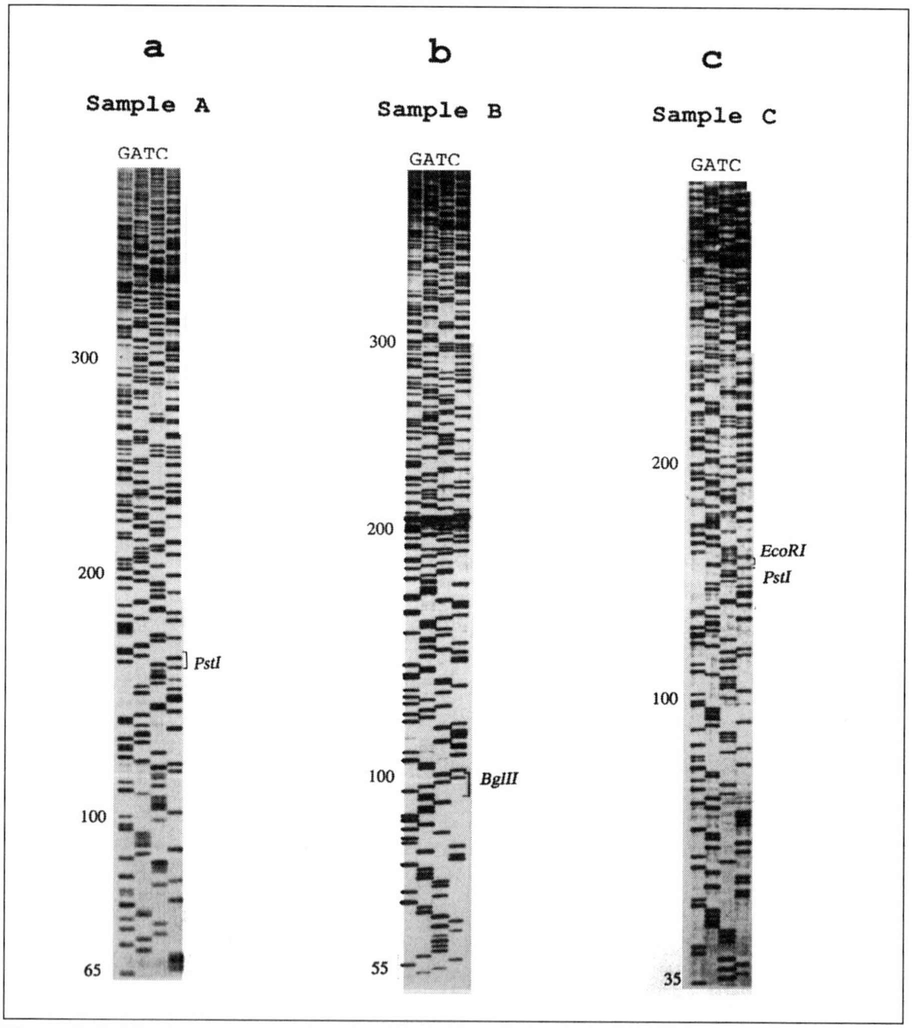

Figure 2. Sequence pattern examples produced by the improved lambda exonuclease digestion method. Samples A and B were amplified with 5′-phosphoryl primers (5′-phosphoryl A1 and A3 for A, 5′-phosphoryl B1 and B2 for B). Sample C was amplified with A2 with the *Bam*HI sequence attached and A3. Amplification conditions were 20 cycles for sample A, 15 cycles for sample B and 20 cycles for sample C. Each amplified sample was digested to single-stranded DNA by the method described in the text and sequenced with 10 pmol of unlabeled A1 for A, B1 for B and A2 for C using ^{35}S-dCTP and Sequenase.

MATERIALS AND METHODS

DNA Sample

The vicinity of the mouse myelin basic protein (MBP) gene promoter region (5) is used as sample. Three samples (A: *Pst*I digested, B: *Bgl*II digested, C: *Pst*I-*Eco*RI digested followed by formation of blunt-ends with T4 DNA polymerase) were prepared from mouse genomic DNA, amplified by inverse PCR, analyzed on 0.7% agarose gel, recovered from the gel slices, and diluted to a concentration of 1 ng/µL. Fragment lengths were approximately 1.4 kb, 2.3 kb and 1.4 kb, respectively. Five oligonucleotide primers were used (A1, A2 and A3 for samples A and C; B1 and B2 for sample B).

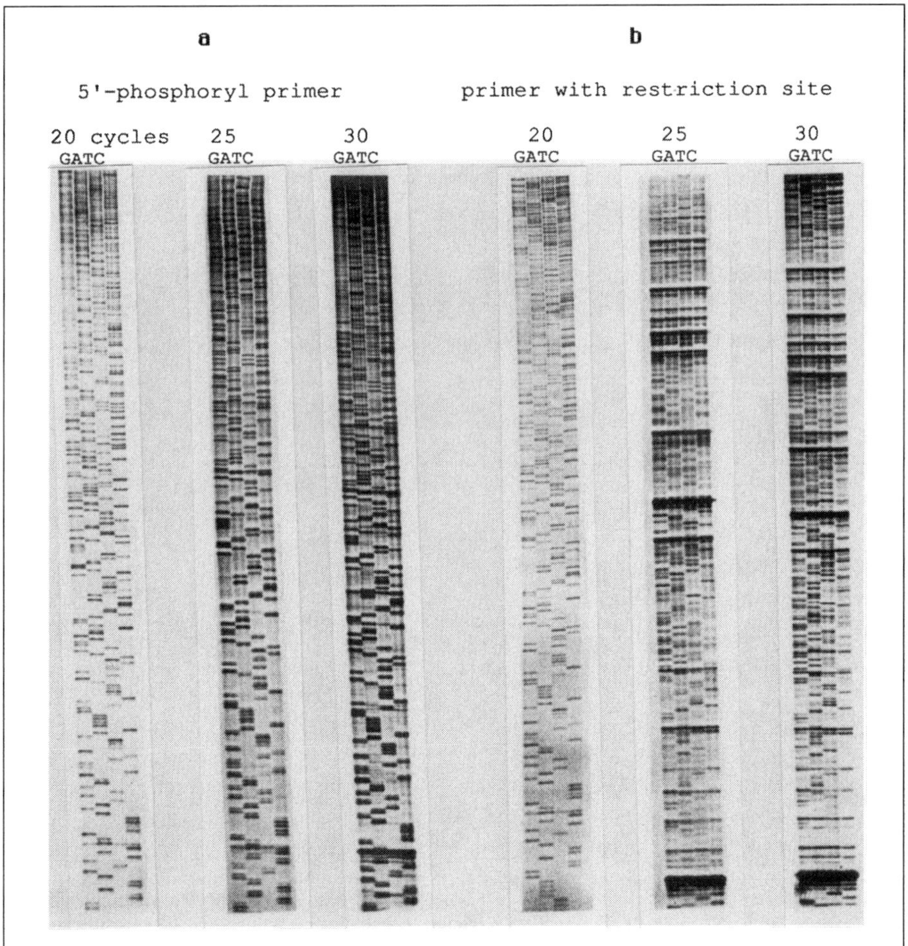

Figure 3. Effects of number of PCR cycles. Panel a: Sample A was amplified with 5′-phosphoryl A1 and A3 for 20, 25 and 30 cycles and sequenced. Panel b: Sample A was amplified with A2 and A3 for 20, 25 and 30 cycles and sequenced.

A2 had a *Bam*HI site and a "CAC" clamp sequence attached at the A1 5′-end (Figure 1).

Method

The 5′ terminus of primer A1 and B1 was phosphorylated for 1 h at 37°C with 1 U of T4 polynucleotide kinase (Toyobo, Tokyo, Japan) per 2 pmol of primer in kinase buffer [50 mM Tris-HCl, pH 7.6, 5 mM $MgCl_2$, 10 mM dithiothreitol (DTT), 1 mM spermidine, 0.1 mM EDTA, 2 mM ATP]. The phosphorylated mixture was diluted with TE (10 mM Tris-HCl, pH 8.0, 1 mM EDTA) to one pmol of primer per μL, and the enzyme was heat-deactivated at 70°C for 15 min.

One nanogram of each DNA sample was amplified in 100 μL of reaction buffer [1× PCR buffer, 20 nmol of each deoxyribonucleoside triphosphate (dNTP), 5 pmol of each primer, 2.5 U of *Taq* DNA polymerase (Stratagene, La Jolla, CA, USA)]. Amplification conditions for one cycle were 96°C for 30 s, 57°C for 30 s and 72°C for x/1000 min (x = fragment length in bp), and there were 15 to 30 cycles. The number of PCR cycles was chosen to avoid producing nonspecific fragments and to avoid extra cycles after reaching the plateau of amplification.

PCR products were desalted using a Centricon®-100 microconcentrator (Amicon Division, W.R. Grace & Co.-Conn, Beverly, MA, USA) with centrifugation at 1000× *g* for 20 min.

Desalted PCR products with a restriction sequence on one 5′ terminus were digested with restriction endonuclease for 1 h at 37°C and desalted again using a Centricon-100.

Desalted DNAs (0.4 to 2 μg) were digested to single-stranded DNAs by 4 U of lambda exonuclease (GIBCO BRL/Life Technologies, Gaithersburg, MD, USA) in 40 μL of reaction buffer (67 mM glycine-KOH, pH 9.4, 2.5 mM $MgCl_2$) for 30 min at 37°C.

Single-stranded DNAs were precipitated with ethanol, rinsed

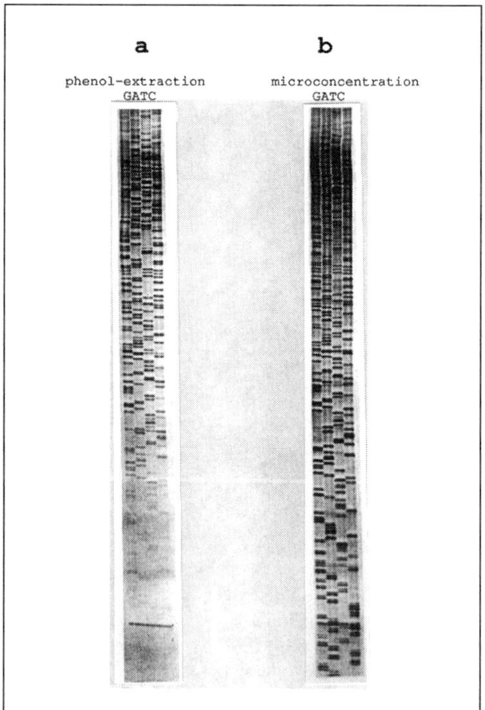

Figure 4. Comparison of the phenol extraction method and microconcentrator method for sample preparation. Sample A was amplified with 5′-phosphoryl A1 and A2 for 30 cycles and both methods are applied to the same amount of products. Panel a: Phenol extraction. Panel b: Microconcentration.

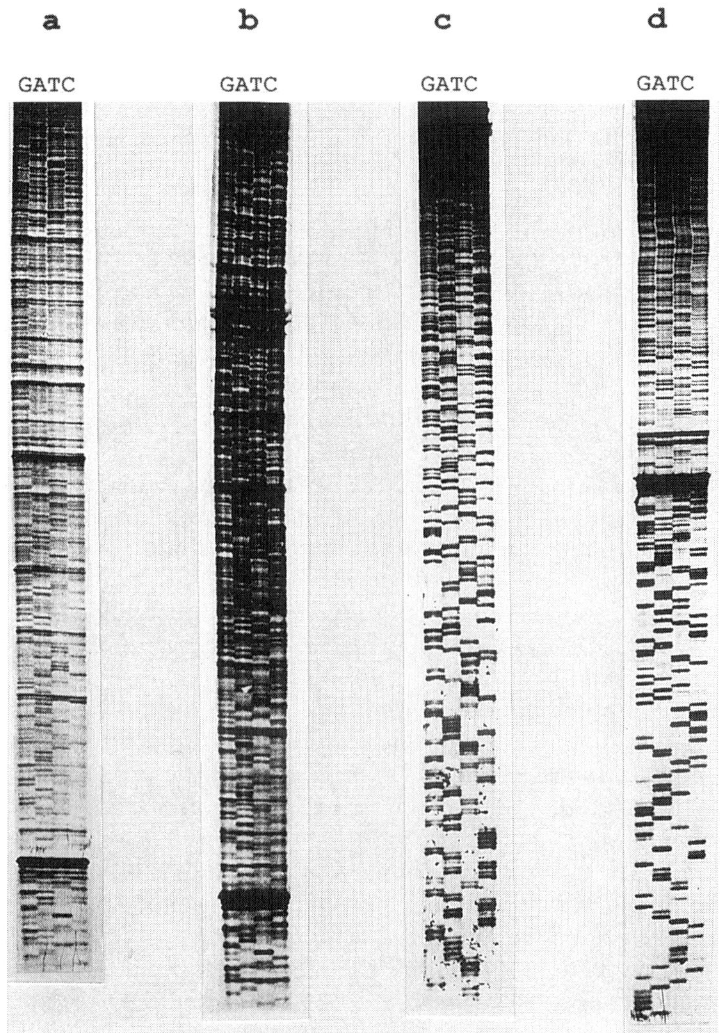

Figure 5. Direct sequencing double-stranded DNA with the snap-cooling procedure and asymmetric PCR. Sample A (Panel a) was amplified with A1 and A3, desalted with Centricon-100 three times, precipitated with ethanol, rinsed with 70% ethanol, dried under vacuum, dissolved in 7 µL of distilled water, mixed with 10 pmol of unlabeled primer (1 µL) and 2 µL of Sequenase buffer, boiled for 3 min, immediately cooled on ice and sequenced with the standard protocol of Sequenase using ^{35}S-dCTP. Sample A was asymmetrically amplified using a 20-base primer pair (A1 and A3) and a 29-base primer pair (A2 and A3+9base) and sequenced with 10 pmol of limiting primers (A1 for Panel b and A2 for Panel c), ^{35}S-dCTP and Sequenase. Asymmetric PCR conditions were 0.5 pmol of limiting primers, 50 pmol of excess primers and 35 PCR cycles. Sample A was amplified with A1 and A3 and sequenced by the snap-cooling procedure using unlabeled A1, ^{35}S-dCTP, and Sequenase. Sample A was asymmetrically amplified using a 20-base primer pair (A1 and A3) and a 29-base primer pair (A2 and A3+9base) and sequenced with Sequenase (Panel b and Panel C). Sample B was asymmetrically amplified using B1 and B2 and sequenced (Panel d).

with 70% ethanol, dried under vacuum and sequenced by the dideoxynucleotide chain-termination method with 10 pmol of unlabeled primer, ^{35}S-dCTP (Amersham, Little Chalfont, Buckinghamshire, England, UK) and Sequenase® (United States Biochemical, Cleveland, OH, USA).

RESULTS AND DISCUSSION

Figure 2 shows sequencing pattern examples. In all cases, more than 300 bases from the 3′ end of the primers were readable. Because of the 5′→3′ exonuclease activity of T4 DNA polymerase, some *Eco*RI digested 5′-protruding termini were clipped by one base and the upper part of the autoradiogram C shows a doublet sequence (unclipped and 1 base clipped).

Figure 3 shows the effect of a different number of PCR cycles on the sequence patterns. For both cases, sequence patterns were the best with 20 amplification cycles. At the initial concentration of 1 ng of target DNA, the plateau of amplification was reached at around 20 cycles and excess PCR cycles did not increase target DNA products but increased backgrounds. Samples amplified using a primer with an attached restriction endonuclease site were sensitive to excess PCR cycles (Panel b). Presumably because excess denaturation and association cycles increased the probability of DNA aggregation or irregular association. This in turn prevented the DNAs from reannealing the restriction sites at their termini.

Figure 4 shows the effect of using phenol extraction (2) instead of a microconcentrator for desalting and enzyme removal. With the phenol extraction method, a marked loss of DNA was observed.

We performed direct sequencing of double-stranded DNA by the typical snap-cooling procedure (1,3) and the typical asymmetric PCR procedure (4) several times. The best patterns obtained by these methods are shown in Figure 5. The snap-cooling procedure could not produce good patterns with unlabeled primers (Panel a). The quality of sequencing patterns by asymmetric PCR depended on samples and primer length (Panels b, c and d). The number of readable bases and pattern quality were inferior to those in Figure 2.

REFERENCES

1. **Casanova, J.-L., C. Pannetier, C. Jaulin and P. Kourilsky.** 1990. Optimal conditions for directly sequencing double-stranded PCR products with Sequenase. Nucleic Acids Res. *18*:4028.
2. **Higuchi, R.G. and H. Ochman.** 1989. Production of single-stranded DNA templates by exonuclease digestion following the polymerase chain reaction. Nucleic Acids Res. *17*:5865.
3. **Kusukawa, N., T. Uemori and I. Kato.** 1990. Rapid and reliable protocol for direct sequencing of material amplified by the polymerase chain reaction. BioTechniques *9*:66-72.
4. **McCabe, P.C.** 1990. Production of single-stranded DNA by asymmetric PCR, p. 76-83. *In* M.A. Innis, D.H. Gelfand, J.J. Sninsky and T.J. White (Eds.), PCR Protocols, Academic Press, New York.
5. **Miura, M., T. Tamura, A. Aoyama and K. Mikoshiba.** 1989. The promoter elements of the mouse myelin basic protein gene function efficiently in HG108-15 neuronal/glial cells. Gene *75*:31-38.

Update to:

Direct Sequencing of PCR Products Using Unlabeled Primers

Minoru Kimura

Division of Molecular Life Science, School of Medicine, Tokai University, Bohseidai, Isehara, Japan

Several groups have used the method presented in the original paper to analyze PCR products. For example, the diversity of the E7 region of human papillomavirus (1) and the sequence of the toxin gene in *Clostridium botulinum* (2) were efficiently analyzed using our method.

REFERENCES

1. **Fujinaga, Y., K. Okazawa, A. Nishikawa, Y. Yamakawa, M. Fukushima, I. Kato and K. Fujinaga.** 1994. Sequence variation of human papillomavirus type 16E7 in preinvasive and invasive cervical neoplasias. Virus Genes *9*:85-92.
2. **Fujinaga, Y., K. Takeshi, K. Inoue, R. Fujita, T. Ohyama, K. Moriishi and K. Oguma.** 1995. Type A and B neurotoxin genes in a *Clostridium botulinum* type AB strain. Biochem. Biophys. Res. Commun. *213*:737-745.

Double-Stranded DNA Sequencing by Linear Amplification with *Taq* DNA Polymerase

Bin Wang[1], Qiong Fang[2], William V. Williams[2] and David B.Weiner[1,2]

[1]The Wistar Institute of Anatomy and Biology, Philadelphia, PA, [2]Department of Medicine, University of Pennsylvania, Philadelphia, PA, USA

Recently, techniques for combining dideoxy sequencing and polymerase chain reaction (PCR) have become popular (2–4). These techniques promise to provide a rapid alternative method to sequencing DNA segments derived from DNA templates at low concentrations. In some protocols, an extra step is required either to generate a single-stranded template or to radioactively label the primer prior to sequencing (3,4). A linear PCR sequencing method has been developed (5,6) to eliminate these inconveniences. However, the currently available protocol for the linear PCR sequencing is somewhat difficult to perform because of the need to concentrate large volumes of PCR product prior to loading onto the sequencing gel (5,6). In addition, signals obtained using these protocols are frequently too weak and lack the level of detection necessary to obtain consistent results (5). To address these problems, we scaled down the volume of PCR to eliminate the concentration step. Additionally, as several chemical reagents have previously been reported to improve specificity (1), we tested the effects of these reagents on our PCR-sequencing reaction. Here we present a step-by-step protocol for the PCR-sequencing reaction:

1. Add 1 µL of **Termination Mixture** into each of four sequencing reaction tubes. [*Termination Mixture* = "G Mixture": 40 µM ddGTP, 15 µM of each dNTP (deoxyribonucleoside triphosphate); "A mixture": 1 mM of ddATP, 15 µM of each dNTP; "T mixture": 1.5 mM ddTTP, 15 µM of each dNTP; "C mixture": 1 mM ddCTP, 15 µM of each dNTP.]
2. Add 5 µL of **Mixture 1** into each of the four reactions. [*Mixture 1* = 2.5 µL of 10× *Taq* buffer, 12.5 pmol of primer, 0.05–0.1 pmol of DNA template, 0.5 µL of ^{35}S-α-dATP (5 µCi) (or ^{32}P-α-dATP) (Amersham, Indianapolis, IN, USA), 1 unit of *Taq* DNA polymerase (Perkin-Elmer, Norwalk, CT, USA); bring to 21 µL total volume with H_2O.]

(Reprinted from BioTechniques 13:527-530, 1992)

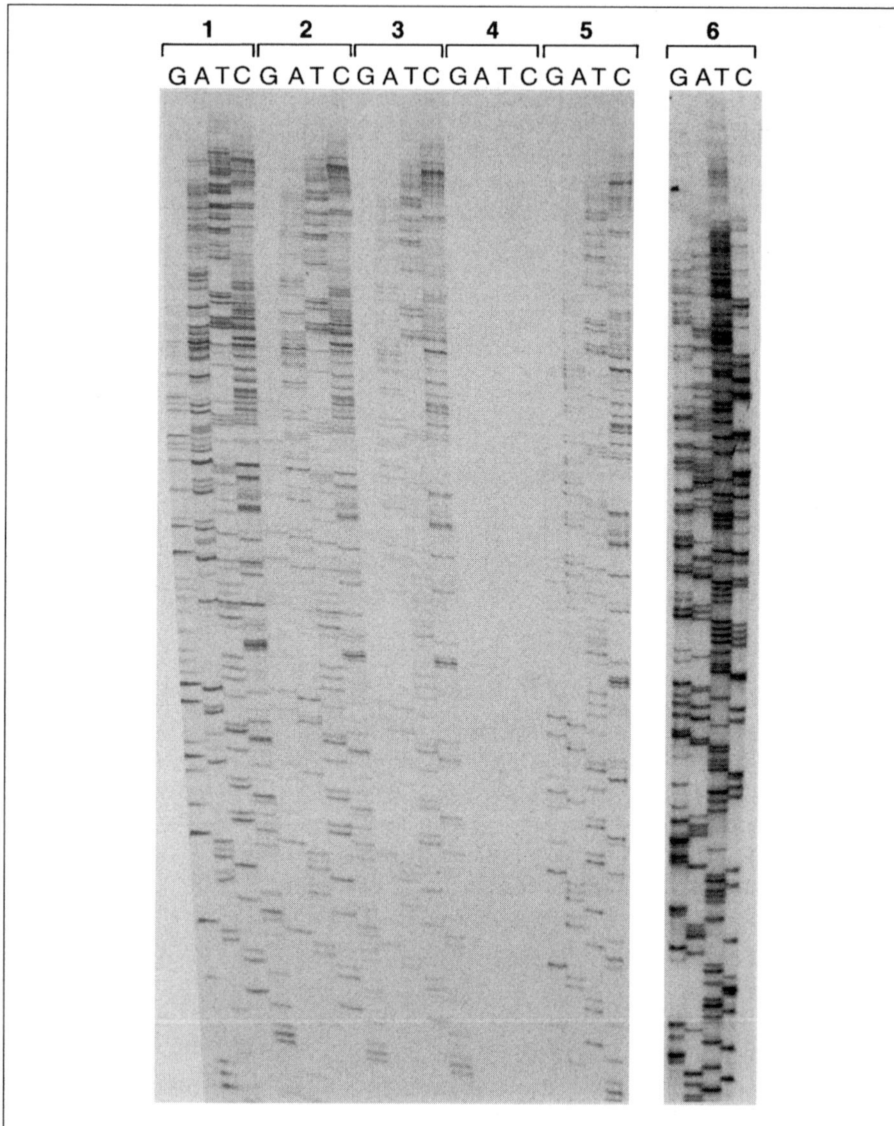

Figure 1. 1.8-kb DNA template (lanes 1–5) from HIV-1 gag segment was subcloned into Bluescript® vector (Stratagene, La Jolla, CA, USA), purified on a $CsCl_2$ gradient. The sequencing reaction was performed according to the protocol described above with ^{35}S-α-dATP. **Lane 1**, reaction with 1× standard *Taq* buffer (10× *Taq* buffer = 500 mM KCl, 100 mM Tis-HCl, pH 8.3, 15 mM $mgCl_2$, 0.1% gelatin, 5% NP-40); **lane 2**, 1× *Taq* buffer with 0.5% of Tween 20; **lane 3**, 1× *Taq* buffer with 5% of DMSO; **lane 4**, 1× *Taq* with 1 M urea; **lane 5**, 1× *Taq* buffer with 0.5% of NP-40. **Lane 6**, a 0.7-kb DNA template obtained from the T cell receptor β17 subfamily (7) was cloned into TA Cloning™ vector (Invitogen, San Diego, CA, USA), purified by an alkaline mini-preparation. The sequencing reaction was performed according to the protocol described above with 1× *Taq* buffer, 0.5% NP-40 and ^{32}P-α-dATP. The primers for these reactions are T7 primer: TGTAATACGACTCACTATAG (lanes 1–5), and cα5: 5′ GTACACGGCA-GGGTCAGGGTTCTGGATATATT 3′ (lane 6). All sequencing reactions wre loaded onto 40 × 40 cm^2 of 5% polyacrylamide gel with 8 M urea, and run at 2000 V for 2–3 h.

3. Add one drop of mineral oil into each tube.

4. The PCR-sequencing reaction is carried out as follows: an initial denaturing step at 94°C for 3 min, then 25 cycles of 93°C for 30 s, 58°C for 45 s.

5. Add 4 µL of **Stop Mixture** (95% formamide, 10 mM EDTA, 10 mM NaOH, 0.1% bromophenol blue, 0.1% xylene cyanol) into each tube; then heat for 4 min at 95–100°C and quickly cool on ice. (Note: It is not necessary to get rid of the mineral oil.)

6. Load 4 µL per lane onto a sequencing gel.

First, we found that using a thermal cycler program which omits the 72°C extension step in fact yields better sequence data (and takes less time) than a program that includes the 72°C step. Second, because the volume of PCR is so small (5 µL), we can add stop mixture directly to the samples with the mineral oil still on the top. By not removing the mineral oil, we can prevent sample condensation during the denaturing treatment. Third, we have found that only 0.5% NP-40 can notably enhance sequencing signals (Figure 1, lane 5); other reagents tested (e.g., 5% dimethyl sulfoxide [DMSO] in lane 2, 0.5% Tween 20 in lane 3) did not improve the quality of this sequencing reaction. Surprisingly, 1 M urea (lane 4), which had previously been thought to increase *Taq* DNA polymerase activity (1), completely inhibited the sequencing reaction. Results obtained with 0.5% NP-40 are comparable to those of the standard single-stranded DNA sequencing reaction. Using this modified protocol, we have obtained good quality sequencing from $CsCl_2$-purified DNA (lanes 1–5), alkaline mini-prepped plasmid (lane 6) and gel-purified PCR products (data not shown). It is important to use no more than 0.1 pmol of DNA template in our protocol to achieve a good sequencing result. Various concentrations of DNA template should be tested if the sequencing result is not satisfactory. We have used more than 10 different kinds of primers to sequence either T cell receptor genes or the HTLV-II envelope region sequence, and so far, results from these reactions were consistent and of satisfactory quality.

In conclusion, we report here a protocol which can improve the quality of the sequencing reaction that results in clearer bands and easier preparation. Furthermore, an error rate of sequencing from this protocol is minimal because all amplification reactions are based solely upon the original DNA template; in contrast, typical PCRs that utilize templates generated by PCR amplification have an increased error rate due to the low fidelity of the *Taq* DNA polymerase.

ACKNOWLEDGMENTS

We thank Alice I. Sato for assistance in preparing this manuscript. This work was supported by grants from NIH to W.V.W. and D.B.W.

REFERENCES

1. **Gelfand, D.H.** 1989. Thermus aquaticus DNA polymerase, p.11-17. *In* H.A. Erlich, R. Gibbs and H.H. Kazazian (Eds.), Current Communications in Molecular Biology. Cold Spring Harbor Press, Cold Spring Harbor, NY.
2. **Gyllenstein, U.B.** 1989. PCR and DNA sequencing. BioTechniques *7*:700-708.
3. **Gyllensten, U.B. and H.A. Erlich.** 1988. Generation of single-stranded DNA by the polymerase chain reaction and its application to direct sequencing of the HLA-DQA locus. Proc. Natl. Acad. Sci. USA 85:7652-7656.
4. **Kaltenboek, B., J.W. Spatafora, X. Zhang, K.G. Kousoulas, M. Blackwell and J. Storz.** 1992. Efficient production of single-stranded 2 DNA as long as 2 kb for sequencing PCR-amplified DNA. BioTechniques *12*:164-171.
5. **Lee, J-S.** 1991. Alternative dideoxy sequencing of double-stranded DNA by cyclic reactions using *Taq* polymerase. DNA Cell Biol. *10*:67-73.
6. **Murray, V.** 1989. Improved double-stranded DNA sequencing using the linear polymerase chain reaction. Nucleic Acids Res. *17*:8889.
7. **William, V.W.** 1992. Restricted heterogeneity of T cell receptor transcripts in rheumatoid synovium. J. Clin. Invest. *90*:326.

Update to:

Double-Stranded DNA Sequencing by Linear Amplification with Taq DNA Polymerase

Bin Wang
Sinogenesis Institute, Haidian, Beijing, China

An additional reference to this article is:

REFERENCES

1. **Murray, V., C. Monchawin and M.J. Cairns.** 1993. Detection of polymorphisms using thermal cycling with a single oligonucleotide on a DNA sequencing gel. Hum. Mutat. *2*:118-127.

Simple Cycle Sequencing by Labeling Primer Using *Taq* DNA Polymerase

Han-Jip Kim and Byung-Kook Kim

Ajou University, Suwon, South Korea

In cycle sequencing, two methods are available for incorporating isotopes into DNA. These are the direct incorporation of labels and the use of end-labeled primers (1,2). The end-labeled primer method uses a polynucleotide kinase to phosphorylate the sequencing primer with the γ-labeled nucleoside triphosphate (NTP) (Figure 1A). This method is widely accepted and several companies supply an end-labeling kit for their cycle sequencing systems. The other method, the direct incorporation of labels, uses the α-labeled deoxyribonucleoside triphosphate (dNTP) to label the extending DNA by *Taq* DNA polymerase in a thermal cycler. This method, however, requires more than ten times the concentration of plasmid templates compared with the end-labeled primer protocol and is not recommended for lambda templates, according to the technical manual of the *fmol*™ DNA Sequencing System (Promega, Madison, WI, USA).

We hypothesized that *Taq* DNA polymerase could be used instead of polynucleotide kinase to label primers with the α-labeled dATP. The sequencing primer, in the presence of the DNA to be sequenced, would be extended by *Taq* DNA polymerase using [α-^{32}P]dATP in a thermal cycler. By omitting other dNTPs, however, the extension would be terminated after the incorporation of the labeled dATP. The omitted nucleotides are the ones that follow the dATP on the extending DNA and are complementary to the plasmid template. Consequently, the extension would be arrested after incorporating α-labeled dATP at the 3′ end of the primer. Although the process is essentially that of "arrested primer extension" in the presence of selected nucleotides (3), primer labeling by *Taq* DNA polymerase has, to our knowledge, never been documented, and no such protocol is available in any commercial cycle sequencing systems. Our method would eliminate the need to purchase either a γ-labeled nucleotide or a polynucleotide kinase. The method would reduce the cost of cycle sequencing. The simplicity of the procedure and the handling of fewer kinds of radioisotopes by utilizing an easily available α-labeled deoxynucleotide would be advantageous over the current methods.

(Reprinted from BioTechniques 16:576-580, 1994)

To evaluate this method, we used pBluescript® II KS+ phagemid vector (GenBank® Accession No. 52327; Stratagene, La Jolla, CA, USA) as the template. The vector contained an unidentified cDNA insert of a rice cDNA library. A T3 primer (20-mer) was provided by Korea Biotech (Daejon, Korea) and was utilized as the sequencing primer. The sequence of the T3 primer was 5′-d(ATTAACCCTCACTAAAGGGA)-3′. According to the template sequence, the primer is followed by 5′-d(ACAAAAGCTG---). To label the primer using *Taq* DNA polymerase, we added 0.5 μL of 800 Ci/mmol [α-^{32}P]dATP (Amersham International, Little Chalfont, Bucks, England, UK), but no other dNTPs were added so that extension after labeling would be arrested. In addition to [α-^{32}P]dATP, 10 fmol (33 ng) of the phagemid template, 3 pmol of the primer, 1 μL of the sequencing grade *Taq* DNA polymerase (5 units/μL) in a storage buffer (Promega) and 5 μL of 5× sequencing buffer (250 mmol Tris-HCl, pH 9.0 at 25°C and 10 mmol $MgCl_2$) were added, and the volume was adjusted to 17 μL with deionized H_2O. This mixture, called the labeling/extension mix, was covered with mineral oil and heated to 95°C for 5 min in a Model N801-0100, DNA Thermal Cycler 480 (Perkin-Elmer, Norwalk, CT, USA). The following profile was used for the labeling/extension cycling: 94°C for 1 min; 55°C for 2 min; 72°C for 3 min. Maximum labeling for sequencing was achieved in 8 cycles (Figure 1D). As a result of the limited extension, the labeled/arrested primer became a 21-mer when thermocycling was completed (Figure 1, B–E).

The labeled/arrested primer was then used in the extension/termination reaction. Four aliquots were prepared by pipetting 4 μL of labeled/arrested primer mix into each tube containing 2 μL of the appropriate dNTP/dideoxyribo (dd)NTP mix (*fmol* DNA Sequencing System). The extension/termination reaction was conducted for 30 cycles using the same temperature profiles as described earlier for the labeling/extension reaction. The temperature profiles were effective for the labeling/extension and the extension/termination but were not necessarily the optimum conditions for either reaction. After completion, 3 μL of the stop solution (10 mmol NaOH, 95% formamide, 0.05% bromophenol blue and 0.05% xylene cyanol) were added. The reaction mixtures were then heated to 70°C for 2 min and loaded onto a sequencing gel.

Direct sequencing from a bacterial colony containing the recombinant pBluescript II KS+ phagemid was also performed using our method (Figure 1E). To prepare the template DNA from a colony, a large (2-mm) colony was transferred into a 0.5-mL microcentrifuge tube containing 10 μL of 10 mmol Tris-HCl (pH 8.0 at 25°C), 1 mmol EDTA and 50 μg/mL proteinase K. The mixture was vortex mixed and briefly spun in a microcentrifuge. The sample was extracted by incubating at 55°C for 15 min then heated to 90°C for 15 min. After cooling on ice for 1 min, the sample was centrifuged at 12 000× *g* for 5 min. An aliquot of 6.5 μL was added to the labeling/extension mixture and served as the template DNA. The concentrations of [α-^{32}P]dATP, the primer, 5× sequencing buffer and the sequencing grade *Taq* DNA polymerase were the same as those described earlier, and the volume was adjusted to 17 µL with deionized H_2O. This mixture was covered with mineral oil, and 8 cycles of labeling/extension were performed. The labeled/extended primer was used in the extension/termination reaction as described earlier for the pure phagemid DNA. The temperature profiles for the labeling/extension and the extension/termination reactions were identical to those described earlier.

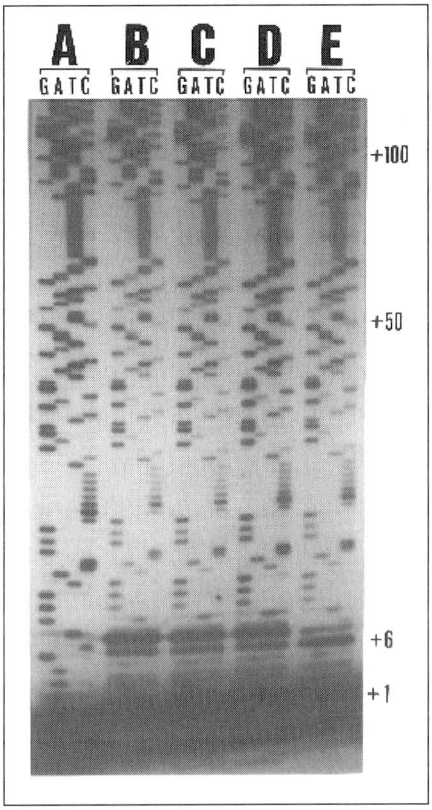

Figure 1. Comparison of the sequencing ladders obtained as follows: A) The primer was labeled with [γ-^{32}P]ATP using the T4 polynucleotide kinase according to the protocol of *fmol* DNA Sequencing System. The template DNA was 40 fmol. B) Four cycles of labeling/extending primer with [α-^{32}P]dATP using the *Taq* DNA polymerase. The template DNA was 10 fmol. C) Six cycles of labeling/extending primer with [α-^{32}P]dATP using the *Taq* DNA polymerase. The template DNA was 10 fmol. D) Eight cycles of labeling/extending primer with [α-^{32}P]dATP using the *Taq* DNA polymerase. The template DNA was 10 fmol. E) Direct sequencing from a bacterial colony using the *Taq* DNA polymerase for the labeling/extending primer with [α-^{32}P]dATP. Eight cycles of labeling/extending primer were performed. The concentration of the template DNA was not determined.

The results showed the phagemid sequences, including the *Eco*RI linker site and the cDNA insert (2 kbp) containing 17 polyA tails. The sequence was easily readable up to 350 bp in the sequence ladders (data not shown). The artifacts at +6 and +7 nucleotide from the primer (20-mer) did not present any difficulty in reading the sequence (Figure 1, B–E). The artifacts were produced during the labeling/extension period since the extension/termination procedure was identical for both the end-labeled primer method (Figure 1A) and our [α-^{32}P] labeled primer method (Figure 1, B–E). For this particular T3 primer and the phagemid clone used for our illustration, more than one labeled dATP could have been incorporated erroneously if AA was added in place of AC. Nonspecific primer extension would explain the artifact noted at +6 since six dATPs could be incorporated consecutively to the primer according to the template sequence. The artifacts and nonspecific extension might be minimized by choosing proper primer sequences to arrest extension.

In addition to the T3 primer used for the illustration, the T7 promoter primer (20-mer; Promega) was utilized for sequencing other recombinant pBluescript phagemids. The sequence of the T7 primer is followed by 5′-d(CGAATTGGAG---) according to the template sequence. To label the primer with [α-^{32}P]dATP, extremely low concentrations of dCTP and dGTP (0.17 μmol) were added along with [α-^{32}P] dATP for the labeling/extension but no dTTP was added. The phagemid clones were sequenced successfully (data not shown). Although this system could be used with other α-labeled dNTPs that are complementary to the template DNA for the labeling/extending primer, our experiment proved that any primer could be labeled and arrested by using only [α-^{32}P]dATP in the presence of selected nucleotides.

This method has enabled us to label primers extensively for sequencing cDNA clones without the use of a polynucleotide kinase and has eliminated the need to purchase a γ-labeled nucleotide for end labeling. To complete both processes of the labeling/extension (primer labeling) and the subsequent extension/termination (sequencing), there was no need to add more *Taq* DNA polymerase. Moreover, only ten fmol of the phagemid DNA was sufficient for template concentration. This method is as efficient as the end-labeled primer protocol and, furthermore, allows the direct incorporation of the label, which would increase the intensity of labeling as the DNA is being extended. In addition, this method could be just as effective for sequencing directly from a bacterial colony and therefore would save more time and materials.

ACKNOWLEDGMENTS

We would like to thank Drs. Won Kim and Neil Howes and Mr. Wayne Mauthe of Agriculture Canada, Winnipeg Research Station for their valuable comments on the manuscript.

REFERENCES

1. **Gyllensten, U.B., M. Allen and A. Josefsson.** 1992. Sequencing of *in vitro* amplified DNA, p. 1-15. *In* J. Ellingboe and U.B. Gyllensten (Eds.), The PCR Technique: DNA Sequencing. Eaton Publishing, Natick, MA.
2. **Kusukawa, N., T. Uemori, K. Asada and I. Kato.** 1990. Rapid and reliable protocol for direct sequencing of material amplified by the polymerase chain reaction. BioTechniques *9*:66-72.
3. **Sandhu, G.S., B.C. Kline, M.E. Bolander and G. Sarkar.** 1993. Gene identification by arrested primer extension. BioTechniques *14*:948-952.

Cycle Sequencing Protocol Using Deep Vent$_R$® (exo$^-$) DNA Polymerase and Reduced dNTP and [α-^{35}S]dATP Concentrations

Bernard Mariamé

Centre d'Immunopathologie et de Génétique Humaine, Toulouse, France

Direct sequencing of polymerase chain reaction (PCR) products is a very powerful technique, especially for clinical investigations where quantities of biological material are often very limited and results are rapidly needed. However, numerous researchers still prefer to sequence cloned PCR fragments, a very time-consuming strategy, as most published protocols for direct sequencing of PCR products suffer from fairly serious drawbacks. Manual sequencing using exonuclease-deficient bacteriophage T7 derived DNA polymerases gives good results but requires large amounts of material: 0.5–1 pmol (corresponding to ±325 to 650 ng of a 1-kb fragment) for one set of reactions, generally providing no more than 200-bp information. In contrast, protocols using PCR technology allow the sequencing of femtomolar quantities of fragment due to the fact that template molecules are released at the beginning of each cycle and thus are again available for the synthesis of new complementary strands. However, most commercial thermostable DNA polymerases for PCR sequencing are supposed to require relatively high nucleotide concentrations, in the range of 100–200 μM. Therefore, the high concentration of nonradioactive dATP strongly competes with the commonly used [α-^{35}S]dATP nucleotide, which consequently is poorly incorporated. The synthesized strands are thus weakly labeled, imposing long autoradiography exposure times. This is often compensated for by increasing the input of expensive radioactivity (up to 20 μCi), which has an adverse effect on waste disposal. It is, therefore, often preferable to label the sequencing primer, but this dictates the use of high-exposure (^{32}P) or very expensive (^{33}P) radionucleotides, and this may be very tedious if different primers are needed for the sequencing of long fragments.

Here, I propose an easily implemented modification that eliminates most drawbacks of current protocols for cycle sequencing with [α-^{35}S]dATP

(Reprinted from BioTechniques 21:18-19, 1996)

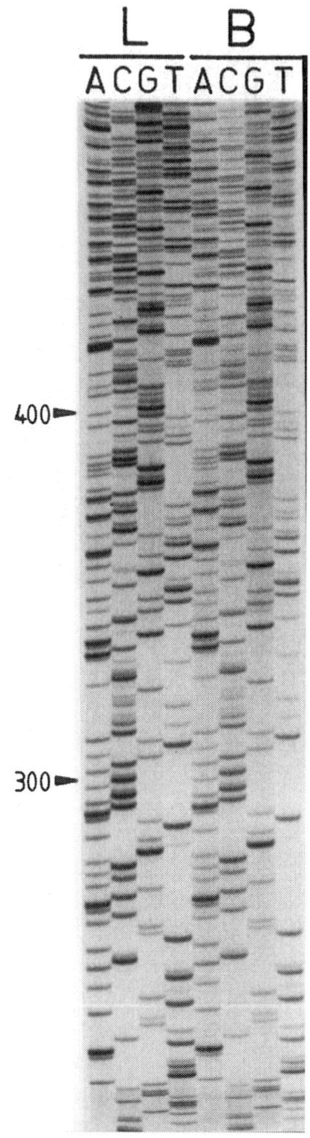

Figure 1. Partial sequences of the third exon of the Epstein-Barr virus BNLF1 gene, in strains associated with T lymphoma (L) and with Burkitt's lymphoma (B). Sequencing reactions were performed either directly on the purified PCR fragment (L) or on the double-stranded plasmid after cloning of the PCR fragment in the pGEM-7 vector. Reactions were loaded on a 6% homogeneous polyacrylamide gel and run for 4.5 h in 0.5× TBE buffer. Under these conditions, the readable sequence easily extends to 400 nucleotides from the primer.

incorporation. This modification is based on the observation that $Vent_R$® (exo^-) or Deep $Vent_R$ (exo^-) DNA polymerases (New England Biolabs, Ozyme, France), at least in this application, do not require very high dNTP concentrations. These two enzymes work perfectly with dNTP concentrations 10 times lower than those currently recommended for sequencing. I have, therefore, determined conditions where, despite a 2.5 times reduction in radioactivity input, the ratio between labeled and nonradioactive dATP is much more favorable to [α-^{35}S]dATP incorporation.

The fragment to be sequenced is purified after agarose gel electrophoresis, using JetSorb™ (Bioprobe Systems, Paris, France) or GENECLEAN® (Bio 101, Ozyme, France) kits. Ten to one hundred fmol of purified PCR fragment (corresponding to 6.5–65 ng of a 1-kb fragment) in 2 µL Tris-EDTA (TE) or water are mixed with 10 µL of the following solution: 6 µL water, 1.2 µL 10× sequencing buffer, 1 µL 2.4% Triton® X-100, 0.5 µL of 10 µM sequencing primer, 0.8 µL [α-^{35}S]dATP (1000 Ci/mmol, 10 mCi/mL; ICN Biomedicals, Orsay, France) and 1.0 U Deep $Vent_R$ (exo^-) DNA polymerase (0.5 µL). After careful homogenization, 2.5 µL of this template-primer-enzyme solution are then added to each of four 0.5-mL microcentrifuge tubes, each containing 2 μL of one 1× dNTP/ddNTP mixture. In these mixtures, dNTP and ddNTP concentrations are about ten times lower than in the mixtures recommended by the enzyme manufacturer. After addition of 10 μL PCR oil and brief centrifugation, tubes are placed in a TR3 Omnigene thermal cycler (Hybaid, Teddington, England, UK) and submitted to the following program: one cycle at 95°C for 2.5 min and then 22 cycles at 95°C for 20 s, 60°C for 5 s and 72°C for 20 s. Of course, the annealing temperature may be modulated according to the melting temperature (T_m) of the primer used. After addition of 3 μL stop dye and a 3-min denaturation step at 85°C, 1.8–2 μL of each sample are loaded on a sequencing polyacrylamide gel, pre-run in 0.5× Tris borate EDTA (TBE) buffer. After electrophoresis, the gel is fixed for 15 min in a bath of 20% ethanol/10% acetic acid, dried and exposed at room temperature on a Kodak BioMax MR Film (Eastman Kodak, Rochester, NY, USA) for 16 h. The dNTP/ddNTP mixtures are prepared and stored as 10× concentrated solutions that contain 45 μM dATP, 100 μM dCTP, 100 μM 7-deaza-GTP and 100 μM dTTP; moreover, the 10× dNTP/ddATP, dNTP/ddCTP, dNTP/ddGTP and dNTP/ddTTP mixtures, respectively, contain 2.5 mM ddATP, 1.2 mM ddCTP, 1.8 mM ddGTP and 2 mM ddTTP. Before use, the dNTP/ddNTP mixtures are made 1× in 1× sequencing buffer: 10 mM KCl, 10 mM $(NH_4)_2SO_4$, 20 mM Tris-HCl, pH 8.8, 7 mM $MgSO_4$. dNTP and ddNTP were purchased from Pharmacia Biotech (Orsay, France) and Boehringer Mannheim (Meylan, France), respectively. In some cases, artefactual bands across the four lanes are observed, which are clearly sequence specific and dependent on fragment purity. When encountered, this problem is solved by treating the samples with deoxynucleotidyl terminal transferase (TdT) as described by Fawcett and Bartlett (2). At the end of the cycle sequencing program (and before addition of the stop dye), 2 μL of a solution containing 1 mM each dNTP and 1 U TdT (Life Technologies, Eragny, France) in 2× TdT buffer are added to the reactions that are incubated 30 min at 37°C. In this case, the quantity of stop dye is increased to 4 μL.

By using this protocol, I routinely read 220–240 nucleotides (and sometimes up to 300), starting at 10–15 nucleotides from the primer, on a single buffer gradient gel (3). However, the sequences may be read up to 400 nucleotides, if the same reactions are loaded on a second gel and run for a longer time (Figure 1).

This protocol strongly improves the efficiency of [α-^{35}S]dATP incorporation, eliminating the most serious drawback of this sequencing technique. The procedure is simple and short, and six series of ten templates may be performed easily in one day, provided the capacity of the thermal cycler is sufficient. The technique is reliable. When sequencing the BNLF1 gene of different Epstein-Barr virus strains whose sequences were already determined, the sequences obtained with this protocol were always identical to those already published.

This protocol is also universal. I have used it without modification for the sequencing of double-stranded PCR fragments in the range between 165 bp and 4.3 kb and of double-stranded plasmids prepared either by centrifugation on CsCl gradients or by the "miniprep" technique of Del Sal et al. (1). Finally, this protocol provides good quality sequences at an appreciably lower cost due to the reduction in nonradioactive and labeled nucleotide concentrations and the use of the relatively inexpensive and efficient Deep $Vent_R$ (exo^-) DNA polymerase.

ACKNOWLEDGMENTS

I thank Dr. John Clayton for his help in the preparation of this manuscript.

REFERENCES

1. **Del Sal, G., G. Manfioletti and C. Schneider.** 1988. A one-tube plasmid DNA mini-preparation suitable for sequencing. Nucleic Acids Res. *16*:9878.
2. **Fawcett, T.W. and S.G. Bartlett.** 1990. An effective method for eliminating "artifact banding" when sequencing double-stranded DNA templates. BioTechniques *9*:46-53.
3. **Sheen, J.Y. and B. Seed.** 1988. Electrolyte gradient gels for DNA sequencing. BioTechniques *6*:943-946.

Update to:

Cycle Sequencing Protocol Using Deep Vent$_R^{®}$(exo$^-$) DNA Polymerase and Reduced dNTP and [α-^{35}S]dATP Concentrations

Bernard Mariamé
Centre d'Immunopathologie et de Génétique Humaine, Toulouse, France

The sequencing protocol described in the original paper has now been used by different people in our group for more than one year, and we have accumulated experience corresponding to about 100 kb and several hundred independent experiments. This has allowed the fine-tuning of some parameters of this protocol, the original idea of which was to reduce the dNTP (and consequently the ddNTP) concentrations in order to favor [α-^{35}S]dATP incorporation in the neosynthesized strand. Here, I describe the latest (although minor) optimizations I am currently using.

First, I tried to define the highest acceptable dilution of the ddNTP/dNTP mixtures and found that using them 0.5× or even 0.25× instead of 1× gives a stronger signal on a clearer background without significant loss in the length of the readable sequence (between 350 and 400 nucleotides from the primer, on two separate—one long and one short—sequencing gels). As stated in the original paper, 10 to 100 femtomoles of template (corresponding to 6.5 to 65 ng of a 1-kb double-stranded fragment) generally gives good results. However, in most cases, little is better than too much, because limited amounts of template (10-20 femtomoles) produce clearer pictures than larger quantities. On the other hand, to easily read close to the primer or further than 250 nucleotides from it (particularly with 0.25× ddNTP/dNTP mixtures), the template input should not be decreased below 20 femtomoles.

The updated protocol I am now using includes the following: to 2 µL of the template (corresponding to 20 femtomoles in TE or water) are added 10 µL of a mixture containing 1.2 µL 10× sequencing buffer, 1.2 µL 2.4% Triton® X-100, 0.5 µL of a 10 µM solution of the sequencing primer, 6–8 µCi [α-^{35}S]dATP (1000 Ci/mmol, 10 mCi/mL; ICN Biomedicals, Orsay, France) and 1 U of Deep Vent$_R$(exo$^-$) DNA polymerase (New England Biolabs, Ozyme, France). After careful homogenization by pipetting up and down, 2.7 µL of this mixture are transferred to each of four 0.5-µL microcentrifuge tubes, each containing 2 µL of one specific 0.25× ddNTP/dNTP mixture. After a brief centrifugation and addition of 10 µL PCR oil, tubes are placed in a TR3 Omnigene thermocycler (Hybaid, Teddington, England, UK) and submitted to the following program: 2 min at 95°C and then 20 cycles at 95°C for 20 s, 60°C for 5 s and 72°C for 20 s. Tubes are then briefly centrifuged again, and 1 U deoxynucleotidyl terminal transferase (TdT; Life Technologies, Eragny, France) in 2 µL of 2× TdT buffer, 1 mM each dNTP, is added to eliminate occasional artifactual bands across the four lanes. Reactions are incubated 30 min at 37°C, blocked by addition of 4 µL stop dye and heated 3 min at 85°C. Three microliters of these reactions are then loaded on the sequencing polyacrylamide gel as previously described. ddNTP/dNTP mixtures and the sequencing buffer are described in the original paper, and it is very important that ddNTP/dNTP mixture dilutions are made in 1× final sequencing buffer in order to maintain a high Mg^{2+} ion concentration, which seems necessary to stabilize the template-polymerase complex at the beginning of the elongation. Decrease in Mg^{2+} concentration to 2 or 3 mM often results in bands across the four lanes in all positions up to 50 nucleotides after the primer.

One drawback of this protocol is that the bands on the autoradiogram are not all of equal intensity, which occasionally renders the reading a little more difficult. As it has been suggested that unequal intensity could result from inorganic pyrophosphate production (1), I have tried to incorporate varying amounts of the Tth thermostable pyrophosphatase (Boehringer Mannheim, Meylan, France) in the sequencing reactions. Unfortunately, and although this had no adverse effect on the sequence quality, no homogenization in band intensity was observed using 20, 10 or 5 mU Tth pyrophosphatase per reaction.

Thus, after extensive testing, this protocol has proved to be very robust and to work quite well on all types of template correctly purified, including PCR fragments and supercoiled double-stranded plasmids obtained by various "miniprep" procedures.

ACKNOWLEDGMENT

I thank Dr. John Clayton for his help in the preparation of this manuscript.

REFERENCES

1. **Tabor, S. and C. Richardson.** 1990. DNA sequence analysis with a modified bacteriophage T7 DNA polymerase. J. Biol. Chem. *265*:8322-8328.

Direct Cycle Sequencing of Mutated Alleles Detected by PCR Single-Strand Conformation Polymorphism Analysis

Samuel Chi-Ho Mok[1], Kwok-Wai Lo[3] and Sai-Wah Tsao[1,2]

[1]Brigham and Women's Hospital, Boston; [2]Dana-Farber Cancer Institute, Harvard Medical School, Boston, MA, USA; [3]Chinese University of Hong Kong, Hong Kong

ABSTRACT

A rapid protocol for direct sequencing of a mutated allele, detected by combining polymerase chain reaction with single-strand conformation polymorphism (PCR–SSCP) analysis and cycle sequencing using a thermal cycler, is described. End-labeled radioactive primers were used both for PCR–SSCP analysis for the detection of p53 gene mutation and for cycle sequencing using ΔTaq™ *Version 2.0 DNA Polymerase. The point mutations along the various exons of the p53 gene can be rapidly determined by this sequencing method. This protocol requires only a small amount of DNA template (0.1 µg) and produces sequencing images with low backgrounds and very uniform band intensity. It has also been used successfully in the study of other gene mutations including* ras *and NF-1 (neurofibromatosis 1) genes.*

INTRODUCTION

Inactivation or activation of a specific gene by point mutations has been shown to be involved in human carcinogenesis and hereditary diseases. Several methods have been developed for the analysis of such mutations. The polymerase chain reaction (PCR) has been shown to be particularly versatile in this aspect. A particular gene fragment of interest is amplified by PCR, and the mutated DNA fragment can be detected by a non-denaturing SSCP (single-strand conformation polymorphism) gel as a distinct DNA band with different mobility shift (7). The mutated sequence in the DNA fragment generated by PCR can be determined either by cloning the PCR fragments into the M13 vector and subsequently sequencing the single-strand template by Sequenase® or by direct sequencing of the PCR product (2,5,10). Direct

(Reprinted from BioTechniques 14:790-794, 1993)

sequencing of the PCR products without an additional cloning step is generally preferable because it does not require an additional cloning step. However, there are inherent difficulties in this method because of the rapid reassociation of the short linear template strands. In order to solve this problem, asymmetric PCR has been developed utilizing an unequal concentration of the two amplification primers to generate single-stranded DNA templates. The single-stranded template generated can then be directly sequenced with an internal primer that would provide additional specificity. Recently, a new method of DNA sequencing using a thermal cycler and end-labeled primers has been developed and applied successfully in the direct sequencing of the yeast genomic DNA (4,8). We have adapted this method for direct sequencing of the PCR products. In addition, we have combined the PCR cycle sequencing with PCR–SSCP analysis for the rapid detection of p53 mutations in human ovarian cancer.

MATERIALS AND METHODS

Specimen Collection and DNA Extraction

Ovarian tumors were obtained from patients at the Brigham and Women's Hospital in Boston according to a protocol approved by the Human Subjects Committee of the hospital. Control tissues consisted of segments of normal fallopian tube, uninvolved round ligament or blood lymphocytes. DNA from these samples were extracted by the phenol/chloroform method as described previously (9).

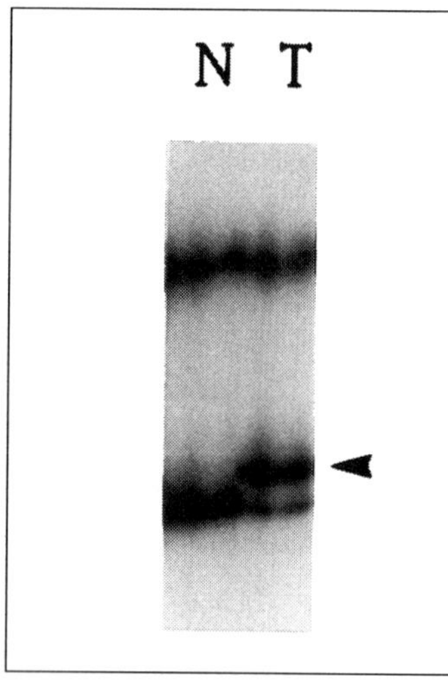

Figure 1. Example of PCR–SSCP analysis of exon 4 of the p53 gene. PCR products from normal tissue (N) and tumor tissue (T) from patient A were electrophoresed on an 8% nondenaturing gel followed by autoradiography. Arrow indicates the variant band in the tumor DNA that has a mutation in exon 4.

PCR–SSCP Analysis

Exons 4 to 8 of the p53 gene were amplified by PCR according to procedures published by Hsu et al. (exons 5 to 8) (3) and Okamoto et al. (exon 4) (6). The oligonucleotide primers used were purchased from Genosys Biotechnologies (Woodlands, TX, USA). The procedures for the PCR–SSCP analysis was as follows: One microliter each of left and right PCR primers (10 µM) flanking the exons of the p53 gene was radiolabeled with 5 units of T4 polynucleotide kinase (Boehringer Mannheim, Indianapolis, IN, USA) and 0.5 µL [γ-^{32}P] ATP (7000 Ci/mmol; ICN Biomedicals, Costa Mesa, CA, USA) in 5 µL of reaction mixture for 30 min at 37°C.

The reaction mixture was then diluted with 255 µL water, 40 µL of 10× PCR buffer (0.1 M Tris-HCl, 0.5 M KCl, pH 8.3, and 20 mM $MgCl_2$), and 20 µL of 1.25 mM deoxyribonucleoside triphosphate (dNTP) mixture. Ten units of *Taq* DNA polymerase (Perkin-Elmer, Norwalk, CT, USA) were added subsequently to the diluted reaction mixture. Four microliters of this primer–PCR mixture were then mixed with 50 ng of genomic DNA in a total volume of 5 µL. The PCR was carried out in a DNA Thermal Cycler (Perkin-Elmer) under the following conditions: The template DNA was first denatured by incubation for 5 min at 95°C, followed by 40 cycles of PCR (1 min incubation at 60°C for annealing; 30 s at 72°C for polymerization; and 30 s at 94°C for denaturation). The PCR products were then mixed with 45 µL loading buffer containing 95% formamide, 20 mM EDTA, 0.05% bromophenol blue and 0.05% xylene cyanol FF (Sigma Chemical, St. Louis, MO, USA). One microliter of the final product was loaded onto an 8% polyacrylamide (49:1, acrylamide:bisacrylamide) gel and electrophoresed at 30 W at 4°C. The gel was dried on a filter paper with a vacuum gel dryer and exposed to x-ray film at -70°C for 6–12 h with an intensifying screen.

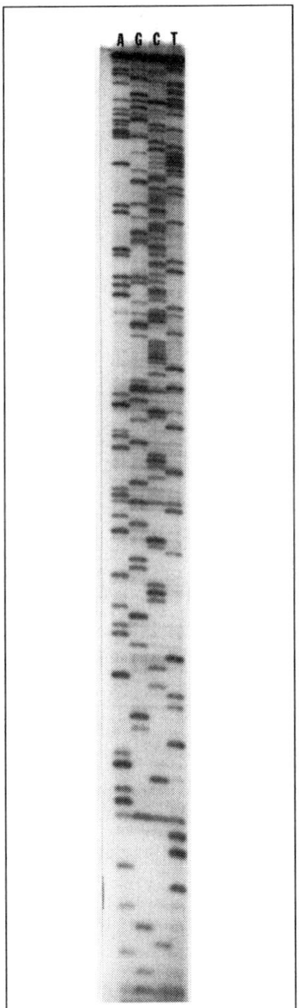

Figure 2. A representative sequence of exon 4 of the p53 gene generated by direct PCR cycle sequencing using *Taq* DNA polymerase.

Direct Cycle Sequencing of PCR Product

Based on the results of the SSCP analysis, DNA from the tumor specimens that carried a point mutation were reamplified for direct cycle sequencing. In case of heavy contamination by the normal cells in the tumor specimen, the DNA bands that displayed a mobility shift in the SSCP gel were cut out, and the DNA was extracted with 50 µL of water at 65°C for 6 h. Ten microliters of the eluted DNA were taken out and reamplified with the corresponding sets of primers using the same conditions as described above in a total volume of 100 µL. The PCR products were purified by running on a 5% nondenaturing polyacrylamide gel, extracted with phenol/chloroform and precipitated with 2 volumes of absolute alcohol. Purified DNA (0.1 µg) was used for direct cycle sequencing. Either the left or the right primer of the corresponding exon of the p53 gene was labeled with [γ-^{32}P]ATP by polynucleotide kinase in a total volume of 10 µL as described. The labeled primer was subsequently diluted fivefold with sterile distilled water.

One microliter of the diluted labeled primer was mixed with 0.1 μg template DNA, 2 μL of reaction buffer (160 mM Tris-HCl, pH 9.5, 65 mM $MgCl_2$) and 8 units of Δ*Taq*™ Version 2.0 DNA Polymerase (United States Biochemical [USB], Cleveland, OH, USA) in a total volume of 17 μL. Then 4 μL of the labeling mixture were transferred to each of the 4 termination tubes containing 4 μL of the termination mixture (A, G, C and T) as described in the *Taq* Cycle Sequencing kit (USB). The termination reactions were incubated in a DNA Thermal Cycler using the same conditions as described above and stopped with 45 μL of stop solution (95% formamide, 20 mM EDTA, 0.05% bromophenol blue, and 0.05% xylene cyanol FF; USB). All samples were boiled for 10 min and 2 μL of each sample were loaded onto an 8% sequencing gel.

RESULTS AND DISCUSSION

By PCR–SSCP analysis, variable patterns of mobility shift were observed in the different p53 exons amplified from the ovarian tumor DNA as compared with the corresponding normal control DNA. Figure 1 shows the SSCP patterns of exon 4 of the p53 gene from the ovarian tumor DNA and the control DNA from patient A. In order to sequence the mutated form of the amplified exon alone, the variant band was cut out from the SSCP gel. DNA was eluted with water and reamplified. The amplified DNA fragment was then purified by polyacrylamide gel before being used for PCR cycle sequencing. The same set of ^{32}P end-labeled primers for the PCR–SSCP analysis was used to sequence the two complementary strands of the reamplified DNA fragments. Figure 2 shows a representative sequencing pattern of exon 4 of the p53 gene generated by using the upstream primer and the cycle sequencing method. Since most of the PCR products for SSCP analysis are in the range of 100 to 500 base pairs, the complete sequence can usually be resolved in two hours from one loading of the sequencing reaction products. Furthermore, the DNA sequence determined by the upstream primer can be confirmed by the complementary sequence determined by the downstream primer. Figure 3 shows a missense mutation involving a C to T transversion at codon 63 of the p53 gene in an ovarian tumor.

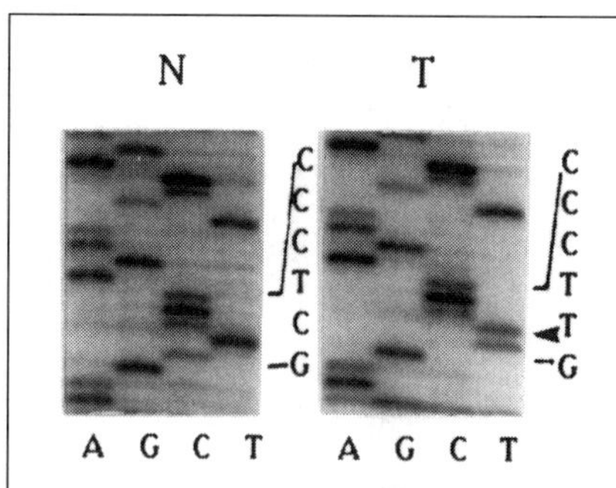

Figure 3. Sequencing of the mutated exon of the p53 gene in ovarian cancer. Portion of the sequence of the sense strand of exon 4 is shown. Arrow indicates a missense mutation (GCT to GTT) in codon 63 in the tumor tissue (T) as compared with the normal control (N).

In this protocol, the mutated form of the exon can be isolated, reamplified and sequenced without the interference of the normal allele. Since the sequencing reaction is performed at a higher temperature, a lot of secondary structures of the DNA fragment are eliminated. Furthermore, the use of end-labeled primers for DNA sequencing generates more uniform band intensities on the sequencing gel than the use of [α-^{32}P]dCTP in the labeling mixture. In addition, the sequencing results can be obtained in as little as 30 min if the gel is exposed at -70°C with an intensifying screen. The only drawback of this sequencing method is that the autoradiograms of the sequencing gels labeled with ^{32}P are generally not as sharp as those labeled with ^{35}S. However, a much sharper image can be generated if the gel is exposed overnight at room temperature. Apart from analyzing p53 mutation, this protocol has also been shown to work successfully in the determination of genetic mutations in the K-*ras* and the neurofibromatosis 1 (NF-1) genes in our laboratory. Therefore, we believe that this method is particularly useful for rapid determination of the sequence of the amplified DNA fragments generated by PCR.

ACKNOWLEDGMENTS

The authors are grateful for the support and advice for this study from Professors Robert C. Knapp and Ross S. Berkowitz, Harvard Medical School. This study is partly supported by the William Graves Ovarian Cancer Fund, Boston, MA, a research grant from Centorcor, Inc., Malvern, PA, and the Croucher Foundation, Hong Kong.

REFERENCES

1. **Gyllensten, U.B. and H.A. Erlich.** 1988. Generation of single-strand DNA by the polymerase chain reaction and its application to direct sequencing of the HLA-DQA locus. Proc. Natl. Acad. Sci. USA *85*:7652-7656.
2. **Higuchi, R., C.H. von Beroldingen, G.F. Sensabaugh and H.A. Erlich.** 1988. DNA typing from single hairs. Nature *332*:543-546.
3. **Hsu, I.C., R.A. Metcalf, T. Sun, J.A. Welsh, N.J. Wang and C.C. Harris.** 1991. Mutational hotspot in the p53 gene in human hepatocellular carcinomas. Nature *350*:427-428.
4. **Huibregtse, J.M. and D.R. Engelke.** 1991. Direct sequencing and footprinting analysis of yeast DNA by primer extension. Methods Enzymol. *194*:550-562.
5. **Newton, C.R., N. Kalsheker, A. Graham, S. Powell, A. Gammack, J. Riley and A.F. Markham.** 1988. Diagnosis of α1-antitrypain deficiency by enzymatic amplification of human genomic DNA and direct sequencing of polymerase chain reaction products. Nucleic Acids Res. *16*:8233-8243.
6. **Okamoto, A., Y. Sameshima, S. Yokoyama, Y. Terashima, T. Sugimura, M. Terada and J. Yokota.** 1991. Frequent allelic losses and mutations of the p53 gene in human ovarian cancer. Cancer Res. *51*:5171-5176.
7. **Orita, M., H. Iwahana, H. Kanazawa, K. Hayashi and T. Sekiya.** 1989. Detection of polymorphisms of human DNA by gel electrophoresis as single strand conformation polymorphisms. Proc. Natl. Acad. Sci. USA *86*:2766-2770.
8. **Smith, D.P., E.M. Johnstone, S.P. Little and H.M. Hsiung.** 1990. Direct DNA sequencing of cDNA inserts from plaques using the linear polymerase chain reaction. BioTechniques *9*:48-54.
9. **Tsao, S.W., C.H. Mok, K. Oike, M. Muto, H.M. Goodman, E.E. Sheets, R.S. Berkowitz, R.C. Knapp and C.C. Lau.** 1991. Involvement of p53 gene in the allelic deletion of chromosome 17p in human ovarian tumors. Anticancer Res. *11*:1975-1982.
10. **Wrischnik, L.A., R.G. Higuchi, M. Stoneking, H.A. Erlich, N. Arnheim and A.C. Wilson.** 1987. Length mutations in human mitochondrial DNA: direct sequencing of enzymatically amplified DNA. Nucleic Acids Res. *15*:529-542.

AmpliTaq® DNA Polymerase, FS Dye-Terminator Sequencing: Analysis of Peak Height Patterns

L.T. Parker, H. Zakeri, Q. Deng[1], S. Spurgeon[2], P.-Y. Kwok and D.A. Nickerson[1]

Washington University School of Medicine, St. Louis, MO; [1]University of Washington School of Medicine, Seattle, WA; [2]Perkin-Elmer/Applied Biosystems Division, Foster City, CA, USA

ABSTRACT

Taq *DNA polymerases in which the phenylalanine is substituted by a tyrosine at position 667 (Taq F667Y) are members of a new class of DNA polymerases that incorporate chain-terminating dideoxyribonucleoside triphosphates (ddNTPs) much more efficiently than the wild-type* Taq *DNA polymerase. Improved incorporation of ddNTPs into DNA during cycle sequencing using AmpliTaq® DNA polymerase, FS (Taq-FS, a member of the Taq F667Y family), and dye-labeled primers results in nearly uniform peak heights in the sequencing trace. This is not the case when dye-labeled ddNTPs are used in Taq-FS cycle sequencing reactions. While the rate of dye-terminator incorporation is more efficient with Taq-FS, the peak pattern is still highly variable and different from that produced by the wild-type enzyme. We have systematically examined pairs of sequence-tagged sites that vary at only a single nucleotide to determine how base changes influence the peak heights of neighboring bases in sequencing traces generated by the Taq-FS dye-terminator chemistry. In 31 of 64 possible 3-base windows (48%), we find that the peak height of a particular base can be predicted by knowing just one or two bases 5′ to the base in question. We have also compared and contrasted the peak patterns produced by the Taq-FS enzyme with those previously identified for the wild-type enzyme. Establishing the patterns in peak heights within local sequence contexts can improve the accuracy of base-calling and the identification of polymorphisms/mutations when using the Taq-FS dye-terminator cycle-sequencing chemistry.*

INTRODUCTION

Direct sequencing of DNA amplified by the polymerase chain reaction (PCR) is being used increasingly in large-scale sequencing projects and in

(Reprinted from BioTechniques 21:694-699, 1996)

scanning specific regions for polymorphisms and mutations (1–3,8). Among the available sequencing chemistries, cycle-sequencing with thermostable DNA polymerases and dye-labeled dideoxy chain terminators is the most versatile because it eliminates the need for specially modified sequencing primers, since the PCR primers can also be used as sequencing primers (6,8). It is also the simplest sequencing chemistry because only one sequencing reaction is needed for each DNA sample.

Recently, a major improvement in this sequencing chemistry has come from the development of mutant *Taq* DNA polymerases where a phenylalanine residue in the active site has been replaced by a tyrosine (F667Y) (7). Like bacteriophage T7 DNA polymerase, these mutant *Taq* DNA polymerases incorporate chain-terminating dideoxyribonucleoside triphosphates (ddNTPs) into DNA much more efficiently than wild-type *Taq* DNA polymerase (7). This property greatly improves the quality of the sequencing data obtained with this chemistry and reduces the amount of ddNTPs required for each sequencing reaction (5,7). Although improved incorporation of ddNTPs into DNA during cycle sequencing using dye-labeled sequencing primers results in nearly uniform peak heights in a sequencing trace (5), this is not the case when dye-labeled ddNTPs are used in the cycle sequencing reactions. While the rate of dye-terminator incorporation is more efficient when a mutant *Taq* DNA polymerase is used, the peak pattern is still uneven and different from that produced by the wild-type enzyme. Uneven peak heights can decrease the accuracy of base-calling and make mutation and polymorphism detection more difficult. The ability to predict peak heights in a given local sequence context can help to compensate for these drawbacks.

Analysis of the sequencing traces in cycle sequencing using the wild-type enzyme and dye-terminators has revealed sequence context-dependent trends in patterns of dye-terminator incorporation as well as a number of reproducible artifacts (4). Knowledge of these trends and artifacts has increased the accuracy and confidence in identifying polymorphisms and mutations in a sequencing trace and in editing sequences obtained using this chemistry. To increase the accuracy of sequence analysis with the new mutant *Taq* DNA polymerase (F667Y), we have analyzed the trends and reproducibility of dye-terminator incorporation in sequencing traces produced by one member of this new family of polymerases, AmpliTaq® DNA Polymerase, FS. These results are compared to the patterns previously established for dye-terminator incorporation with the wild-type enzyme (4).

MATERIALS AND METHODS

PCR Amplification and Purification of PCR Products

Human genomic DNA (16 ng) from individuals with known genotype were amplified in 40-µL reactions, gel-purified and eluted into 50 µL of wa-

Table 1. Effect of 5′ Bases on Peak Heights of the 3′ Base

Base String[a]	Mean Peak Height (mm)[b]	Range (mm)	Percent <10 mm (small)[c]	Percent 10–20 mm (average)[d]	Percent >20 mm (large)[e]
1. CA**A**	7.0 ± 2.1	3–10	39/39 (100%)		
2. TA**A**	6.9 ± 2.1	3–13	60/65 (92%)		
3. AC**A**	7.3 ± 2.0	5–13	38/41 (93%)		
4. NG**A**	29.1 ± 3.0	16–30			117/120 (98%)
5. AA**C**	14.2 ± 2.3	11–20		38/38 (100%)	
6. GA**C**	13.7 ± 4.4	8–29		26/30 (87%)	
7. TA**C**	15.1 ± 4.0	10–24		27/30 (90%)	
8. TG**C**	16.8 ± 3.5	10–24		26/30 (87%)	
9. CT**C**	25.0 ± 4.5	17–30			30/35 (86%)
10. TT**C**	27.1 ± 4.6	13–30			29/31 (94%)
11. AA**G**	7.6 ± 3.2	2–14	28/30 (93%)		
12. CA**G**	6.1 ± 2.8	3–15	49/51 (96%)		
13. TA**G**	5.5 ± 2.4	2–11	29/30 (97%)		
14. AC**G**	6.0 ± 2.7	2–13	27/30 (90%)		
15. GC**G**	7.4 ± 2.7	3–17	36/41 (88%)		
16. TC**G**	7.1 ± 4.2	3–23	28/32 (88%)		
17. NG**G**	28.1 ± 3.6	16–30			117/120 (98%)
18. AC**T**	12.8 ± 4.0	6–25		28/33 (85%)	
19. GC**T**	11.9 ± 2.3	7–17		27/30 (90%)	
20. TC**T**	12.3 ± 2.7	9–20		29/31 (94%)	
21. TG**T**	26.5 ± 5.5	10–30			35/41 (85%)
22. NT**T**	8.3 ± 2.0	3–13	143/155 (92%)		

[a]The 3′ base is in bold type.
[b]Mean peak height followed by standard deviation.
[c]The proportion of cases for a particular window where the 3′ base has a peak height of ≤1/3 of full scale (≤10 mm, designated as "small").
[d]The proportion of cases for a particular window where the 3′ base has a peak height of 1/3–2/3 of full scale (10–20 mm, designated as "average").
[e]The proportion of cases for a particular window where the 3′ base has a peak height of ≥2/3 of full scale (≥20 mm, designated as "large").

ter as previously described in detail (2,4). The purified DNA was used as sequencing template without further characterization.

Taq-FS Cycle-Sequencing Using Dye-Labeled Terminators

Cycle sequencing was performed on the GeneAmp® PCR System 9600 (Perkin-Elmer, Norwalk, CT, USA) utilizing the PRISM™ Ready Dye Terminator Cycle Sequencing kit with AmpliTaq DNA polymerase, FS (Taq-FS; Perkin-Elmer/Applied Biosystems Division [PE/ABI], Foster City, CA, USA), according to the manufacturer's directions. Briefly, the gel-purified DNA (10.4 μL) was added to a MicroAmp™ reaction tube (Perkin-Elmer)

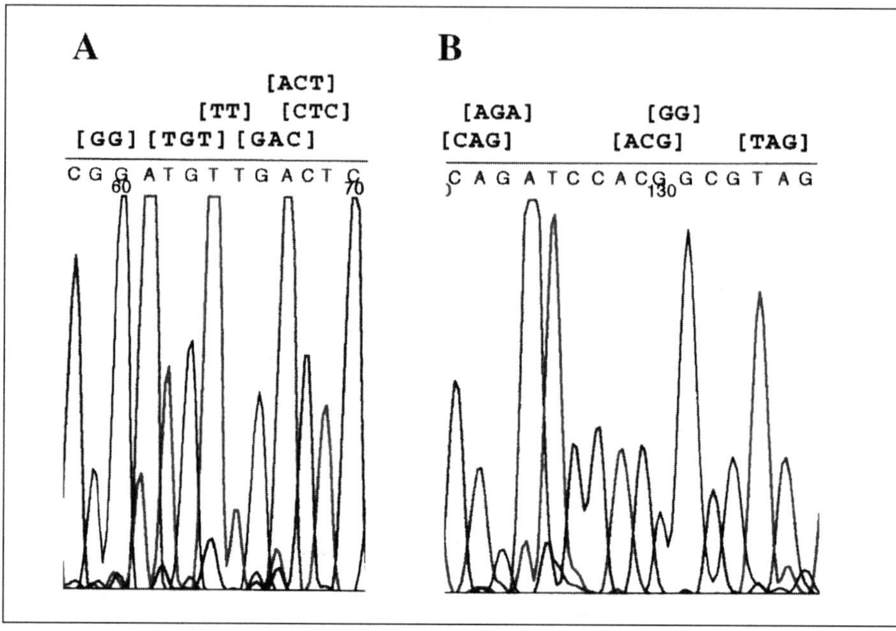

Figure 1. Effects of 5′ base(s) on peak heights. Panels A and B show the patterns of peak height variation in Taq-FS cycle sequencing using dye-labeled dideoxy chain terminators. (A) The G peak following G is large, the T peak following TG is large, the T peak following T is small, the C peak following GA is average, the T peak following AC is average and the C peak following CT is large. (B) The G peak following CA is small, the A peak following AG is large, the G peak following AC is small, the G peak following G is large and the G peak after TA is small.

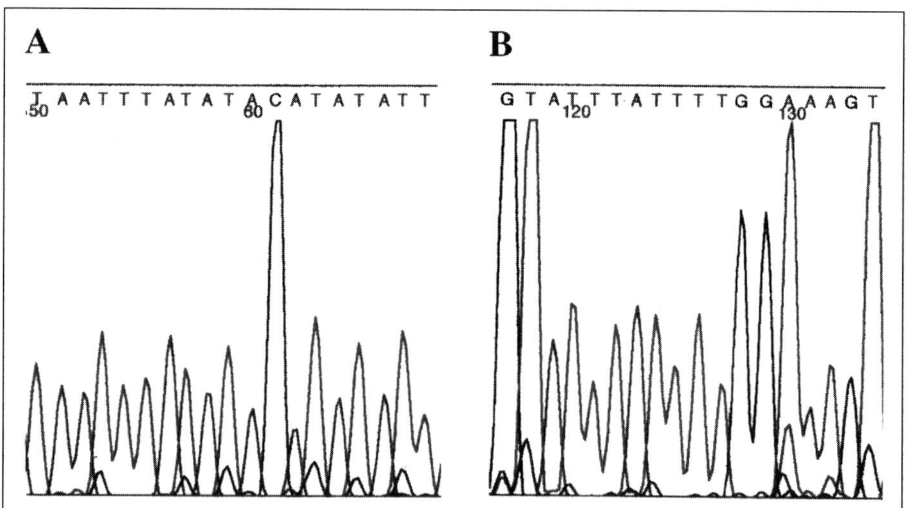

Figure 2. Sequencing artifacts. (A) T peaks after A's contain artifactual C signals. (B) T peaks after G's contain artifactual C signals and A peaks after G's contain artifactual T signals. These artifacts occur frequently, but are not always present.

Table 2. Comparison of Peak Patterns in Taq-FS and AmpliTaq Sequencing Traces

	Base string[a]	Peak Height of 3′ Base: Taq-FS	Peak Height of 3′ Base: AmpliTaq
1.	NT**T**	small	small
2.	CA**A**	small	small
3.	TA**A**	small	small
4.	CT**C**	large	large
5.	TT**C**	large	large
6.	AC**A**	small	large
7.	TC**T**	average	large
8.	NG**A**	large	small/average
9.	NG**G**	large	variable
10.	AA**G**	small	average/large
11.	CA**G**	small	average/large
12.	TA**G**	small	average/large
13.	AC**G**	small	average/large
14.	GC**G**	small	average/large
15.	TC**G**	small	average/large
16.	AC**T**	average	variable
17.	GC**T**	average	variable
18.	AA**C**	average	variable
19.	TG**T**	large	small/average
20.	NG**C**	average/large	small
21.	AA**A**	small/average	small
22.	GA**A**	small/average	small
23.	AT**A**	small/average	small
24.	GT**A**	small/average	small
25.	GC**A**	small/average	large
26.	TC**A**	small/average	large
27.	AC**C**	average/large	large
28.	GC**C**	average/large	large
29.	AT**C**	average/large	large
30.	TA**T**	small/average	large

[a]The 3′ base is in bold type.

containing 3.2 pmol of sequencing primer and 8.0 μL of premix (containing buffer, dNTPs, dye-labeled ddNTPs and Taq-FS/pyrophosphatase). After the initial denaturation at 96°C for 2 min, the reaction mixture was incubated for 25 cycles at 96°C for 15 s, 50°C for 1 s and 60°C for 4 min. Excess dye-labeled terminators were removed from the extension products by spin column purification (CentriSep™ spin columns; Princeton Separations, Adelphia, NJ, USA) according to the manufacturer's directions. [Note: The excess dyes can also be removed successfully by ethanol precipitation, which avoids the labor and expense associated with spin-column purification.] Once separated, the extension products were evaporated to dryness under reduced pressure (SpeedVac®; Savant Instruments, Farmingdale, NY, USA). Each sample was resuspended in 3 µL of loading buffer (5:1, 1% deionized formamide/50 mM EDTA, pH 8.0), heated for 2 min at 90°C and loaded onto an Applied Biosystems 373A sequencer (PE/ABI) according to the manufacturer's instructions.

Analysis of Sequencing Traces

Homozygous individuals containing each of the alternative alleles for 33 different polymorphic sequence-tagged sites (pSTSs: 66 sequencing traces altogether) were sequenced using the Taq-FS enzyme. These traces represent all of the possible three-base windows (64 in total) where the middle base in the window is the only variant base in each pair of sequences. In addition, 20 control sequencing traces, produced using the Taq-FS dye-terminator chemistry, were also analyzed to extend and confirm these trends observed in the allelic pairs of sequencing traces to provide a significantly larger sample of each of the three-base windows. Control sequencing traces were normalized by reanalyzing 3600 data points of the sequencing read using the ABI 373A Analysis Software to produce identical spacing between the traces. To ensure that there was as much uniformity as possible, measurements were taken only from the middle portion (bases 100–200) of the trace. This strategy avoids

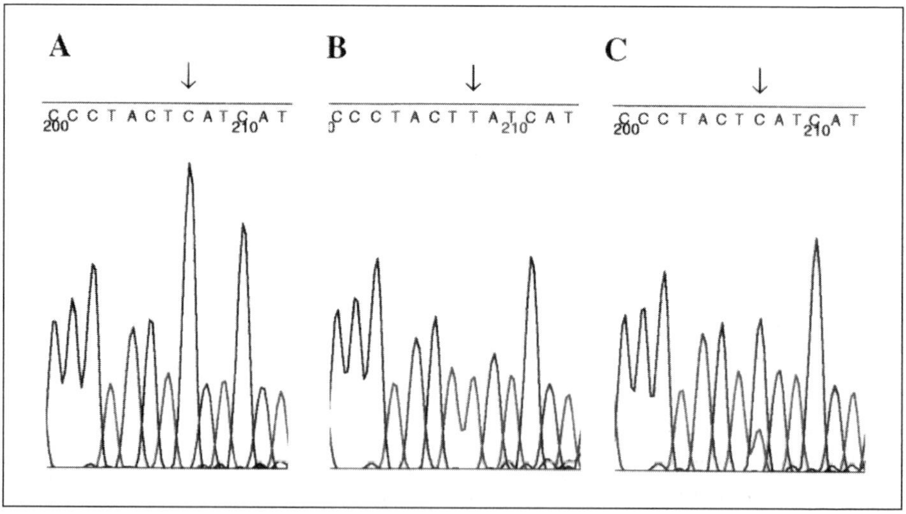

Figure 3. Disparity in peak heights at the polymorphic site in a heterozygote. (A) Sequencing trace of the homozygote with the allele CTC. The 3′ C peak is large and is about twice the size of the C peak in a heterozygote. (B) Sequencing trace of the homozygote with the allele CTT. The 3′ T peak is small but is still twice the size of the T peak in a heterozygote. (C) Sequencing trace of the heterozygote with CT(**C/T**). The base-calling program called this heterozygous base a "C" rather than an "N".

potential problems that could arise by comparing the weaker peaks toward the end of the sequencing run to the stronger peaks at the beginning of the sequencing run. For each of the 64 possible three-base windows, the base most 3′ in each window was measured in the context of the two bases 5′ to it. When a 3′-base peak was found to be equal to or less than 1/3 of full scale, it was defined as a "small" peak. Similarly, when a 3′-base peak was found to be equal to or greater than 2/3 of full scale, it was defined as a "large" peak. When a 3′-base peak was found to be between 1/3 to 2/3 of full scale, it was defined as an "average" peak. For example, if the full height of the analyzed sequencing traces was 30 mm, the "small" peaks measured 10 mm or less in height, "large" peaks measured 20 mm or more in height and average peaks ranged between 10 and 20 mm.

RESULTS AND DISCUSSION

Trends in Dye-Terminator Incorporation by Taq-FS

Analysis of all three-base windows within the polymorphic STSs (64 windows altogether) revealed several trends that were confirmed by analyzing similar three-base windows (>30 independent observations of each window) in 20 control sequencing traces. First, the peak patterns produced with Taq-FS were highly reproducible and predictable in specific sequence contexts. In cases where the two homozygotes of a substitution polymorphism were analyzed (see Table 1 for specific examples), we found that the only affected peak height in these three-base windows was the base immediately 3′ to the substituted nucleotide. Second, only a small number of bases 5′ to a particular base significantly influenced its peak height. For example, among the 64 three-base windows, 31 (48%) consistently showed small, average or large peak heights for the 3′ base. In fact, 12 of these 31 cases were represented by NGA, NGG or NTT (see Table 1). In cases where base-strings of NGA or NGG were found, the 3′ base was consistently "large" (>2/3 full scale in height, see Figure 1). In the base-string of NTT, the 3′ base was found to be consistently "small" (<1/3 full scale in height, see Figure 1A). In each of these instances, knowing only one base immediately 5′ to a particular base was sufficient to predict its peak height.

Several other reproducible trends in the pattern of dye-terminator incorporation by Taq-FS were also apparent. In three-base strings of AAG, CAG, TAG, CAA, TAA, ACG, GCG, TCG or ACA, the 3′ base was found to be consistently "small". In strings of ACT, GCT, TCT, AAC, GAC, TAC or TGC, the 3′ base was consistently "average" in size (1/3 to 2/3 of full scale). For base-strings of CTC, TTC and TGT, the 3′ base was consistently "large". See Figure 1 for selected examples of these trends.

It is interesting to note that we did not find any consistent sequencing artifacts in sequencing traces produced by Taq-FS. In a significant number of cases, however, small C peaks were found under the T peaks in AT strings (90%, Figure 2A), small T peaks were found under the A peak in GA strings (75%, Figure 2B) and small C peaks were found under the T peak in GT strings (90%, Figure 2B) even in high-quality sequencing traces.

When the peak patterns produced by Taq-FS were compared to those obtained using the wild-type enzyme, AmpliTaq, the majority (ca. 80%) of the three-base windows showed patterns (explained below) that differed considerably depending on which enzyme was used in the sequence analysis.

I. Peak patterns that are similar using Taq-FS and AmpliTaq. There are eight 3-base strings that give similar peak patterns with both enzymes. The NTT, CAA and TAA windows all result in small peaks at the 3′ base with either enzyme, while the windows CTC and TTC result in large 3′ C peaks (Table 2).

II. Peak patterns that are different but predictable with both enzymes. Two 3-base strings, namely, ACA and TCT, give different but predictable 3′-peak heights for each enzyme. In sequencing traces produced with AmpliTaq, one finds large 3′ A and 3′ T peaks, whereas in Taq-FS sequencing traces one finds small 3′ A and average 3′ T peaks for ACA and TCT, respectively (Table 2).

III. Peak patterns predictable with Taq-FS but not with AmpliTaq. Eighteen 3-base strings are predictable when Taq-FS is used, but are variable when the wild-type enzyme is used. For example, in Taq-FS sequencing traces, the 3′ base peak is consistently small with base strings AAG, CAG, TAG, ACG, GCG and TCG; large with NGA, NGG and TGT; and average with AAC, ACT and GCT. These 3-base strings produce no consistent patterns in sequencing traces generated by AmpliTaq (Table 2).

IV. Peak patterns variable with Taq-FS but predictable with AmpliTaq. Fourteen 3-base strings have predictable peak heights for the 3′ base in sequencing traces generated with AmpliTaq. These include NGC, AAA, GAA, ATA and GTA (always small); GCA, TCA, ACC, GCC, ATC and TAT (always large). However, there were no consistent trends when Taq-FS was used (Table 2).

Polymorphism/Mutation Detection by Taq-FS Sequencing

In sequencing pSTSs with the wild-type enzyme, we previously found that changes in the peak heights of bases immediately 3′ to polymorphic bases were useful in providing additional criteria for identifying heterozygous bases in a sequencing trace (2,4). In the current study, we have found that the trends in 3′ bases can be used to predict which heterozygous base combinations may be missed by existing base-calling software when Taq-FS sequencing chemistry is used. An example of this situation is shown in Figure 3. In Figure 3A, the window CTC shows the characteristically large 3′ C peak. In Figure 3B, the allelic CTT window shows the characteristically small 3′ T peak. When a heterozygote C/T is present at the 3′ base (Figure 3C), the C peak at the polymorphic site is about one-half the size of the C peak in the homozygous CTC window and the T-peak at this location is about one-half the size of that in the homozygous CTT. Because of the large disparity in peak heights for these two bases, one could predict that this would be a difficult combination for the base-calling program to recognize in the heterozygote. Indeed, the base-calling software assigned a "C" for the polymorphic base in the heterozygote instead of an "N" (Figure 3C).

Although Taq-FS incorporates dye-terminators much more efficiently than does the wild-type enzyme, the patterns produced by this improved DNA polymerase are still quite variable. As our results show, the pattern of dye-terminator incorporation in many instances is highly reproducible and dependent on the local sequence context. For example, ddGTP is incorporated much slower than average by Taq-FS when it is following an A base. This leads to the production of a very small G peak following an A peak. Conversely, ddGTP is incorporated much faster than average when it follows another G peak, therefore producing a very large G peak.

By examining sequences of PCR products (STSs) containing more than 80 single base-pair substitutions, we have identified several general trends among the variable peak heights produced by Taq-FS cycle sequencing using dye-labeled terminators. Our analysis reveals that, in 12 out of the possible 64 three-base windows, the peak height of the 3′ base could be accurately predicted by knowing only the base immediately 5′ to the base in question (NGA, NGG and NTT, see Table 1). In addition, in 19 out of the remaining 52 three-base combinations, the peak height of the 3′ base can be predicted by knowing the 2 bases immediately 5′ to it. In short, in almost half of the cases (48%), one can predict the peak height of a base just by knowing the 2 bases immediately 5′ to it. In the other half of the cases where the pattern of the 3′ base does not display a consistent trend, the height of the 3′ base is usually in the average or higher range (>1/3 full scale). Therefore, these windows are rarely problematic in base assignment. It is possible that the peak heights of 3′ bases in these combinations might be predictable if larger windows of bases were considered. Further elucidation of trends involving more than a string of 3 bases requires a much larger data set and a more rigorous algorithm, which is currently under development.

Some of the patterns found for Taq-FS sequencing are similar to those identified with the wild-type enzyme (AmpliTaq). In this regard, both DNA polymerases give sequencing traces with some troublesome small peaks and these would be excellent targets to address when developing improvements in the sequencing chemistry, i.e., new mutant enzymes or improved reaction conditions. Unlike the extremely small peaks (usually much less than 1/6 of full scale) found in sequencing traces produced by the wild-type enzyme (such as the T peak following another T, the A peak following another A or the C peak following a G), the small peaks found in sequencing traces generated by Taq-FS are usually between 1/6 to 1/3 of full scale, making it relatively easy to assign the bases accurately. The one exception in sequencing with Taq-FS and dye-labeled terminators is the G peak following an A. Here, the G peaks can be extremely small, with a significant proportion of them at <1/6 of full scale.

The ability to predict the expected peak height in sequencing traces can improve base-calling accuracy and help one to recognize heterozygotes with greater confidence. Furthermore, it is possible to predict which overlapping heterozygous bases will be difficult to recognize because of the potential disparity of peak heights for those windows with reproducible trends (Table 1 and Figure 3). Incorporating these trends into base-calling programs will enhance their accuracy and allow for greater automation in the identification of polymorphisms/mutations by DNA sequence analysis.

ACKNOWLEDGMENTS

We thank I. Bauer-Sardina for her technical assistance. This work was supported by grants from the Department of Energy to D.A.N. and P.Y.K. (D.E.-FG06-94ER-61909), the National Science Foundation to D.A.N. (DIR 8809710) and a National Institutes of Health Training Grant to L.T.P. (5T32AR07284).

REFERENCES

1. **Gibbs, R.A., P.N. Nguyen, L.J. McBride, S.M. Koepf and C.T. Caskey.** 1989. Identification of mutations leading to the Lesch-Nyhan syndrome by automated direct DNA sequencing of in vitro amplified cDNA. Proc. Natl. Acad. Sci. USA *86*:1919-1923.
2. **Kwok, P.Y., C. Carlson, T.D. Yager, W. Ankener and D.A. Nickerson.** 1994. Comparative analysis of human DNA variations by fluorescence-based sequencing of PCR products. Genomics *23*:138-144.
3. **Lee, L.G., C.R. Connell, S.L. Woo, R.D. Cheng, B.F. McArdle, C.W. Fuller, N.D. Halloran and R.K. Wilson.** 1992. DNA sequencing with dye-labeled terminators and T7 DNA polymerase: effect of dyes and dNTPs on incorporation of dye-terminators and probability analysis of termination fragments. Nucleic Acids Res. *20*:2471-2483.
4. **Parker, L.T., Q. Deng., H. Zakeri, C. Carlson, D.A. Nickerson and P.-Y. Kwok.** 1995. Peak height variations in automated sequencing of PCR products using *s* dye-terminator chemistry. BioTechniques *19*:116-121.
5. **Reeve, M.A. and C.W. Fuller.** 1995. A novel thermostable polymerase for DNA sequencing. Nature *376*:796-797.
6. **Rosenthal, A. and D.S. Charnock-Jones.** 1992. New protocols for DNA sequencing with dye terminators. DNA Seq. *3*:61-64.
7. **Tabor, S. and C.C. Richardson.** 1995. A single residue in DNA polymerases of the Escherichia coli DNA polymerase I family is critical for distinguishing between deoxy- and dideoxyribonucleotides. Proc. Natl. Acad. Sci. USA *92*:6339-6343.
8. **Tracy, T.E. and L.S. Mulcahy.** 1991. A simple method for direct automated sequencing of PCR fragments. BioTechniques *11*:68-75.

Efficient, Automatic Detection of Heterozygous Bases During Large-Scale DNA Sequence Screening

R.S. Phelps, R.B. Chadwick[1], M.P. Conrad[1], M.N. Kronick[1] and A. Kamb

Myriad Genetics, Inc., Salt Lake City, UT and [1]Perkin-Elmer, Applied Biosystems Division, Foster City, CA, USA

ABSTRACT

A crucial factor in the success of positional cloning efforts is the ability to screen rapidly many different candidate genes for mutations. By modifying standard software, we have improved the detection of heterozygous base positions in PCR products sequenced by cycle sequencing. A key element of the method is the incorporation of a modified heterozygote detection algorithm that permits the use of DNA sequence data derived from PCR and sequencing reactions that have not been fully optimized. This allows sequencing runs of average quality to be used. We demonstrate that the sensitivity and specificity of the method are well suited to mutation detection applications such as positional cloning.

INTRODUCTION

The vast amount of DNA sequence information derived from the Human Genome Intitiative can be exploited to explore human genetic diversity—neutral polymorphism and mutation—on a scale heretofore unthinkable. Integration of disease gene isolation with genetic diversity studies facilitates risk analysis and presymptomatic disease testing. But to collect and compare DNA sequences rapidly, it is essential to detect sequence variants efficiently. In the general case, this involves detection of heterozygous point mutations.

Recently, several strategies for detection of heterozygous point mutations have been developed (Reference 2, for review). These include techniques that do not involve direct DNA sequence determination as well as sequencing methods that depend on hybridization rather than gel electrophoresis. However, for de novo detection of previously unknown sequence variants, gel-based DNA sequencing remains the best and most widely used method.

(Reprinted from BioTechniques 19:984-989, 1995)

Software to automatically detect heterozygous nucleotide positions has been developed by a number of commercial and academic groups (6,9). In general, the algorithms for calling heterozygous base positions involve peak height comparisons (4,6,8,9). At each base position on the chromatogram, the peaks of all 4 bases are located and the intensity (I_1) of the highest peak is compared to the intensity (I_2) of the second highest peak. If the ratio of the two intensities (I_2/I_1) exceeds a certain threshold, a heterozygote is called at that position. This simple approach works well when the DNA sequence is of high quality and the peak heights are fairly uniform. However, in many instances it is difficult and time-consuming to optimize polymerase chain reaction (PCR) and sequencing conditions so as to generate DNA sequence of the highest quality. For example, during positional gene cloning, a multitude of fragments must be screened for mutations to identify linked sequence variants. It is prohibitive in terms of time and expense to optimize each fragment for DNA sequence determination using conventional methods that employ T7 DNA Polymerase (Sequenase®) and biotin capture (5). Under such circumstances, a strategy for detection that incorporates cycle sequencing is preferred. Here we present a method to screen PCR fragments rapidly using cycle sequencing. This method includes a heterozygote detection algorithm that involves a second peak height comparison to eliminate a large fraction of false-positive heterozygote calls (noise) that otherwise obscure the true heterozygotes. The method is general and could be implemented with a variety of machines and software. We have used standard equipment and software to show that the method has excellent signal-to-noise properties even when the sequence quality is mediocre.

MATERIALS AND METHODS

PCR Conditions

PCR products were amplified using genomic DNA from human placenta (Sigma Chemical, St. Louis, MO, USA) or from various pedigree individuals. Two PCR amplicons derived from the human BRCA1 gene were analyzed: ALG27 and ALG35 (7). In some individuals who carry genetic predisposition to breast cancer, ALG27 contains a heterozygous A→C transversion in exon 21, while in others, ALG35 contains an 11-bp heterozygous deletion in exon 2 (7). The primers used for amplification were: ALG26, FM13F (TGT AAA ACG ACG GCC AGT ATT CCC TGC TCG GGT TCA GAT); RM13R (CAG GAA ACA GCT ATG ACC CAT CGT GGG ACC TTG CTT ATA); ALG35, FT7 (TAA TAC GAC TCA CTA TAG GGG TTC ATT GGA ACA GAA AGA AAT GG); and RT3 (ATT AAC CCT CAC TAA AGG GAC TTC CCT AGT ATG TAA GGT CAA). A variety of $MgCl_2$ concentrations (1.0–3.5 mM) and buffer pH values (8.0–10.0) were tested for each fragment using a PCR optimization kit according to the manufacturer's instructions

(Invitrogen, San Diego, CA, USA). Reaction mixtures contained 10 ng template DNA, 50 nM tailed primers (each primer was a 38-mer: 18 bases of a universal primer at the 5′-end followed by 20 bases of gene-specific sequence at the 3′-end), 100 µM dNTPs and buffer from the PCR optimization kit in a 100-µL reaction volume. For both ALG35 and ALG26, the buffer conditions were 3.5 mM $MgCl_2$ and pH 8.5. Two cocktails were prepared: *(i)* a lower mixture containing water, dNTPs and primers; and *(ii)* an upper mixture containing water, buffer, enzyme and template. Fifty microliters of lower mixture were placed in a tube with one AmpliWax® PCR Gem 100 bead (Perkin-Elmer, Norwalk, CT, USA), heated to 80°C for 2 min and cooled to 4°C for 1 min. Fifty microliters of upper mixture were added to the cooled lower mixture and the thermal cycling process begun. The cycling conditions on a GeneAmp® PCR System 9600 (Perkin-Elmer) were: one cycle at 98°C (20 s); 8 cycles at 98°C (5 s), 62°C (30 s), 70°C (1 min); 32 cycles at 96°C (10 s), 70°C (1 min).

Note that the primer and dNTP concentrations were lower compared to normal PCR conditions so as to minimize interference in subsequent DNA sequencing reactions. This permitted dilution of the amplified product into the sequencing reactions without any purification steps. It is critical to avoid primer dimer if this process is to be successful.

DNA Sequencing Chemistry

The DNA sequencing reactions were carried out using cycle sequencing with dye-labeled primers and the Prism™ Ready Mix kit based upon AmpliTaq® CS+ Polymerase (currently Perkin-Elmer/Applied Biosystems Division [PE/ABI], Foster City, CA, USA). Briefly, after a 1:10 dilution in water, the PCR template was mixed with the standard Ready Mix cycle sequencing kit (PE/ABI), dye-labeled M13 forward or reverse primers and cycled in a PCR System 9600 for 15 cycles of 95°C for 30 s, 55°C for 30 s and 70°C for 1 min. An additional 15 cycles were then used of 95°C for 30 s and 70°C for 1 min. After pooling and ethanol-precipitation, the individual reactions were loaded on an ABI Model 373 Sequencer (PE/ABI).

Software Analysis

Data were collected on a Model 373A Sequencer and analyzed with ABI Sequence Analysis software Version 2 (PE/ABI). Sequence files were analyzed for heterozygosity with the commercially available Version 1.2 of the ABI Factura™ application (PE/ABI) (included in the ABI Sequence Navigator™ software) or with a specially modified version of the software (not available commercially; licensed users of Factura Version 1.2 should contact Morgan Conrad at PE/ABI, Foster City, CA, USA, for information on obtaining a version of Factura that incorporates the improvements discussed in this article). This output was used for sequence alignment and analysis of double-stranded sequence. The read length used for the analysis was 198 bases for ALG35 and 250 bases for ALG26. For ALG26, false-positive rates from 3 forward and 4 reverse runs were totaled. For ALG35, false-negative rates from 2 forward and 2 reverse runs were totaled. For the numbers plotted in Figure 4, the denominators for the percentage calculation were 198 for each ALG 35 (graph A) run and 250 minus the number of positions flagged as stops by the modified software (on average, 18 for 0.3 threshold) for each ALG 26 run (graph B). This approach was conservative, since it tended to favor the unmodified software performance by inflating false-positive rates for the modified version.

RESULTS AND DISCUSSION

Rapid Screening for Mutations

At present, the most accurate DNA sequence determination of PCR products involves Sequenase DNA Polymerase (United States Biochemical, Cleveland, OH, USA) in conjunction with dye-labeled primers and streptavidin capture of biotinylated, single-stranded template (5). This protocol and its variants produce DNA sequence that can be read automatically by machine. In addition, commercially available software can detect heterozygous positions within such a sequence with great reliability (6). However, the procedure is time-consuming relative to cycle sequencing, and it requires at least

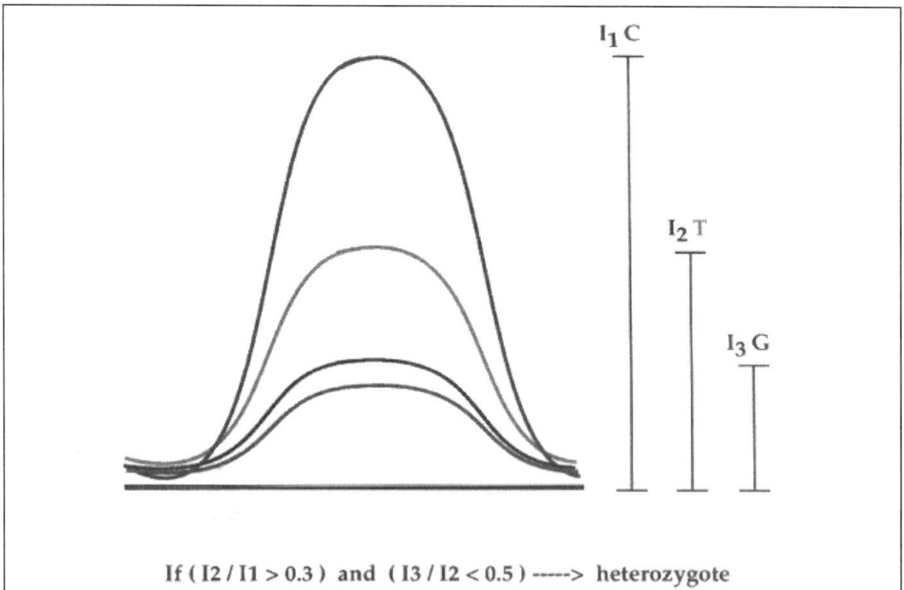

Figure 1. Diagram of the principle behind heterozygote detection. Two peak-height ratio comparisons are made to minimize false-positive heterozygote calls.

50-fold more sequencing template.

In certain situations, for instance, positional cloning projects, it is advantageous to screen hundreds of distinct DNA fragments for sequence variants (7). The speed and accuracy with which this can be accomplished is a crucial factor in the success of the enterprise. There is a trade-off between speed and accuracy, but in these cases it is reasonable to sacrifice some accuracy for a large gain in speed. Thus, it may be desirable to obtain imperfect sequence rapidly while still maintaining an acceptable rate of mutation detection. We have modified the standard ABI mutation screening software to detect mutations and polymorphisms under conditions where the DNA sequence is suboptimal. The modification involves incorporation of a second comparative step in the detection algorithm (Figure 1). Not only must the base position display a ratio of the top 2 intensities (I_2/I_1), which is above a specified threshold, but also the ratio of the second and third highest intensities (I_2/I_3) must fall below another threshold. In the analysis reported here, this second threshold was set at 0.5. Only when both conditions are satisfied is a heterozygote position called. If the I_2/I_3 intensity ratio exceeds a specified threshold, the position is flagged as a strong stop. This extra condition enforces a level of specificity and selectivity on the algorithm that eliminates much of the background noise (false-positives) that complicates detection.

The entire procedure for automatic detection involves the following (Figure 2): *(i)* PCR of the fragment of interest using gene-specific primers with 18-base generic "tails"; *(ii)* cycle sequencing biochemistry using the generic, dye-labeled primers to prime synthesis on both strands (separately); *(iii)* electrophoresis; *(iv)* automated heterozygote analysis by a modified version of the Factura software; and *(v)* comparison of the two strands and elimination of residual false-positive base positions that are not called as heterozygotes in both directions. During the development stages, the PCR is optimized in a single step by testing 16–32 reaction conditions (see Materials and Methods). If no set of conditions is found to achieve satisfactory sequence quality, the primers are redesigned, and the optimization process is repeated.

Sensitivity and Selectivity of Method

We quantified the performance of the procedure by testing the false-positive rate (frequency with which base positions are flagged incorrectly as heterozygotes) and the false-negative rate (frequency with which true heterozygote positions are not flagged). I_2/I_1 thresholds were examined using several forward and reverse sequencing reactions performed on 2 different amplified fragments from parts of the human BRCA1 gene (see Materials and Methods; Reference 6). Each of these runs involved an independently amplified product.

False-positive rates were estimated using PCR products that contained a single heterozygous base. In all cases, this base was correctly called. However, depending on the thresholds set and the particular sequence being analyzed, different numbers of false-positives were also called (Figure 3A and Figure 4A). The false-positive rates using the modified Factura software were dramatically lower than the rates using the unmodified version. For example, at an I_2/I_1 threshold of 0.3, the modified software had an average false-positive rate nearly 20 times less than the unmodified version. This is because most of the false-positive calls made by the original software involved positions where there were moderate signals for all bases (Figure 3). These positions were easily identified by the second peak-height comparison. Therefore, the selectivity of the method was dramatically improved by the software modification so that imperfect sequences did not generate an avalanche of false-positive calls.

To estimate false-negative rates, PCR fragments that contained frameshift mutations were used (1). Similar frameshift mutations have been analyzed previously by ABI software (3). Fragments that contain frameshift mutations were selected to determine false-negative rates for two reasons: *(i)* this test generated large numbers of artificial heterozygote positions and made it possible to quantify the performance of the method in this context (Figure 3B and

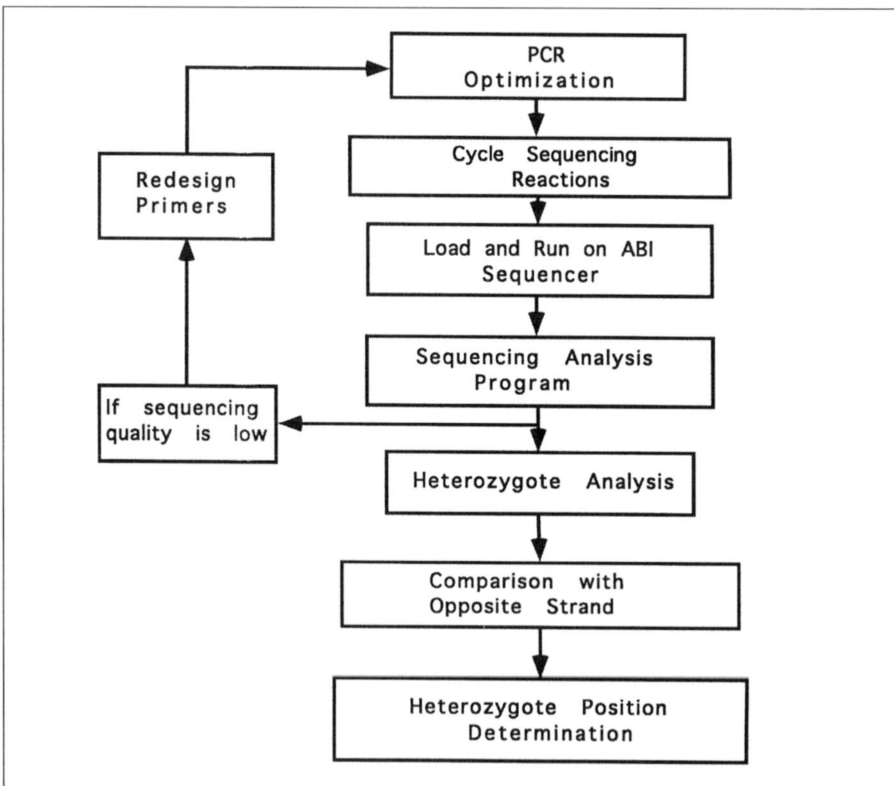

Figure 2. Work-flow diagram for detection of heterozygotes in cycle-sequenced PCR products.

Table 1. Signal-to-Noise Ratios for the Analyses

		I_2/I_1 Threshold 0.2	0.3	0.4	0.5
Signal/Noise	Modified	26.3	162.5	462.7	640
	Unmodified	7.8	9.3	38.6	64.7
	Mod./Unmod.	3.4	17.5	12	9.9

The ratio is the fraction of true positives called over the fraction of false-positives called (total # of true positives - total # false-negatives / total # false-positives). The last row shows ratios of the 2 signal-to-noise ratios in the above rows. "Modified" refers to the modified Factura software; "Unmodified" refers to the original, commercially available software.

Figure 4B) and *(ii)* the test was judged to be more stringent than other tests because heterozygote detection using frameshift mutations may be confounded due to the variable peak heights of 2 sequences superimposed beyond the frameshift position. Thus, the false-negative rates of this test may be higher than those using fragments that contain point mutations. The false-negative rates for single-stranded sequence of frameshift mutants using the modified software were roughly equivalent to rates using the original version. For example, at an I_2/I_1 threshold of 0.3, the average false-negative rate was only 1.2 times higher for the modified software than for the unmodified version. Therefore, the added noise filter did not significantly impair the heterozygote-calling capacity of the software. The overall improvement in performance at the 0.3 threshold as measured by the signal-to-noise ratio (number of true positives over number of false positives) was greater than 17-fold (Table 1).

Our goal is to reduce human involvement in the process of heterozygote detection. Thus, it is preferable to avoid visual inspection of sequence chromatograms. The results presented here suggest that an I_2/I_1 threshold of 0.3 is optimal for automatic heterozygote detection under the conditions that we tested. At this threshold, the signal-to-noise ratio is high and the false-negative percentage is reasonable (about 10%). The false-positive rate of 1% at an I_2/I_1 threshold of 0.3 achieved by the methods described here is still too high if a single strand is surveyed. However, if the additional constraint was placed on the call that the heterozygous position must be present on both strands, the rate dropped to about 0.01%—roughly the square of the rate on one strand. If this constraint was applied to true positives as well, the rate at which heterozygotes would be missed may increase to roughly 20% $(1–0.1)^2$. This expectation is more difficult to test explicitly since the frameshift mutations generate artificial heterozygotes that do not overlap on the 2 strands. It is possible that the performance might be improved using other threshold values, depending on the circumstances and sequence quality. One can envision, for example, statistical analysis of peak-height ratios or sequence-specific threshold optimization to further improve the software performance.

Comparison with Manual Radioactive Sequencing

Manual, radioactive DNA sequencing is an effective method for detecting sequence variants. In a previous effort to find BRCA1, roughly 15 000 bp were screened for mutations and polymorphisms in 8 individuals (3,6). Nineteen different sequence changes were identified (D. Shattuck-Eidens and A. Kamb, unpublished). The manual method, though difficult to quantify, is believed to be very sensitive and selective, largely because the human eye is adept at pattern recognition. If the A reactions from several samples are loaded side by side, followed by the C, G and T reactions, differences are readily apparent even when the autoradiographic image of the fragments is mediocre. The read length of such experiments under normal conditions is about 250 bp. We consider manual, radioactive sequencing to be the standard against which other methods should be compared.

The advantages of automated heterozygote detection using the procedure described above compared to manual, radioactive sequencing are: *(i)* it is relatively straightforward to quantify the performance of the method; *(ii)* all sequencing reactions of reasonable quality can be used to improve a consensus sequence; *(iii)* the procedure minimizes human involvement in the detection;

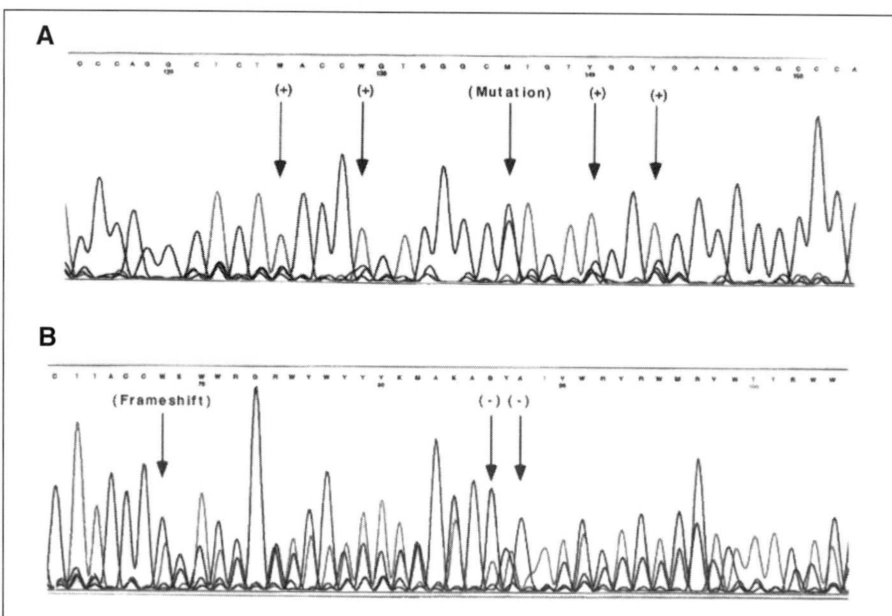

Figure 3. Portions of representative chromatograms from cycle sequencing experiments using as template (A) a BRCA1 PCR amplicon (ALG26) with a single heterozyous mutation; or (B) a BRCA1 amplicon (ALG35) with a frameshift mutation at a position marked on the chromatogram (7). "(+)" symbols mark the positions where the original Factura software called a heterozygote at a 0.3 I_2/I_1 threshold, but where the modified software did not (false positives). "(-)" symbols show heterozygous positions that were not called by the software (false negatives).

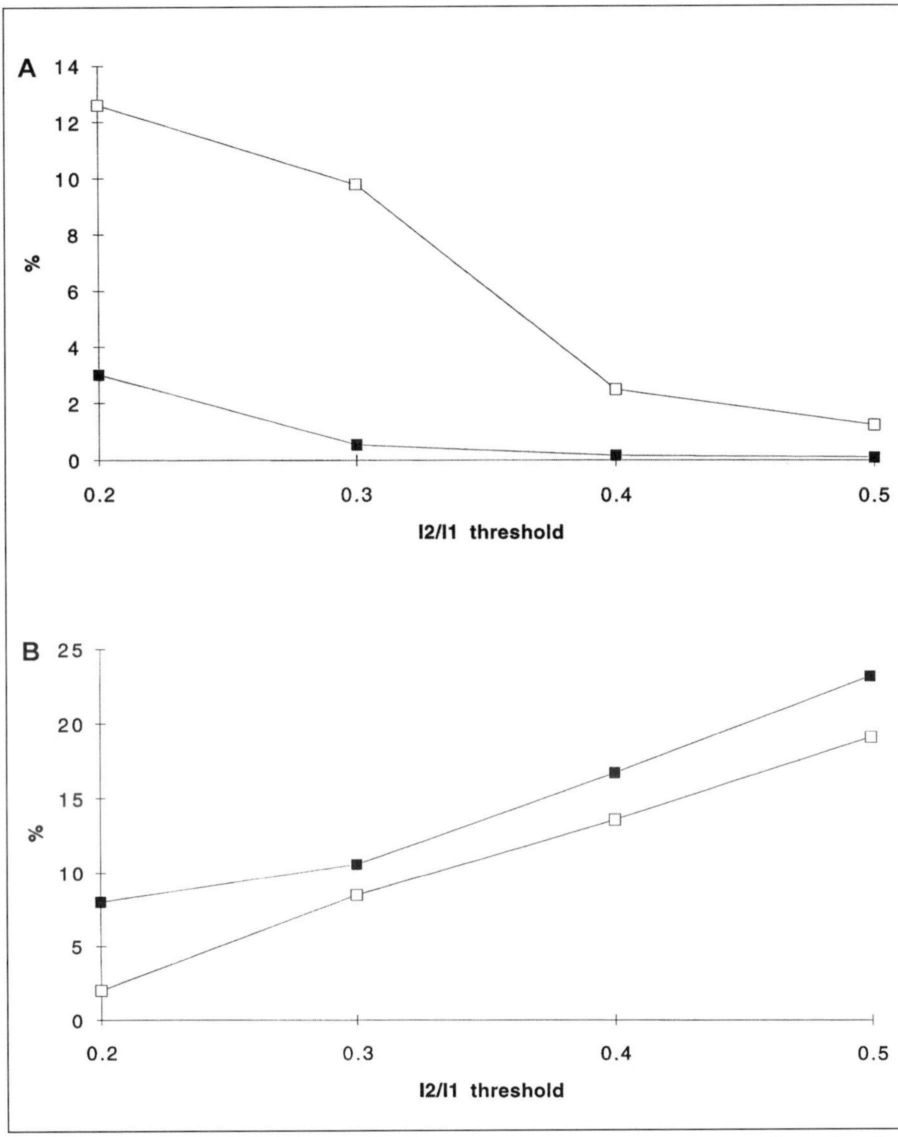

Figure 4. Graphs plotting the percentage of miscalls against the I_2/I_1 threshold using amplicons that allow estimation of (A) false-positive rates. For example, for the 0.3-threshold value, the false-positive percentages were (100)(9/1621) for the modified software (9 = total # of false positives; 1621 = total # of bases analyzed, see Materials and Methods) and (100)(172/1747) for the unmodified software. (B) False-negative rates. For the 0.3-threshold value, false-negative percentages were (100)(21/198) for the modified software (21 = total # of false negatives; 198 = total # of true positives) and (100)(17/198) for the unmodified software. The 2 sequence sets used to estimate false-negative and false-positive rates (ALG35 and ALG26, respectively) were of comparable quality as judged by using the unmodified Factura software to determine the average false-positive rate at an I_2/I_1 threshold of 0.3 for each sequence set in a region unaffected by the frameshift mutation (upstream of base 67) in all sequences. The false-positive rate for the ALG35 set in this region was 9.3% compared to 8.7% for the ALG26 set. The open squares indicate the unmodified Factura software, while the filled squares indicate the modified version.

(iv) radioisotopes are eliminated; *(v)* the sequencing chemistry is simple; *(vi)* avoidance of Sequenase, magnetic beads and large quantities of PCR template reduces cost; and *(vii)* we expect an increased speed of detection as measured by bases screened/time.

CONCLUSION

It is important to match the method of heterozygote detection to the specific task. In many cases, for example, in a diagnostic setting, nearly perfect detection is required. Thus, the sequence quality is crucial, and investment in sequence optimization is well worthwhile. However, when a multitude of different sequences must be screened rapidly, it is preferable to use methods such as those discussed here, which, in the absence of extensive optimization, still perform well and are executed simply and inexpensively. The procedures we describe, which combine the use of direct cycle sequencing of PCR products and improved software, allow accurate, automatic detection of heterozygotes under such circumstances where the sequence quality may be suboptimal. The detection algorithm is not dependent on specific sequencing chemistry. It should perform better with higher quality sequence and with more uniform peak heights such as those produced by Sequenase-like enzymes. This work represents a step in the development of more sophisticated schemes for automated heterozygote detection, which incorporate information in addition to a single peak-height comparison, such as contextual information or interlane comparative information.

REFERENCES

1. **Grompe, M.** 1993. The rapid detection of unknown mutations in nucleic acids. Nat. Genet. *5*:111-117.
2. **Harshman, K., R. Bell, J. Rosenthal, H. Katcher, Z. Gholami, C. Frye, W. Ding, P. Dayananth, et al.** 1995. Comparison of the positional cloning methods used to isolate the BRCA1 gene. Hum. Mol. Genet. *4*:1259-1266.
3. **Kolodner, R.D., N.R. Hall, J. Lipford, M.F. Kane, R.R.S. Rao, P. Morrison, L. Wirth, P.J. Finan, J. Burn, P. Chapman, C. Earabino, E. Merchant and D.T. Bishop.** 1994. Structure of the human MSH2 locus and analysis of two Muir-Torre kindreds for msh2 mutations. Genomics *24*:516-526.
4. **Kwok, P., C. Carlson, T.D. Yeager, W. Ankener and D.A. Nickerson.** 1994. Comparative analysis of human DNA variations by fluorescence-based sequencing of PCR products. Genomics *23*:138-144.
5. **Leren, T.P., O.K. Rodningen, O. Rosby, K. Solberg and K. Berg**. 1993. Screening for point mutations by semi-automated DNA sequencing using Sequenase® and magnetic beads. BioTechniques *14*:618-623.
6. **McGinnis, M.D., M.P. Conrad, A.G.M. Bouwens, M.G.J. Tilanus and M.N. Kronick.** 1995. Automated, solid phase sequencing of DRB region genes using T7 sequencing chemistry and dye-labeled primers. Tissue Antigens *46*:173-179.
7. **Miki, Y., J. Swensen, D. Shattuck-Eidens, P.A. Futreal, K. Harshman, S. Tavtigian, Q. Liu, C. Cochran, et al.** 1994. A strong candidate for the breast and ovarian cancer susceptibility gene BRCA1. Science *266*:66-71.
8. **Rozemuller, E., A.G.M. Bouwens, B.E.J.E.G. Bast and M.G.J. Tilanus.** 1993. Assignment of HL-DPB alleles by computerized matching based upon sequence data. Hum. Immunol. *37*:207-212.
9. **Versluis, L.F., E. Rozemuller, S. Tonks, S.G. Marsh, A.G.M. Bouwens, J.G. Bodmer and M.G.J. Tilanus.** 1993. High resolution HLA-DPB typing based upon computerized analysis of data obtained by fluorescent sequencing of the amplified polymorphic exon 2. Hum. Immunol. *38*:277-283.

Update to:

Efficient, Automatic Detection of Heterozygous Bases During Large-Scale DNA Sequence Screening

Mel N. Kronick
Perkin-Elmer, Applied Biosystems Division, Foster City, CA, USA

The approach described in this article was motivated by a desire to improve the accuracy with which heterozygote or mixed base positions appeared in data generated by automated sequencers. The approach is especially valuable when the application is screening and the PCR used to generate sequencing template is only crudely optimized. The presence of third or fourth peaks at a particular base position usually indicates the existence of noise which, more often than not, can be mistaken for a true heterozygote. By insisting that a true second base is at least a defined percentage greater than the third most intense peak, a noticeable improvement in accuracy of heterozygote detection is achieved. This improvement has now been incorporated into the commercially available software used for heterozygote identification so that interested researchers need not contact the authors (2). The problem of signal to noise in automated sequencer data can be aggravated by the peak-to-peak variability in intensity that results from the fact that the incorporation efficiency of sequencing enzyme is dependent on sequence context. In the paper by Phelps et al., the sequencing enzyme employed was the AmpliTaq® CS+ enzyme, which does exhibit noticeable base-to-base variation in dideoxynucleotide incorporation. Since the publication of this paper, significantly improved thermally stable sequencing enzymes have become available (3,5,6). The significant increase in uniformity of dideoxynucleotide incorporation has the effect of making the second most intense base in a heterozygote almost always above the noise. The advantages of the application of one such enzyme, AmpliTaq® FS enzyme, to sequencing heterozygote samples has already been demonstrated (1). Furthermore, as noted above with regard to Phelps et al., more extensive optimization of the PCR will always help reduce noise. Also, sequencing both strands and comparison of reference to wild-type sequence provide significant additional degrees of control and cross-checking. Finally, the use of pattern comparison algorithms in addition to chemistry improvements will provide additional means to determine with increasing accuracy the position of mixed bases in a sequence (4).

REFERENCES

1. **Chadwick, R.B., M.P. Conrad, M.D. McGinnis, L. Johnston-Dow, S.L. Spurgeon and M.N. Kronick.** 1996. Heterozygote and mutation detection by direct automated fluorescent DNA sequencing using a mutant *Taq* DNA polymerase. BioTechniques *20*:676-683.
2. Factura™ 2.0, PE Applied Biosystems, Foster City, CA.
3. **Kalman, L.V., R.D. Abramson and D.H. Gelfand.** 1995. Thermostable DNA polymerases with altered discrimination properties. Genome Sci. Technol. *1*:42.
4. **Parker, L.T., H. Zakeri, Q. Deng, S. Spurgeon, P.-Y. Kwok and D.A. Nickerson.** 1996. AmpliTaq® DNA Polymerase, FS dye-terminator sequencing: analysis of peak height patterns. BioTechniques *21*:694-699.
5. **Spurgeon, S.L., F.L. Reichert, S.M. Koepf, S.M. Chen, R.D. Abramson, J. Brandis, L.V. Kalman and A.M. Wang.** 1995. Improved fluorescent dye primer and dye terminator sequencing using a new mutant of DNA polymerase. Genome Sci. Technol. *1*:48.
6. **Tabor, S. and C.C. Richardson.** 1995. A single residue in DNA polymerases of the Escherichia coli family is critical for distinguishing between deoxy- and dideoxyribonucleotides. Proc. Natl. Acad. Sci. USA *92*:6339-6343.

Rapid Cycle Sequencing in an Air Thermal Cycler

Harold Swerdlow, Kerry Dew-Jager and Raymond F. Gesteland

University of Utah, Salt Lake City, UT, USA

ABSTRACT

Cycle sequencing using Taq *DNA polymerase has gained popularity recently due to reduced template requirements, improved signal and its ability to directly sequence PCR fragments. A major drawback to the technique is the time required for performing reactions in a block-based thermal cycler. To help cycle sequencing compete with other methods, we have modified the protocol to be performed in capillaries using an air-based thermal cycling instrument. This instrument has been developed and optimized for rapid, specific amplification of DNA by PCR. The resulting cycle sequencing methodology is faster than block-based approaches; a reaction can be completed in 25 min, compared with about 2 h in a conventional instrument. Thus, the speed of the technique is competitive with standard uncycled T7 or* Taq *reactions. Accuracy of the sequencing data is improved; two problem areas in the sequence obtained with a block cycler are ameliorated by the capillary methodology. This technique represents a novel approach to cycle sequencing that will further the development of capillary-based analytical methods.*

INTRODUCTION

With the advent of the Human Genome Initiative, pressure has been placed upon the scientific community to develop DNA sequencing methods that are simultaneously more rapid, accurate and cost-efficient than current methodologies (21,23). The polymerase chain reaction (PCR) has allowed researchers to easily amplify high-purity templates from complex mixtures of nucleic acids (15,16). Direct sequencing of PCR products has been hampered to a varying extent by the presence of competing primers remaining from the reaction, by spurious DNA amplification products and by the double-stranded nature of the templates produced (6,7,10). Cloning of PCR products for sequencing purposes is both time-consuming and error-prone (19). Asymmetric PCR has been used to generate single-stranded templates for sequencing by forcing linear amplification with an excess of one primer over the other

(Reprinted from BioTechniques 15:512-519)

(8,12). Magnetic beads coated with streptavidin have been used to purify a biotinylated template strand from its complementary strand. This allows PCR products to be sequenced accurately by single-strand techniques (9).

An alternative approach, cycle sequencing, takes advantage of the thermostability of *Taq* DNA polymerase to generate multiple, dideoxynucleotide-terminated sequence fragments from a single-template molecule (1,2,4,17). This method of sequencing requires only a few hundred nanograms of template DNA to produce ample quantities of sequence product for manual or automated, radioactive or fluorescent techniques. Cycle sequencing works well as a method to directly sequence PCR products, probably due to the lower amounts of template DNA required and an associated reduction of primer carryover. Numerous commercial kits are now available to simplify the routine use of this technique in the laboratory (18). Although cycling provides several advantages over conventional sequencing protocols, it is far slower. At least 2 h of cycle time are required on a block thermal cycler compared with a 5- to 10-min incubation period in standard methods.

The attraction of cycle sequencing could be improved by a radical reduction in the total cycling time. To accomplish this, we have adapted cycle sequencing protocols to a rapid air thermal cycler. This commercial instrument, developed by Wittwer and collaborators for PCR, employs thin-walled capillary tubes (500-µm i.d., 1000-µm o.d.), with a low thermal mass, and air, with its low specific heat, to accomplish very rapid temperature transitions (25). Consequently, it has been shown that the kinetics of denaturation, annealing and elongation are much faster than previously suspected and allow 30 PCR cycles to be completed in 10–25 min (24). Furthermore, minimizing the time spent at or near optimal annealing temperatures can improve specificity in some amplifications (26). Using this rapid thermal cycling technology has allowed completion of a standard cycle-sequencing protocol in only 25 min.

MATERIALS AND METHODS

Sequencing Templates

Single-stranded control and experimental cycle sequencing reactions used M13mp18 DNA (United States Biochemical, Cleveland, OH, USA) as template. Double-stranded reactions were performed with PCR products from a cDNA clone derived from the zebra fish (*Brachydanio rerio*) *goosecoid* gene (20). PCR products were purified from a 1% agarose gel and eluted using standard procedures with Geneclean® (Intermountain Scientific, Bountiful, UT, USA) before use as templates.

Reaction Mixtures

Fluorescent primers with *Taq* DNA polymerase. All reactions were performed in 10 μL of a buffer containing 80 mM Tris-HCl pH 8.9, 20 mM KCl, 5 mM $MgCl_2$, with 1.3 units of AmpliTaq® DNA Polymerase (Perkin-Elmer, Norwalk, CT, USA) added to each. Fifty micrograms per milliliter bovine serum albumin (BSA) (Molecular Biology Grade; Boehringer Mannheim, Indianapolis, IN, USA) were added to all reactions, including block controls, to stabilize the enzyme in the glass capillary tubes (26). Reactions contained 0.4 pmol of FAM (C) and JOE (A) and 0.8 pmol of TAMRA (G) and ROX (T) dye primers (-21 M13 fluorescent 18-mers; Applied Biosystems, Foster City, CA, USA). Approximately 100 ng of template DNA were used for each of the C and A reactions and 200 ng for each of the G and T reactions, whether single- or double-stranded. Deoxy- and dideoxynucleotides were obtained from Pharmacia Biotech (Piscataway, NJ, USA) and their final concentrations were as follows:

C reactions—12.5 μM dCTP, 50 μM dATP, 75 μM 7-deaza dGTP, 50 μM dTTP and 150 μM ddCTP.

A reactions—50 μM dCTP, 12.5 μM dATP, 75 μM 7-deaza dGTP, 50 μM dTTP and 300 μM ddATP.

G reactions—50 μM dCTP, 50 μM dATP, 18.8 μM 7-deaza dGTP, 50 μM dTTP and 25 μM ddGTP.

T reactions—50 μM dCTP, 50 μM dATP, 75 μM 7-deaza dGTP, 12.5 μM dTTP and 360 μM ddTTP.

Radioactive primers with *Taq* DNA polymerase. M13 -21 universal primer was 5′-end labeled with [γ-^{32}P]ATP (14). All 4 base-specific reactions contained 0.6 pmol of primer and 10 ng of template DNA and the same buffer, enzyme and nucleotide concentrations as for the fluorescent primers.

$Vent_R$® DNA Polymerase. Cycle sequencing reactions using $Vent_R$ (exo$^-$) polymerase (New England Biolabs, Beverly, MA, USA) were performed with a CircumVent™ Thermal Cycle DNA sequencing kit mostly according to the manufacturer's instructions, with the following change. For fluorescent reactions only, deoxynucleotide concentrations were increased 8-fold from the recommended values, and dideoxynucleotide concentrations were reduced by half.

Thermal Cycling

Air thermal cycler. Capillaries (Idaho Technology #1705, Idaho Falls, ID, USA) were siliconized by placing them in a vacuum desiccator along with 2 mL of Sigmacote® (Sigma Chemical, St. Louis, MO, USA) (14). Vacuum was applied for 5 min, and the system was left overnight to coat the capillaries. The next day, capillaries were heated at 150°C for 2 h. Each reaction was flame-sealed in a siliconized capillary and placed into the air thermal cycler (Idaho Technology, Model 1605). After a 3-min initial denaturation at 95°C, 15 amplification cycles were performed with a denaturation temperature of 95°C for 0 s, an annealing temperature of 50°C for 0 s and an elongation temperature of 70°C for 20 s. Fifteen more chase cycles were performed with a denaturation temperature of 95°C for 0 s, an annealing temperature of 69°C for 0 s and an elongation temperature of 70°C for 20 s. Capillaries were cut open with a sapphire cutter and the contents pooled. The capillary manipulation steps (filling, sealing and recovery) take about 3 min for a single template, which compares favorably to standard fluorescent sequencing protocols.

Radioactive samples were prepared in a manner similar to the fluorescent reactions, except they were not pooled after cycling.

Cycle parameters for the $Vent_R$ DNA polymerase varied from those used with *Taq* only in the omission of the chase cycles.

Block thermal cycler. Cycling was performed as per instructions included with Applied Biosystems Taq Dye Primer Sequencing Kit, using Perkin-Elmer GeneAmp® tubes in a Perkin-Elmer DNA Thermal Cycler with a drop of mineral oil covering each sample. After a 3-min initial denaturation at 95°C, 15 amplification cycles were performed with denaturation temperature of 95°C for 30 s, an annealing temperature of 55°C for 30 s and an elongation temperature of 70°C for 1 min. Fifteen more two-stage chase cycles were performed with a denaturation temperature of 95°C for 30 s, followed by an elongation temperature of 70°C for 2 min. Samples were pooled after removing them from under the oil.

Preparation for Electrophoresis

Fluorescent reaction pools (40 µL) were precipitated by the addition of 2 µL of 3 M sodium acetate (NaOAc) and 105 µL of EtOH for 15 min at 4°C or overnight at -20°C. The dried pellets were resuspended in 4 µL of loading buffer (98% formamide, 10 mM EDTA pH 8.0), heated at 95°C for 2 min and cooled on ice before gel electrophoresis.

After cycling the radioactive samples, 4 µL of loading dye (0.3% xylene cyanol, 0.3% bromophenol blue, 0.4% EDTA in formamide) were added to each. The reactions were heated at 95°C for 2 min and cooled on ice before gel electrophoresis.

Gel Electrophoresis

All sequencing gels consisted of 6% wt/vol total acrylamide, prepared using deionized ultrapure acrylamide/ bisacrylamide (pre-mixed 19:1 ratio; Bio-Rad, Hercules, CA, USA), 7 M urea (ultrapure; GIBCO BRL/Life Technologies, Gaithersburg, MD, USA), 1× TBE (pre-mixed 10× TBE concentrate, ultrapure; Boehringer Mannheim), HPLC grade H_2O, 0.04% vol/vol TEMED and 0.05% wt/vol ammonium persulfate (electrophoresis purity; Bio-Rad). Fluorescent-labeled reactions were run on an Applied Biosystems Model 373A DNA Sequencer at 30 W constant power for 10 h, according to manufacturer's instructions. Radioactive samples were run on gels at 30 W for 2 h and autoradiographed with Kodak XAR film (Eastman Kodak, Rochester, NY, USA).

RESULTS AND DISCUSSION

Reaction Conditions

The development of a protocol for rapid cycle sequencing was aided substantially by experience gained in rapid PCR (24–26). Both techniques employ the thermostable DNA polymerase derived from *Thermus aquaticus*, *Taq* DNA polymerase. Thus, some of the optimal reaction and cycling parameters were already known. The air cycle sequencing protocol was optimized for use with the Model 373A fluorescent dye-primer DNA sequencing system, using the Taq Dye Primer Sequencing Kit protocol as a guide.

Previous reports on rapid capillary-based PCR had employed high concentrations (500 μg/mL) of BSA to stabilize the polymerase against the denaturing effect of the capillaries' glass surface (26). Adding BSA to sequencing samples prepared on a block thermal cycler reduces signal strength. Even at concentrations as low as 100 μg/mL, signal strength is 50% of a control reaction, as measured on the Model 373A instrument; phenol extraction reverses this effect (data not shown). To offset the requirements for high BSA concentrations in the reactions, siliconized capillaries were employed. These capillaries gave results far superior to untreated capillaries but could not be used without some added BSA. Siliconized capillaries and an added BSA concentration of 50 μg/mL were found to be both necessary and sufficient to produce high-quality sequence data. Additionally, at this lower BSA concentration, phenol extraction of the samples did not improve the signal strength or quality of the data (not shown).

Taq DNA polymerase is reported to have a fairly broad pH optimum, with high activity between pH 7.0 and 9.0 (3). An 80 mM Tris-HCl, pH 8.9, buffer was used with varying amounts of salt and $MgCl_2$. Best results were obtained with 20 mM KCl (26), while Applied Biosystems' ammonium sulfate buffer showed poorer results as did a buffer containing no salt (10). A magnesium titration from 2–5 mM produced only subtle differences; 5 mM $MgCl_2$ yielded higher signal values and slightly higher fidelity than the other concentrations (data not shown).

Cycling Conditions

Air cycle sequencing reactions were performed with annealing temperatures between 45° and 60°C. The optimal annealing temperature was found to be 50° in the air cycler for M13 universal primer. At 45°, 55°and 60°C annealing temperatures, signal strength was reduced compared with 50°C. The 55° and 60°C annealing temperatures resulted in higher noise in certain regions of the sequence. Extension time was found to be optimal at 20 s. Increasing the extension time to 60 s did not improve the quality of the data, while decreasing the extension time to 15 s resulted in increased false stops and peak height variation.

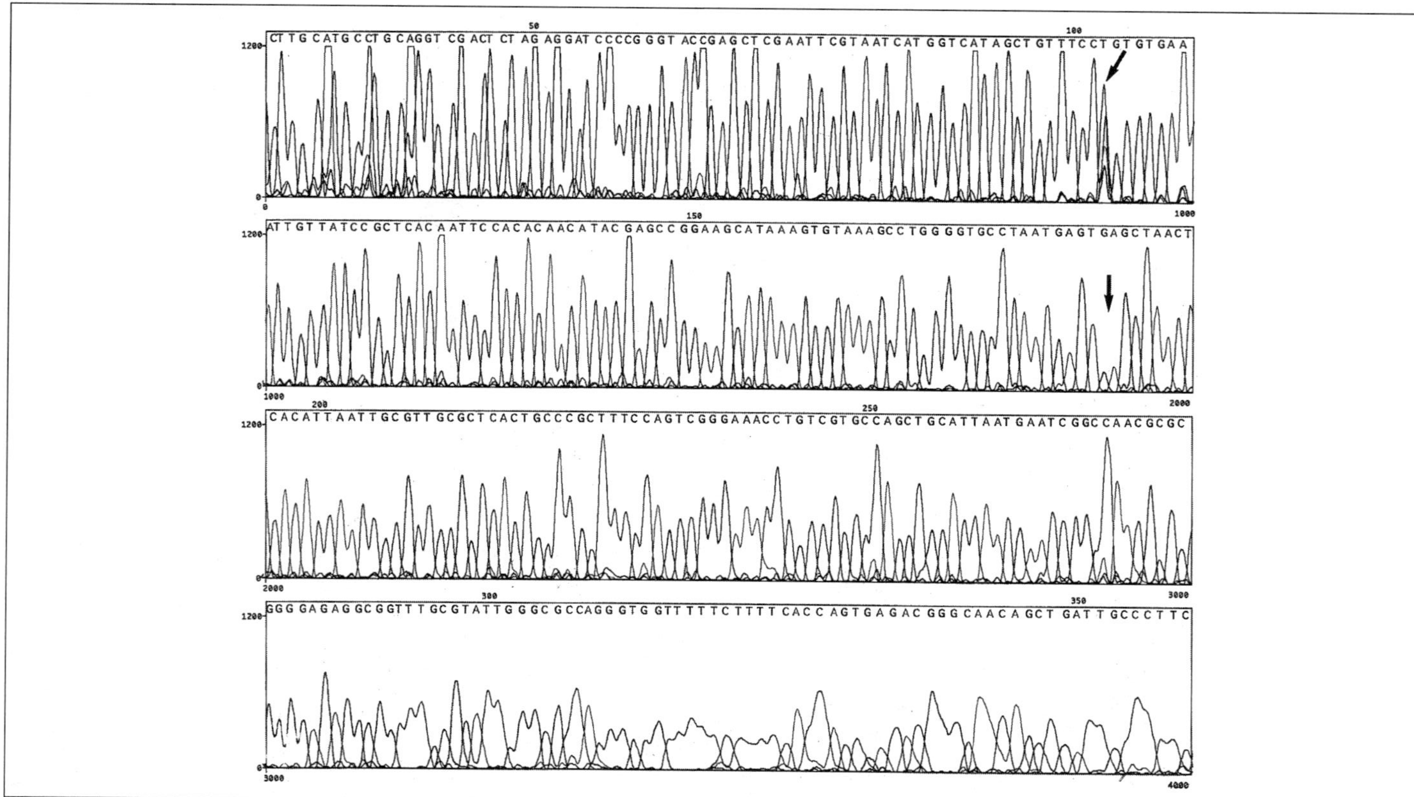

Figure 1. Rapid air-cycled M13mp18 fluorescent control reaction. *Taq* reactions were run and analyzed on an Applied Biosystems Model 373A DNA Sequencer at 30 W for 10 h. Analyzed signal values for this run were C:1289, A:743,G:350, T:253. Reliable sequence was obtained out to about 510 nucleotides. For the region displayed, all the called bases agree with the known sequence. Numbering represents the total length of the terminated fragments, including the 18-mer primer. Arrows indicate locations of problematic nucleotides referred to in the text.

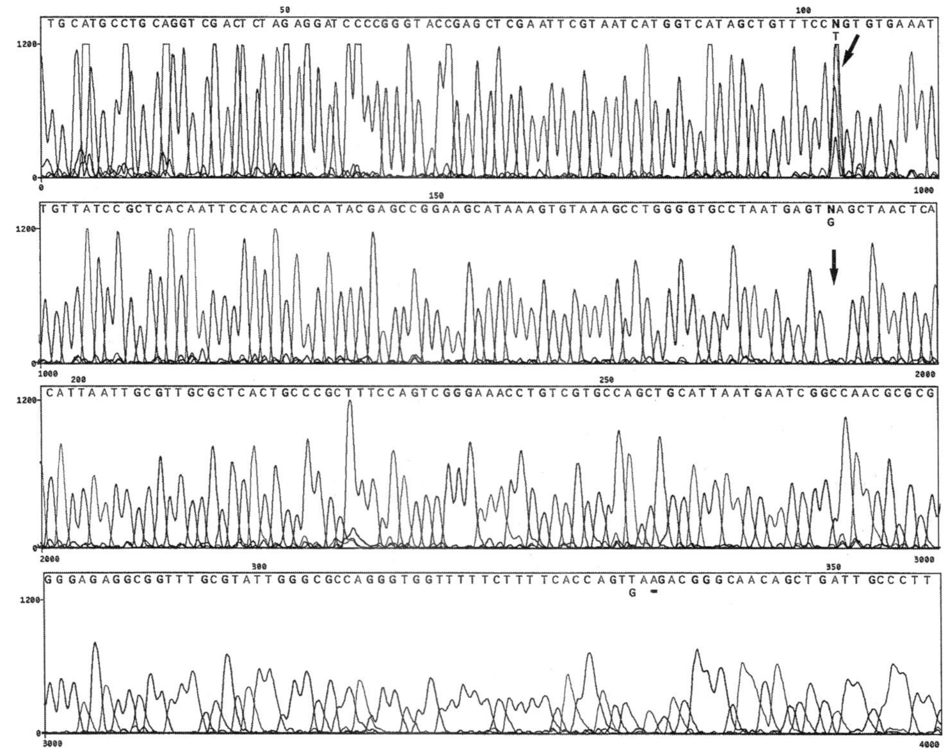

Figure 2. Block thermal-cycled M13mp18 fluorescent control reaction. Running conditions and numbering are described in Figure 1. Analyzed signal values for this run were C:879, A:541, G:238 and T:244. Reliable sequence was obtained out to about 470 nucleotide. Four errors are visible in the displayed data; two problematic regions are indicated by arrows and are discussed in the text. The correct sequence is indicated below the called sequence "-" indicates a base that is called but not present in the known sequence.

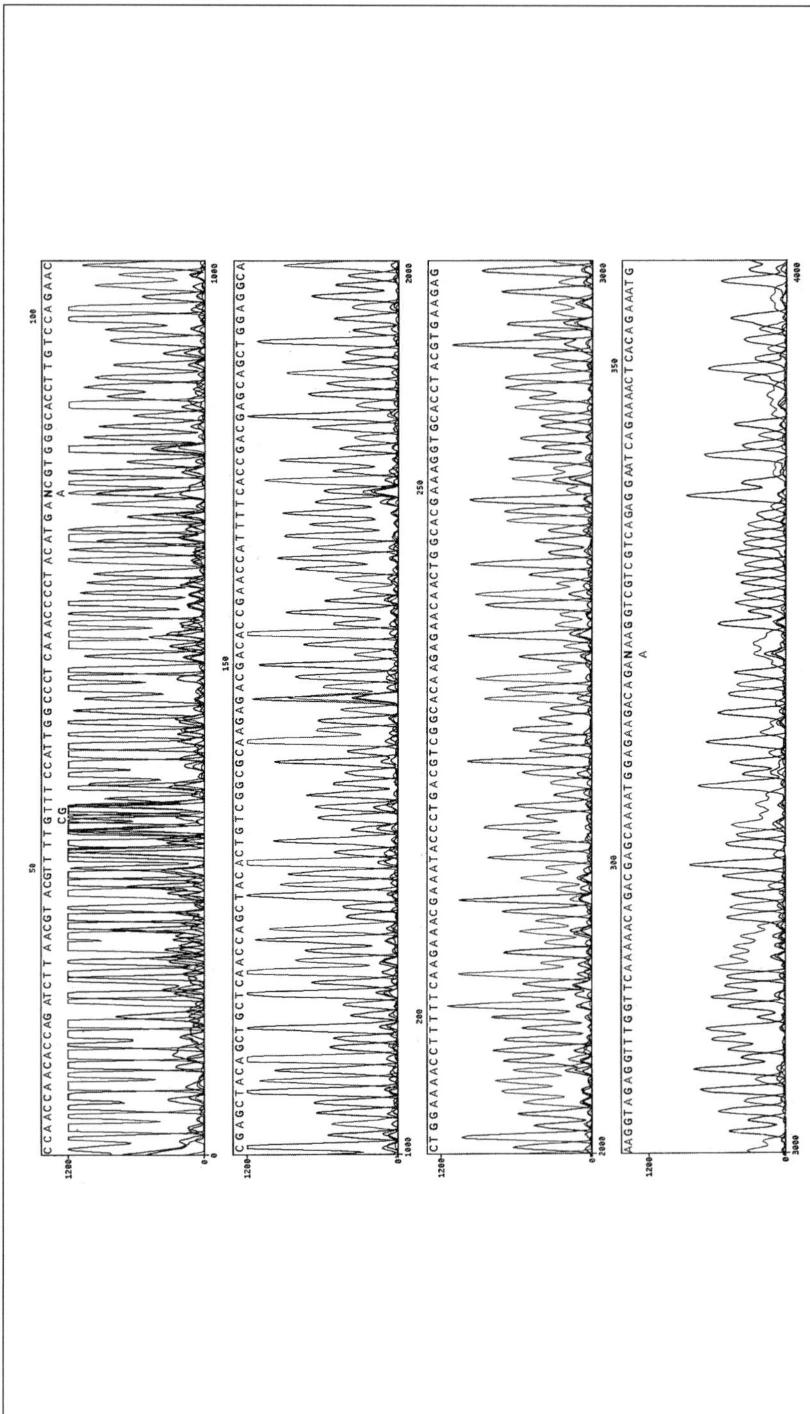

Figure 3. Direct sequencing of a double-stranded PCR product using rapid air cycling. Analyzed signal values for this run were C:314, A:260, G: 130 and T:131. Reliable sequence was obtained out to about 350 bp. Four miscalled bases are indicated. Numbering represents the total length of the terminated fragments, including the 18-mer primer.

The number of cycles was varied. Based on the ratio of primer to template, eight cycles should be sufficient for reaction completion. Fifteen amplification and 15 chase cycles were necessary for single-stranded templates, while 20 amplification and chase cycles were necessary for double-stranded templates. This result implies that the reaction is not 100% efficient, especially for the later cycles. The chase cycles are necessary to extend fragments, prematurely terminated with deoxynucleotides, thereby lowering the number of false stops. An alternative chase strategy proposed by Chissoe et al. did not improve the quality of our sequencing data (4). A few comments about the air-cycling protocol are warranted. As stated in Materials and Methods, 0 s was set for the duration of the annealing and denaturation steps. These nominal settings have been shown to work well for PCRs; the time spent ramping to and from the set temperatures is more than sufficient to achieve both annealing and denaturation (26). To circumvent a programming limitation of the commercial air thermal cycler, a three-temperature chase cycle was performed with 69°C as the annealing temperature. A newer version of the firmware should allow two-temperature thermal cycling.

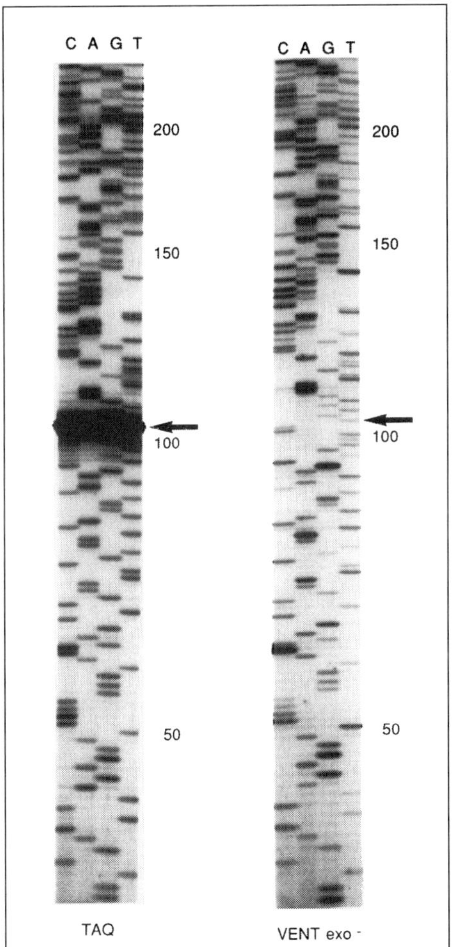

Figure 4. Radioactively labeled M13mp18 DNA sequenced with AmpliTaq and $Vent_R$ (exo⁻) polymerase. The reactions were performed using a ^{32}P-labeled primer (-21M13 19-mer) in the air thermal cycler. Sequence products were run on a 6% polyacrylamide gel at 30 watts for 2 h. Arrows indicate the position of T103, a problematic nucleotide referred to in the text.

Representative Runs

Typical results applying the optimized air-cycling protocol to an M13mp18 single-stranded sequencing reaction are shown in Figure 1. A block-cycled control reaction is shown in Figure 2 for comparison. Analyzed signal values for the data shown are given in their legends. Duplicate runs on the same gel gave averaged signal values of C:1180, A:689, G:350 and T:253 for the air-cycled reactions and C:802, A:551, G:240 and T:239 for

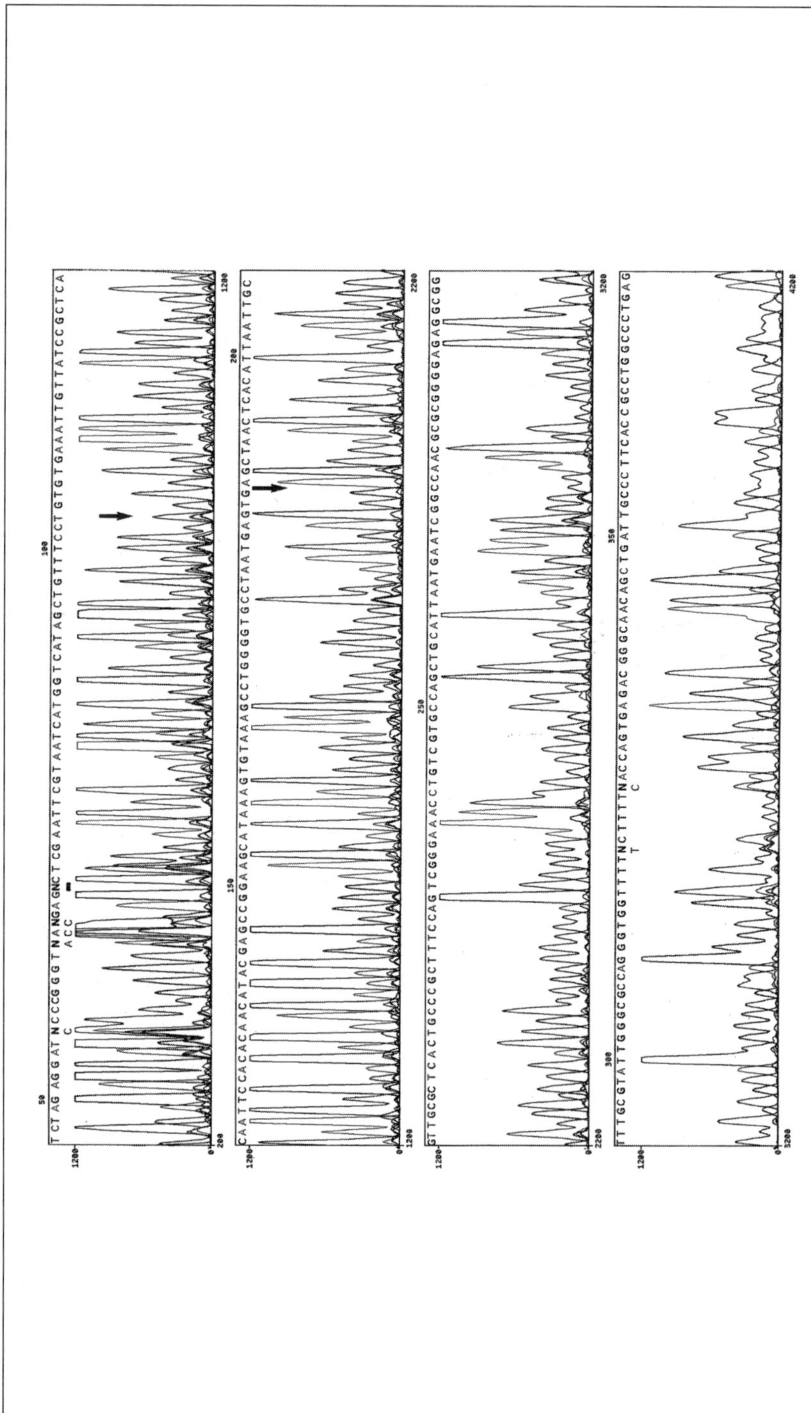

Figure 5. M13mp18 DNA sequenced with Vent$_R$ (exo$^-$) polymerase. Analyzed signal values for this run were C:997, A:759, G:289 and T:153. Reliable sequence was obtained out to about 360 bp. Arrows indicate the positions of T103 and G187-A188, which are normally problematic when sequenced with *Taq* DNA polymerase (see Figure 2).

those performed in the block cycler. Reliable sequence was obtained out to 473 nucleotides with a mean error rate of 0.5% for the air-cycled reactions and 1.2% for the block-cycled reactions ($n = 2$). In the displayed runs, T103 (arrow) is correctly identified when air-cycled, but not when block-cycled. In 12 direct comparisons, the software read T103 4 times in the air-cycled reactions and failed to interpret T103 correctly in any of the block reactions. In the displayed runs, the sequence G187-A188 (arrow) gives readable signal in the rapid air-cycled reactions but not in the block reactions. In 12 direct comparisons, this sequence was correctly interpreted in 11 of the air-cycled reactions, but only in 2 of the block-cycled reactions. Perhaps the improvements we see in air cycling could also be realized by optimization of the block-cycling protocol. However, improved reaction quality using air thermal cycling has also been reported for PCRs (26).

Major interest in cycle sequencing has been fueled by the desire to sequence double-stranded DNA fragments, especially PCR fragments. Using the rapid protocol, we have sequenced 2 kb from two cDNA clones from the zebra fish *goosecoid* gene expressed during early embryogenesis. Direct fluorescent cycle sequencing of double-stranded PCR products derived from the cDNA clones was performed using the rapid air protocol. The finished protein sequence will be published elsewhere (20). A typical sequence is shown in Figure 3. On average, reliable sequence was obtained out to 360 nucleotides. This method thus provides a rapid, viable alternative to conventional thermal block methods for sequencing PCR fragments.

The air-cycling protocol has been adapted to radioactively labeled primers. The results are shown in Figure 4a for *Taq* DNA polymerase. This makes rapid air cycling available to those labs using conventional sequencing methods. Handling of glass capillaries with radioactive contents should be performed with caution. The rapid air cycle sequencing protocol was also adapted to a chemiluminescent sequencing strategy (11).

The use of other thermostable DNA polymerases may be helpful in resolving ambiguities occurring in *Taq* sequencing reactions (18). $Vent_R$ (exo^-) polymerase results are shown in Figure 4b. A definite improvement is seen in the false stop at T103 (arrow), when compared with the *Taq* reactions of Figure 4a. Using $Vent_R$ DNA polymerase with fluorescently tagged primers has proven more problematic. By increasing the deoxynucleotide concentrations and decreasing the dideoxynucleotide concentrations, sequence products out to 350 nucleotides were obtained. A typical fluorescent Vent M13mp18 sequence is shown in Figure 5. The noise is more prevalent than comparable *Taq* reactions (Figure 1), and the 2% error rate is higher. Note, however, that the ambiguous T103 and weak G187-A188 are read with no difficulty by the software (arrows).

CONCLUSION

Significantly faster cycle sequencing has been accomplished using a capillary tube format in an air-based thermal cycle instrument. Thermal cycling can be completed in 25 min in this method compared with 2 h in a conventional block thermal cycler. As in heated cover cyclers, no oil overlays are required; this increases the recovery of sequenced fragments, eliminates a possible source of contamination and prevents problems due to the inadequate removal of oil. Additionally, rapid cycling produced an improvement in the readability of certain false stops and weak bands. The method is simple, fast and may provide a more direct path to automation of cycle sequencing technology.

PCRs performed in a rapid air thermal cycler have shown increased priming specificity when compared to slower cycling protocols (26). This is presumably due to the shorter annealing times employed. In this study, priming specificity did not vary significantly for the M13 universal primer; nonspecific priming would result in higher background noise in the analyzed traces. Other sequencing primers might display higher priming specificity with rapid cycling. This technique could enhance the success rate of primer selection for primer walking sequencing strategies (21).

Recent developments in capillary gel electrophoresis have allowed DNA sequencing to be performed in 50- to 100-µm i.d. fused-silica capillaries using laser fluorescence detection (5,13,22). Capillary-based PCR, followed by capillary cycle sequencing may provide a unique route to automating sample preparation for capillary gel electrophoresis. Such a system would be simple to operate, cost-efficient and fast enough to sequence entire genomes.

ACKNOWLEDGMENTS

The authors wish to express their gratitude to Carl Wittwer for numerous helpful suggestions and support, as well as for critical reading of the manuscript. We also thank Scott Stachel for providing cDNA PCR products. This work was supported in part by Grant 5 P30HG00199 from the National Institutes of Health (R.F.G.). This research was supported in part by an appointment to the Human Genome Distinguished Postdoctoral Fellowships sponsored by the U.S .Department of Energy, Office of Health and Environmental Research, and administered by the Oak Ridge Institute for Science and Education (H.S.). R.F.G. is an investigator of the Howard Hughes Medical Institute. A detailed protocol is available upon request.

REFERENCES

1. **Carothers, A.M., G. Urlaub, J. Mucha, D. Grunberger and L.A. Chasin.** 1989. Point mutation in a mammalian gene: rapid preparation of total RNA, PCR amplification of cDNA, and *Taq* sequencing by a novel method. BioTechniques. *7*:494-499.
2. **Cathcart, R.** 1990. Advances in automated DNA sequencing. Nature *347*:310.
3. **Chien, A., D.B. Edgar and J.M. Trela.** 1976. Deoxyribonucleic acid polymerase from the extreme thermophile Thermus aquaticus. J. Bacteriol. *127*:1550-1557.

4. **Chissoe, S.L., Y.F. Wang, S.W. Clifton, N. Ma, H.J. Sun, J.S. Lobsinger, S.M. Kenton, J.D. White and B.A. Roe**. 1991. Strategies for rapid and accurate DNA sequencing. Methods: A Companion to Methods in Enzymology *3*:55-65.
5. **Cohen, A.S., D.R. Najarian and B.L. Karger.** 1990. Separation and analysis of DNA sequence reaction products by capillary gel electrophoresis. J. Chromatogr. *516*:49-60.
6. **Gibbs, R.A.** 1990. DNA amplification by the polymerase chain reaction. Anal. Chem. *62*:1202-1214.
7. **Gyllensten, U.B.** 1989. PCR and DNA sequencing. BioTechniques *7*:700-708.
8. **Gyllensten, U.B. and H.A. Erlich.** 1988. Generation of single-stranded DNA by the polymerase chain reaction and its application to direct sequencing of the HLA-DQA locus. Proc. Natl. Acad. Sci. USA *85*:7652-7656.
9. **Hultman, T., S. Stahl, E. Hornes and M. Uhlen.** 1989. Direct solid phase sequencing of genomic and plasmid DNA using magnetic beads as solid support. Nucleic Acids Res. *17*:4937-4946.
10. **Innis, M.A., K.B. Myambo, D.H. Gelfand and M.A.D. Brow.** 1988. DNA sequencing with Thermus aquaticus polymerase and direct sequencing of polymerase chain reaction-amplified DNA. Proc. Natl. Acad. Sci. USA *85*:9436-9440.
11. **Karger, A.E., R. Weiss and R.F. Gesteland.** 1993. Line scanning system for direct digital chemiluminescence imaging of DNA sequencing blots. Anal. Chem. *65*:1785-1793
12. **Kreitman, M. and L.F. Landweber.** 1989. A strategy for producing single-stranded DNA in the polymerase chain reaction. Gene Anal. Tech. *6*:84-88.
13. **Luckey, J.A., H. Drossman, A.J. Kostichka, D.A. Mead, J. D'Cunha, T.B. Norris and L.M. Smith.** 1990. High speed DNA sequencing by capillary gel electrophoresis. Nucleic Acids Res. *18*:4417-4421.
14. **Maniatis, T., E.F. Fritsch and J. Sambrook.** 1982. Molecular Cloning: A Laboratory Manual. Cold Spring Harbor Laboratory, Cold Spring Harbor, NY.
15. **Mullis, K. and F.A. Faloona.** 1987. Specific synthesis of DNA in vitro via a polymerase-catalyzed chain reaction. Methods Enzymol. *155*:335-350.
16. **Mullis, K., F. Faloona, S. Scharf, R. Saiki, G. Horn and H. Erlich.** 1986. Specific enzymatic amplification of DNA in vitro; the polymerase chain reaction, p. 263-273. *In* Cold Spring Harbor Symposia on Quantitative Biology, Vol. 51. Cold Spring Harbor Laboratory, Cold Spring Harbor, NY.
17. **Murray, V**. 1989. Improved double-stranded DNA sequencing using the linear polymerase chain reaction. Nucleic Acids Res. *17*:8889.
18. **Roberts, S.S.** 1992. Thermostable DNA polymerases heat up DNA sequencing. J. NIH Res. *4*:89-99.
19. **Saiki, R.K., D.H. Gelfand, S. Stoffel, S.J. Scharf, R. Higuchi, G.T. Horn, K.B. Mullis and H.A. Erlich.** 1988. Primer-directed enzymatic amplification of DNA with a thermostable DNA polymerase. Science *239*:487-491.
20. **Stachel, S.E., D.J. Grunwald and P.Z. Myers.** 1993. Lithium perturbation and *goosecoid* expression identify a dorsal specification pathway in the pre-gastrula zebrafish. Development *117*:1261-1274.
21. **Studier, F.W.** 1989. A strategy for high-volume sequencing of cosmid DNAs: random and directed priming with a library of oligonucleotides. Proc. Natl. Acad. Sci. USA *86*:6917-6921.
22. **Swerdlow, H., J.Z. Zhang, D.Y. Chen, H.R. Harke, R. Grey, S. Wu, N.J. Dovichi and C. Fuller.** 1991. Three DNA sequencing methods using capillary gel electrophoresis and laser-induced fluorescence. Anal. Chem. *63*:2835-2841.
23. **Watson, J.** 1990. The Human Genome Project: past, present and future. Science *248*:44-49.
24. **Wittwer, C.T., G.C. Fillmore and D.J. Garling.** 1990. Minimizing the time required for DNA amplification by efficient heat transfer to small samples. Anal. Biochem. *186*:328-331.
25. **Wittwer, C.T., G.C. Fillmore and D.R. Hill-yard.** 1989. Automated polymerase chain reaction in capillary tubes with hot air. Nucleic Acids Res. *17*:4353-4357.
26. **Wittwer, C.T. and D.J. Garling.** 1991. Rapid cycle DNA amplification: time and temperature optimization. BioTechniques *10*:76-83.

Update to:

Rapid Cycle Sequencing in an Air Thermal Cycler

Robert Sinclair, Heather Lin and Harold Swerdlow
University of Utah, Salt Lake City, UT, USA

We have continued to use the Idaho Technology air thermal cycler for sequencing reactions for the three years since the original article was published. The methods we presented, which we use routinely to prepare a variety of sequencing samples, have proven to be robust and reliable.

To take advantage of the improved DNA sequencing enzymes that became available in 1996, we have modified our protocols. We have been using AmpliTaq™ DNA polymerase FS from Perkin-Elmer/Applied Biosystems (Foster City, CA, USA) for most of our routine DNA sequencing work. Our standard buffer for AmpliTaq-FS contains 50 mM Tris-HCl (pH 8.0), 5 mM $MgCl_2$, and 62.5 μg/mL BSA. We use 1.7 units of enzyme for A and C reactions, and 3.4 units for G and T. Due to the improved properties of AmpliTaq-FS, a uniform ratio of dideoxy to deoxynucleotide triphosphates (about 1:320) works well for all four termination mixes. These mixes are significantly less complex than the mixes we previously presented, which had been optimized to overcome AmpliTaq's uneven discrimination against dideoxynucleotide triphosphates. We have found that the conditions recommended in Perkin-Elmer's ABI PRISM™ Dye Primer Cycle Sequencing Core Kit protocol (Revision C March 1996) work well using our thermal cycling conditions in glass capillaries if the buffer [final concentrations of 80 mM Tris-HCl (pH 9.0), 2 mM $MgCl_2$] is supplemented with 62.5 μg/mL BSA.

Using the slightly higher BSA concentration (62.5 μg/mL versus 50 μg/mL from the original protocol), we have produced good quality sequence ladders in non-siliconized capillaries using Vent (exo^-), AmpliTaq, and AmpliTaq-FS Polymerases. However, the sequence produced is superior when the reactions are supplemented with BSA and performed in siliconized capillaries. We could not detect any adverse effects on sequence quality from capillaries that had been stored at room temperature for over three years after siliconizing, compared with freshly siliconized capillaries. It is therefore acceptable to siliconize capillaries in bulk and store them at room temperature indefinitely.

Direct Microtiter Plate Sequencing of PCR-Amplified M13 Clones from Plaques Using Dried Reagents

Susan J. Dale

Amersham International plc, White Lion Road, Little Chalfont, Bucks, England, UK

ABSTRACT

A method is described using dried reagents in microtiter plates for direct PCR and sequencing of M13 plaques. The method has advantages of reliability and high throughput that make it suitable for medium to large-scale sequencing projects.

INTRODUCTION

Direct sequencing of PCR DNA has been the focus of much effort over the last few years (1,2,8). Several techniques have been suggested; however, the quality of results obtained has not always been sufficient for reliable use in a major sequencing project. By combining the best features of these techniques and minimizing the steps needed, a more efficient direct sequencing protocol has been developed.

To increase convenience and throughput, the PCR sequencing has been combined with the use of dried reagents in microtiter plates. The other advantages of this approach are that PCR reagents can be prepared in areas where targets are not handled; also the yields of amplification are more reproducible when replicate reactions are used.

The sequencing method chosen is a nested asymmetric PCR approach (Figure 1). We have found that annealing of primers to double-stranded PCR products is inefficient due to competition with target reannealing. Therefore, we have selected a method that produces single-stranded target directly.

MATERIALS AND METHODS

Plating Out and Picking Plaques

E. coli TG1 cells were infected with an M13 clone containing a 600-bp fragment of the chicken lysozyme gene. The cells were plated out in molten

(Reprinted from BioTechniques 12:194-197, 1992)

top agar onto LB plates and incubated overnight at 37°C. Individual plaques were lifted using Sealon film (Fuji Photo, Tokyo, Japan) (7). Small triangles of the film were cut out and pressed onto the plaque using a sterile needle. The plaque lift was added directly to a PCR. The film was melted during the

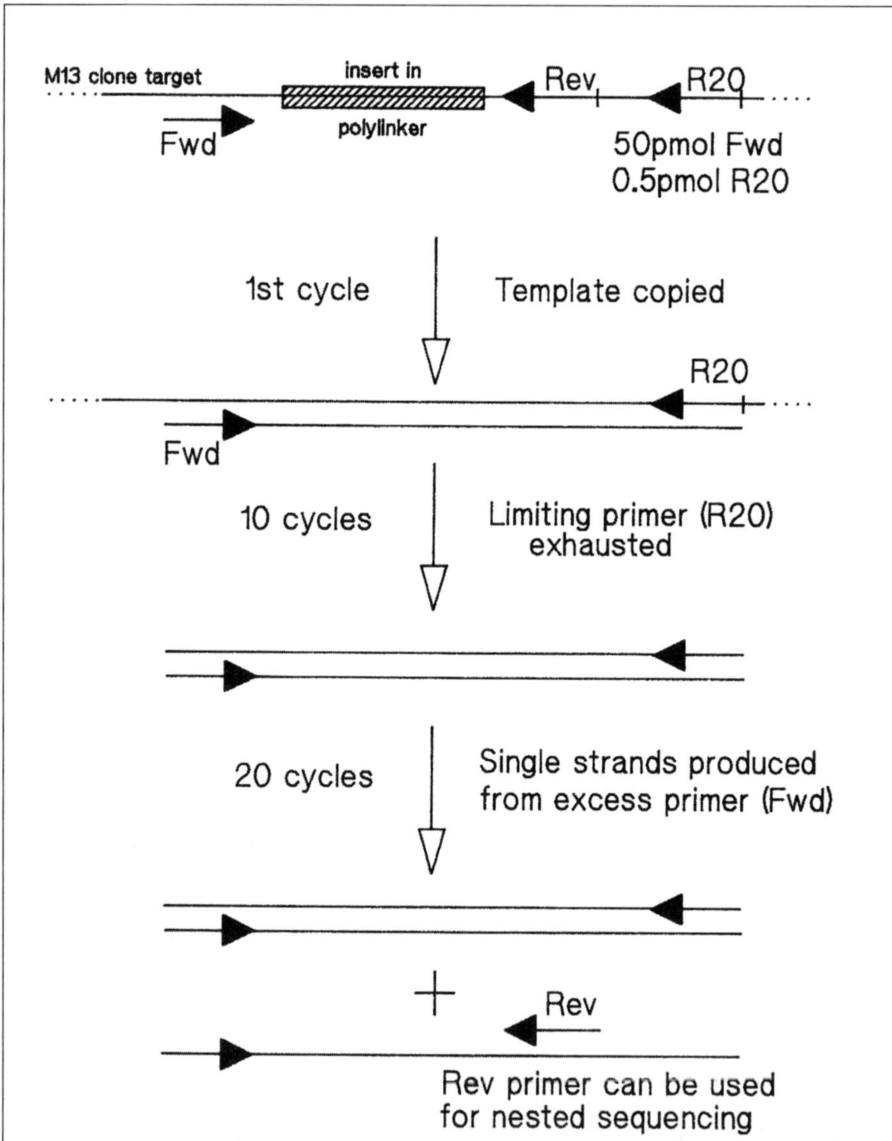

Figure 1. Scheme of nested asymmetric PCR sequencing of M13 clones. Fwd is the M13 universal sequencing primer (5′-GTAAAACGACGGCCAGT-3′); Rev is the M13 reverse primer (5′-ACAGGAAACAGCTATGAC-3′); R20 is a primer approximately 20 bases 5′ to Rev (5′-TGTGGAATTGCGAGCGG-3′).

first denaturation step and had no adverse affect on the yield of amplified DNA.

Preparation and Use of Dried PCRs

A bulk PCR was prepared containing the following reagents per reaction: 2.5 units *Taq* polymerase (Amersham International plc, Bucks, England, UK), 5 µL PCR buffer (as supplied with *Taq* polymerase), 1 µL 10 mM solution of 4 deoxynucleoside triphosphates (dNTPs), 4 µL 2 O.D./mL (50 pmol) M13 forward primer (5′-GTAAAACGACGGCCA GT-3′), 4 µL 0.02 O.D./mL (0.5 pmol) M13R20 primer (5′-TGTGGAATTGCGAGCGG-3′) (Figure 1). A volume equal to one reaction was dispensed into each well of polystyrene microtiter well strips (ICN Biomedicals, Costa Mesa, CA, USA). The wells were covered with sealing sheets (ICN Biomedicals) to prevent contamination with dust, etc. The seals were pierced several times with a sterile needle. The plates were dried in a vacuum oven at room temperature for 1 h.

Just prior to use, the reactions were resuspended thoroughly in 50 µL sterile water and transferred to 0.5-mL microcentrifuge tubes. The target (100 pg control DNA or a plaque lift on Sealon) was added, and the reactions were overlayed with light mineral oil.

The reactions were subjected to 30 cycles of amplification (95°C for 1 min, 55°C for 1 min, 72°C for 2 min) and a final polymerization step of 72°C for 10 min in a PREM II thermocycler (LEP Scientific, Milton Keynes, England, UK).

Removal of Excess Nucleotides and Primers

PCRs were removed from under the oil. An equal volume of 4 M ammonium acetate and 2 volumes of propan-2-ol were added and mixed. The tubes were centrifuged at room temperature for 10 min. The supernatants were discarded and the pellets (which may be invisible) were washed with 100 µL ice-cold 70% ethanol. The tubes were centrifuged for 5 min, and the washes were discarded. The pellets were dried briefly in a SpeedVac (Savant Instruments, Farmingdale, NY, USA) and resuspended by vortex mixing in a volume equal to the original.

Sequencing PCR DNA

Sequencing by dideoxy chain termination (5) was performed in the Multiwells system (Amersham™) (4). The enzyme used was *Taq* polymerase (3), with dried ^{35}S dCTP (Amersham) as the label. The first row of wells (annealing buffer and M13 forward primer) was removed and replaced with the third row (annealing buffer only). The dried ^{35}S dCTP wells were then placed in the empty third row. The layout of the plate was, therefore, as follows: Row 1, annealing reaction; Row 2, *Taq* polymerase; Row 3, ^{35}S label; Row 4, deoxynucleotide labeling mixture; Rows 5–8, A-T dideoxynucleotide termination reactions. Row 1 (annealing reactions) was removed to a separate plate and received 1 µL 2 O.D./mL M13 reverse primer (5′-ACAGGAAACAGCTATGAC-3′) and 29 µL PCR DNA solution. The annealing reactions were covered with a sealing sheet, incubated at 75°C for 10 min and then followed by room temperature for 10 min. The mixtures were transferred to the second row (*Taq* polymerase), the third row (dried ^{35}S) and then the fourth row (labeling mixture), resuspending the well contents thoroughly at each stage. The labeling reaction was incubated at 75°C for 2 min. Then 6 µL labeling mixture was added to each termination mixture (rows 5–8). These termination reactions were incubated at 75°C for 5 min. Formamide stop solution was added (4 µL per well), and the reactions were heated to 90°C for 3 min just prior to loading onto a 6% polyacrylamide gel (Gel-Mix® 6, GIBCO BRL/Life Technologies, Paisley, Scotland, UK).

RESULTS AND DISCUSSION

The sequences produced by this protocol using picked plaques as the target for amplification (Figure 2) are of a similar quality to those derived from purified DNA target. By employing an internal nested primer for sequencing, multiple stops caused by sequencing primer dimers are avoided. The method produces at least 250 readable bases per clone, and the sequence quality is limited

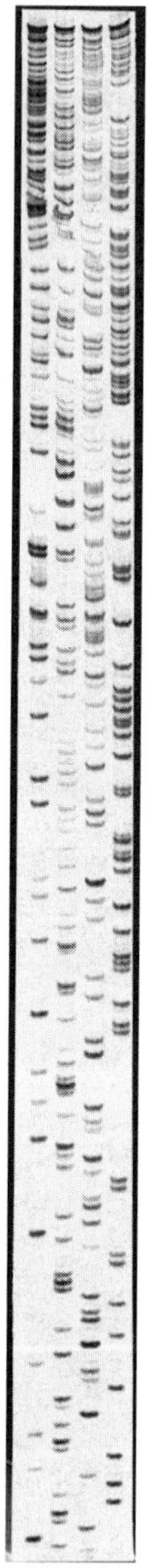

Figure 2. Results of sequencing a clone of chicken lysozyme in M13mp10. A plaque was picked into reconstituted PCR reagents that had been stored dried for 12 weeks. After 30 cycles of PCR, the product was sequenced using dried reagents as described. The autoradiogram was exposed for 36 h using Hyperfilm™ Bmax (Amersham). The sequencing reactions are in the order A, C, G, T from left to right.

by the polymerase, not by the target.

We have found that the method of plaque picking is also important in determining the quality of sequencing results obtained. Contamination of the PCR with agar from the plate interferes with amplification, reducing the yield and enhancing production of nonspecific bands. These cause artifacts that make the sequence unreadable in some regions.

Reagents for PCR can be made in large quantities for drying in microtiter plates because the reactions have an extended stability at -20°C in this format. Amplifications performed using dried reagents that have been stored for 16 weeks give an equivalent yield to fresh reagents. The sequencing results shown (Figure 2) were produced from PCR carried out using 12-week-old reagents.

The use of dried reagents for PCR and sequencing has significantly reduced the hands-on time and increased the reproducibility of both techniques. Direct PCR sequencing can now be used more reliably for medium- to large-scale sequencing projects. For example, the products of an in vitro mutagenesis experiment can be analyzed rapidly without the need for multiple M13 phage preparations. The microtiter plate approach to PCR and sequencing is also adaptable to automation (6) that would make the technique powerful for major projects.

We have successfully employed this technique with plasmid colonies and extracted human genomic DNA as targets (data not shown). It has been necessary in some cases to optimize the asymmetric ratio for particular primer pairs; however, good sequences can be achieved even when no single-stranded DNA is visible on ethidium bromide-stained agarose gels.

REFERENCES

1. **Gyllensten, U.B. and H.A. Erlich.** 1988. Generation of single-stranded DNA by the polymerase chain reaction and its application to direct sequencing of the HLA-DQA locus. Proc. Natl. Acad. Sci. USA *85*:7652-7656.
2. **Higuchi, R.G. and H. Ochman.** 1989. Production of single-stranded DNA templates by exonuclease digestion following the polymerase chain reaction. Nucleic Acids Res. *17*:5865.
3. **Innis, M.A., K.B. Myambo, D.H. Gelfand and M.A.B. Brow.** 1988. DNA sequencing with Thermus aquaticus DNA polymerase and direct sequencing of polymerase chain reaction-amplified DNA. Proc. Natl. Acad. Sci. USA *85*:9436-9440.
4. **Ortlepp, S.A. and I.A. McKay.** 1989. Performing nucleic acid reactions using predispensed lyophilized reaction mixtures. BioTechniques *7*:1110-1115.
5. **Sanger, F., S. Nicklen and A.R. Coulson.** 1977. DNA sequencing with chain-terminating inhibitors. Proc. Natl. Acad. Sci. USA *74*:5463-5467.
6. **Schofield, J.P., M. Vaudin, S. Kettle and D.S.C. Jones.** 1989. A rapid semi-automated microtitre plate method for analysis and sequencing by PCR from bacterial stocks. Nucleic Acids Res. *17*:9498.
7. **Stoddart, M., P. Pierre, C. Spencer and D.W. Harris.** 1991. An improved method for the preparation of M13 plaques for amplification by the polymerase chain reaction. Nucleic Acids Res. *19*:3152.
8. **Wrischnik, L.A., R.G. Higuchi, M. Stoneking, H.A. Erlich, N. Arnheim and A.C. Wilson.** 1987. Length mutations in human mitochondrial DNA: direct sequencing of enzymatically amplified DNA. Nucleic Acids Res. *15*:529-542.

Update to:

Direct Microtiter Plate Sequencing of PCR-Amplified M13 Clones from Plaques Using Dried Reagents

Susan J. Dale

Amersham International plc, White Lion Road, Little Chalfont, Bucks, England, UK

Since the original publication of this article, some of the products used have been superseded. The reader is directed to the manufacturers' catalogs for suitable alternatives.

Robotic Automation of Dideoxyribonucleotide Sequencing Reactions

James J. Earley, Helena Kuivaniemi, Darwin J. Prockop and Gerard Tromp

Thomas Jefferson University, Philadelphia, PA, USA

ABSTRACT

We developed a robot to carry out standard Sanger dideoxyribonucleotide sequencing reactions efficiently and with minimal human intervention. A commercial robot was adapted to our design and specifications, and we programmed it to perform up to 240 sequencing reactions in a single unattended run of 7 h. The robot configuration can be easily altered to allow 480 reactions to be performed in an unattended run of 14 h. The special features of our robot include cooled reagent reservoirs and cooled chambers for storage of DNA templates and completed reactions as well as reproducible aspiration of small volumes by using a sensing algorithm. The robot has successfully performed over 3500 DNA reactions in about 30 separate runs in our DNA core facility.

INTRODUCTION

New DNA polymerases and protocols have considerably improved the original Sanger dideoxyribonucleotide sequencing protocols (9). Nonetheless, DNA sequencing reactions require reproducible pipetting of small volumes and are time-consuming, labor-intensive and subject to fluctuation in reproducibility and efficiency from person to person.

Automation of DNA sequencing reactions can alleviate the tedium, reduce the error rate and provide greater reproducibility. Use of pipetting workstations has demonstrated the advantages of automation (1–3,5–7,13,14,16,17). At least one robotic system designed specifically to perform DNA sequencing reactions is commercially available (CATALYST™; Applied Biosystems [ABI], Foster City, CA, USA) (1). However, the number of reactions that can be performed by these systems without human intervention is relatively small. To increase the number of reactions that can be performed per day with little or no human intervention and to obtain reactions of uniform quality, we

(Reprinted from BioTechniques 17:156-165, 1994)

developed the specifications for a robot that can perform dideoxyribonucleotide sequencing reactions. We contracted the robotic company (Zymark, Hopkinton, MA, USA) to custom build a robot for us, and we programmed it to perform sequencing reactions efficiently. Here we describe the robot and the protocols that we developed.

MATERIALS AND METHODS

The Robot

A custom-built robot was purchased from the Zymark Corporation. The robot consisted of a proprietary robot arm (Zymate XP) and controller (System V controller), standard Pytechnology® components and custom-designed components that conformed to our specifications (Figure 1). The robot arm had a cylindrical work space defined by rotation through 375 degrees, a 35-cm range of movement vertically and a radial extension of about 65 cm. The

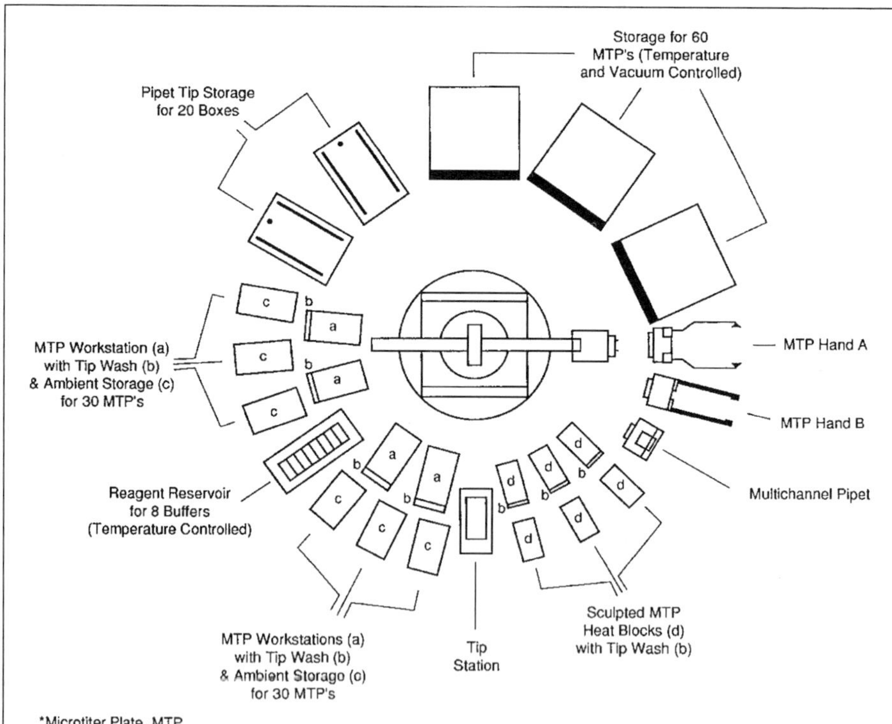

Figure 1. Plan view of the robot. Shown is the maximal configuration that allows 60 microplates (MTP) to be stored at ambient and controlled temperatures. MTP Hand A is necessary to retrieve microplates and pipet tips from ambient storage. MTP Hand B is necessary to place microplates into temperature-controlled storage.

arm exchanged functional components, called hands, to perform different operations. Standard Pysection components included pipet-tip storage racks, microplate storage racks, a pipet-tip workstation, microplate workstations called locators and two hands: the general-purpose hand to move objects and the 8-channel micropipet hand. Nonstandard components were a water-cooled reagent reservoir with an electronically controlled, pneumatically operated lid, water-cooled storage chambers for microplates with electronically controlled, pneumatically operated doors, a more robust hand for moving objects (hand B, Figure 1) and sculpted heat blocks (USA/Scientific Plastics, Ocala, FL, USA). Other nonstandard components that were added during the development of programs and protocols included reservoirs for washing the microtips (components labeled b in Figure 1). An IBM®-compatible PC was used to interface with and program the robot controller.

Sequencing Reactions

Sequenase® Version 2.0 T7 DNA polymerase (United States Biochemical, Cleveland, OH, USA) and *Bst* DNA polymerase (Bio-Rad, Hercules, CA, USA) were used for dideoxyribonucleotide DNA sequencing. Reaction conditions were based on the two-step reaction where the initial extension of the primer and incorporation of radiolabeled nucleotide is performed as a separate step from the termination step (10). Conditions for reactions using Sequenase were similar to those recommended by the manufacturer. Reagents from the Sequenase Version 2.0 kit were used as supplied except that nonionic detergents (see below) were added to the enzyme-labeling and termination solutions to a final concentration of 0.001%. Also, the enzyme was diluted in an enzyme dilution buffer that was one-third the concentration of the standard protocol.

The volumes of the reactions were increased because of constraints on pipetting described below. In addition, the microplate format was prone to evaporation even though the plates were covered with lids at all times except during pipetting. The volumes of the reactions were, therefore, adjusted at the end of the extension labeling step to compensate for evaporation. The annealing volume was 16 µL as compared with the standard 10 µL. After annealing, the effective volume was about 10 µL. The buffer in the annealing solution was adjusted to be correct for the effective volume (see Table 1). After the labeling-extension reaction was complete, the volume was increased by the addition of 9 µL of water to about 25 µL. Five microliters of the labeling-extension reaction were dispensed to each of the 5-µL termination mixtures, and the reactions were terminated by the addition of 4 µL stop solution (Sequenase Kit). The robot performed the reactions by retrieving a microplate with lid, called the source plate, containing up to 96 samples (12 sets of 8) from a cooled storage chamber and placing it on a locator (Figure 1). The lid was removed and placed on an adjacent locator. An empty microplate with lid, called the reaction plate, was retrieved from ambient storage and placed on a

Table 1. Volumes (in µL) of Sequencing Reactions During Different Segments of the Protocol

	Sequenase		*Bst* DNA polymerase
	Standard	Robot	Robot
Denaturing and Annealing			
Annealing solution	10	16	22
Starting volume	10	16	22
[Evaporation]	--	-6	-6
	10	10	16
Labeling-Extension Reaction			
Enzyme-labeling solution added	+5.5	+6	+7
Starting volume	15.5	16	23
[Evaporation]	--	--	-7
	15.5	16	16
Volume added to compensate for evaporation	--	+9	+20
	15.5	25	36
Volume Dispensed (total)	14	20	28
per termination (4 times)	3.5	5	7
Termination Reaction			
Termination solution	2.5	5	7
Dispensed elongation reaction	+3.5	+5	+7
	6.0	10	14
[Evaporation]	--	-2	-7
	6.0	8	7
Addition of Stop Solution			
Stop solution	+4.0	+4	+4
Final volume per termination reaction	10.0	12	11

sculpted block at ambient temperature. (Four of the blocks were at ambient temperature, one was at 37°C and one at 65°C.) The lid was removed and placed on an adjacent block. Samples of template DNA and primer in buffer were pipetted from the source plate to the first column of wells in the reaction plate. The lid of the reaction plate was replaced and the plate was moved to the 65°C block. The lid of the source plate was replaced and it was returned to the cooled storage chamber. After 5 min at 65°C, the reaction plate was moved to a locator at ambient temperature, its lid removed and placed on an adjacent locator. Solutions from the inserts in the cooled reservoir were then dispensed to the plate. The four termination solutions, 5 µL per well, were dispensed to columns 2 through 5 of the target plate. Stop solution and water, 35 µL per well, were dispensed to columns 6 and 7, respectively. Labeling and extension were initiated by pipetting 5.5 µL of enzyme-labeling solution to each template-primer well in column 1. The lid was replaced and the reaction incubated at room temperature. After 4 min, the reaction plate was moved to a 37°C block, the lid removed to an adjacent block and 9 µL of water from column 7 were pipetted to the labeling-extension reaction. Five microliters of labeling-extension reaction were dispensed into the wells containing the termination solutions in columns 2 through 5. The lid was replaced and the termination reactions incubated for 5 min at 37°C. To terminate the reactions, the lid was removed to an adjacent block, and 4 µL of stop solution from column 6 were dispensed to the termination reactions. The lid was replaced and the reaction plate moved to a cooled storage chamber.

For *Bst* DNA polymerase, the sequencing conditions were similar to those developed for the use of Sequenase except that all steps were performed for about 5 min instead of 4 min and 5 min, and the elongation-labeling and termination steps were performed at 65°C instead of at ambient temperature and 37°C, respectively. It was therefore necessary to compensate for a greater evaporative loss of volume (see Table 1). Samples of template DNA and sequencing primer in 22 µL of 19 mM $MgCl_2$ and 64 mM Tris-HCl, pH 8.8, were placed in the source plate. The robot manipulated the microplates as described above. The 22-µL samples were transferred, eight at a time, to a reaction plate. The template DNA was denatured at 65°C for 5 min. Primer was annealed to the template by cooling to room temperature. The water, stop solution and termination solutions were transferred to the reaction plate as described above during annealing. For sequencing with *Bst* DNA polymerase, 7 µL of the termination solutions were dispensed to the reaction plate. The termination solutions consisted of 27.4 µM each of dATP, dCTP, 7-deaza-dGTP and dTTP, 0.025% nonionic detergents, 0.25 µg/mL bovine serum albumin (BSA), and 0.1 mM EDTA in 1 mM Tris-HCl, pH 7.0. The concentrations of dideoxyribonucleoside triphosphates (ddNTPs) in the respective termination solutions that caused termination over a useful range for sequencing were determined empirically and were 428 µM ddATP, 114 µM dCTP, 170 µM ddGTP and 370 µM ddTTP. After annealing, the volume of the sample was about 16 µL because of evaporation. The extension reaction was started by the addition of 7 µL of enzyme-labeling solution that consisted of the following: 0.15 U/µL *Bst* DNA polymerase, 0.46 µM each of dCTP, 7-deaza-dGTP and dTTP, 0.46 µM ^{33}P-dATP with a specific activity of 500 Ci/mmol, 0.02% nonionic detergents, 0.1 mg/mL BSA, 0.075 mM EDTA in 0.75 mM Tris-HCl pH 8.0. The reaction was incubated at 65°C for 5 min. Evaporative loss of volume during extension was about 7 µL. Twenty microliters of water were added to the extension reaction prior to dispensing 7 µL to each of the termination solutions. The termination reaction was incubated at 65°C for 5 min. Evaporative loss of volume reduced the final volume to about 7 µL for each termination reaction. The reaction was stopped by the addition of 4 μL stop solution.

Single-stranded bacteriophage M13 DNA (United States Biochemical) and single-stranded DNA generated by asymmetrical amplification with the polymerase chain reaction (PCR) were used as template. Nonionic detergents were added to most reagents to reduce surface tension. A 10% stock solution of equal parts of Brij®-35, Nonidet® P-40 (NP40), Triton®X-100 and Tween® 20 (Surfact-Amps™; Pierce, Rockford, IL, USA) was used. Radiolabeled dATP (^{35}S-dATP; ^{33}P-dATP) was purchased from Du Pont NEN (Boston, MA, USA) and solutions of deoxyribonucleotides and dideoxyribonucleotides were purchased from Boehringer Mannheim (Indianapolis, IN, USA). Optimal ratios of dideoxyribonucleotide-to-deoxyribonucleotide for *Bst* DNA polymerase were determined empirically. Products of sequencing reactions were separated on 6% polyacrylamide DNA sequencing gels (Sequagel 6%; National Diagnostics, Atlanta, GA, USA) that were fixed in 10% methanol-10% acetic acid, dried under vacuum and exposed to x-ray film (Hyperfilm™-MP; Amersham, Arlington Heights, IL, USA) to produce autoradiographs.

The robot used standard laboratory plasticware: U-bottom microplates and lids (Nunc® brand from Thomas Scientific, Swedesboro, NJ, USA), micropipet tips (USA/Scientific Plastics) and plastic inserts for the cooled reagent reservoir (Beckman Instruments, Fullerton, CA, USA).

RESULTS

The Robot

Specifications for the system included the following features: (a) operation without intervention after the initial setup, (b) cooled storage of reagents with a robotically controllable lid for the reagents, (c) cooled storage for microplates, (d) six or more heat blocks sculpted to fit microplates, (e) a capacity to manipulate microplates and (f) a capacity to pipet volumes as small as 5 μL from microplate wells containing as little as 10 μL with an eight-channel pipet. The most significant specification with regard to pipetting was the ability to recover all but a few microliters from each well in a row of eight wells.

The current configuration of the robot allows 30 microplates and lids to be stored at room temperature prior to sequencing and a maximum of 40 plates with completed sequencing reactions to be stored in cooled chambers. The system can, however, be configured to allow ambient as well as cooled storage of a maximum of 60 microplates (Figure 1). The system incorporates a series of standard robot components for microplate handling and multiwell pipetting (these components are typically used for ELISA assays). The nonstandard components included to comply with our specifications are the cooled reagent reservoir, the heat-blocks that are sculpted to accommodate

U-bottom microplates, the cooled storage chambers and a more robust hand (hand B in Figure 1) that allows for manipulation of microplates and lids on the sculpted heat blocks.

Reproducible Aspiration of Small Volumes

Programming of some basic functions that involved the coordination of power-and-event controller switches, such as the opening and closing of incubator doors, was provided by the Zymark systems engineers. As delivered, the robot was unable to recover 7 µL from a total of 12 µL in microplate wells; many of the channels in the multichannel micropipet were only partially filled. It was evident that the pipet tips had to reach to just above the bottoms of the microplate wells to ensure that they would be immersed in the liquid for the last few microliters without blocking the orifices. Because the Zymark robot has a reach of up to 70 cm, there is some sway or movement at the extremity. As a result, it is not possible to program it using absolute coordinates to within sub-millimeter precision. Therefore, it was necessary to program the robot so that it would "sense" the bottoms of the wells.

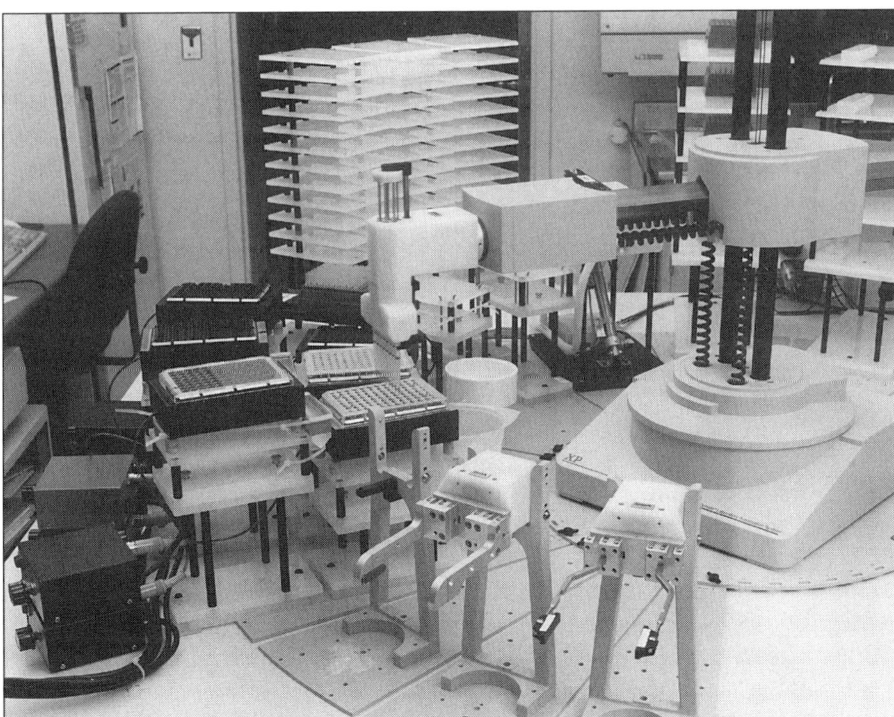

Figure 2. Robot pipetting reagents on microplate. Clockwise starting in foreground: MTP Hand A in stand, MTP Hand B in stand, stand for multichannel pipet, sculpted heat blocks and controllers, tip station with blue box of pipet tips, ambient storage racks for multititer plates (multi-tiered structure behind multichannel pipet). Pipet tip storage racks can be seen directly behind vertical supports and arm housing of robot.

The robot is equipped with force sensors, which made it possible to devise an algorithm to position the tips correctly. The algorithm first determined that the tips exerted force against the bottom of the vessel(s), and then the height was increased to the pipetting height where the orifices were not occluded. Both the position where the tips exerted force and the pipetting height were approached by small changes in position followed by monitoring of the exerted force. To ensure that the final position was reached as rapidly as possible, initial movements were made in large and rapid steps. The large, rapid steps were followed by a series of steps that became smaller in a graded manner. (A detailed description and schematic diagram of the algorithm is available to interested readers through mail, EMail or BioTechNet®.) The algorithm significantly improved the reproducibility of the pipetting. The pipetting height was determined once for each microplate for the set of wells containing the extended primer and template, just prior to dispensing the reaction to the four termination mixtures.

Three other factors influenced the pipetting. First, small volumes of solutions in the microplates had a tendency to bead up into droplets that frequently prevented recovery of the desired volume from a well. Inclusion of a mixture of nonionic detergents to a final concentration of 0.02% was adequate to prevent beading on the polystyrene microplates. Second, the microplates were improperly seated on the sculpted heat blocks in that the plates pivoted on the rounded U bottoms. Shims were added to the heat blocks to ensure that contact was made with the outer edge of the microplates and with the U bottoms. Third, it was necessary for the robot to touch the sides of the wells to ensure that all the dispensed volume reached the target well.

With a combination of stably seated microplates, wiping of tips, nonionic detergents and the sensing algorithm, the robot was able to reproducibly aspirate and dispense small volumes.

Optimization of Throughput

To increase the number of reactions performed in a set time, the reaction conditions for *Bst* DNA polymerase (4,10) were modified so that all steps (heating, annealing, extension and termination) were performed in blocks of about 5 min. Blocks of equal length allowed two sets of reactions to be performed out-of-phase by one 5-min block. Two sets of reactions were completed in 27 min.

One consequence of performing parallel reactions staggered by 5 min was that the time required to change micropipet tips became rate limiting. Therefore, we added reservoirs containing water (components labeled b in Figure 1) at positions where the robot performed pipetting operations. We used plastic inserts similar to those used in the cooled reagent reservoir except that they had a nominal volume of 38 mL. Tips were washed by aspirating and expelling 50 µL several times. The last expulsion of the washing solution was performed slowly while the tips were wiped.

Sequencing with Sequenase

Sequenase (10–12) is currently a standard for sequencing in many laboratories. To compare the sequencing reactions performed by the robot with those performed manually in our and other laboratories, the reaction conditions for Sequenase were modified for use with the robot. The sequences generated (Figure 3) were comparable to those generated manually by an individual experienced at sequencing. Furthermore, there was little, if any, variation from well to well and from plate to plate for a single template (Figure 3).

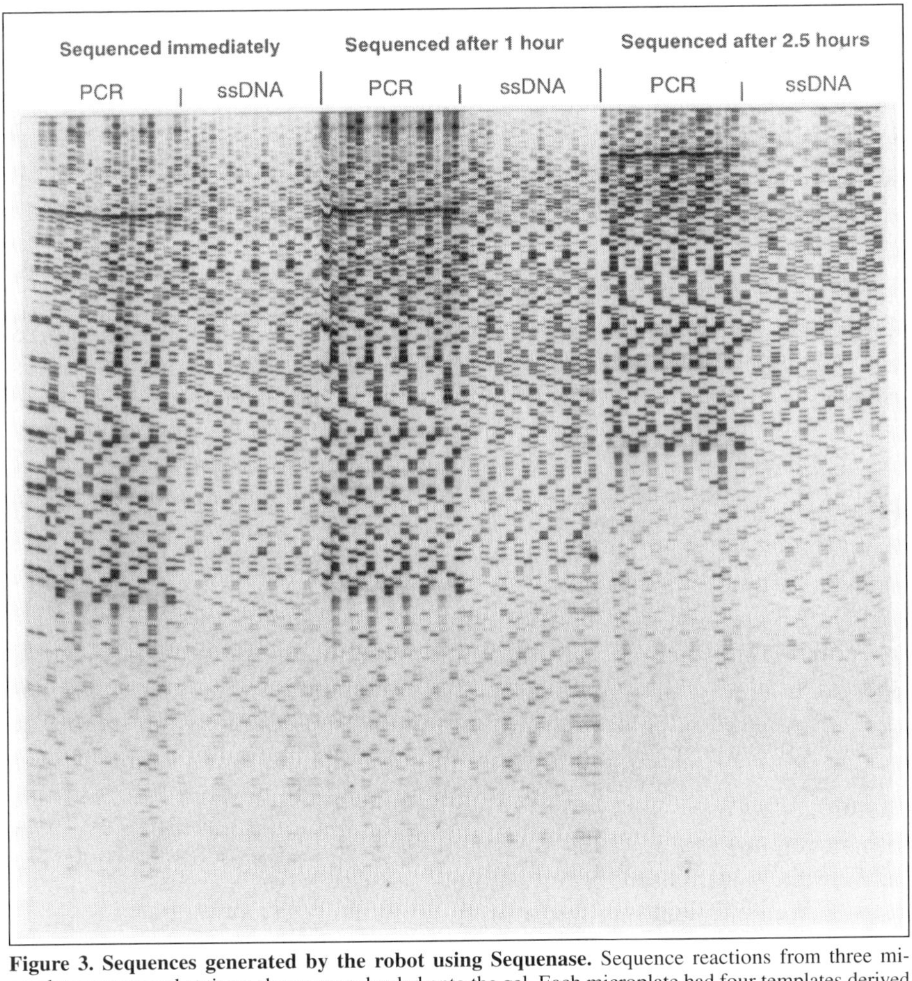

Figure 3. Sequences generated by the robot using Sequenase. Sequence reactions from three microplates sequenced at times shown were loaded onto the gel. Each microplate had four templates derived from asymmetrical amplification (PCR) and four templates that were purified single-stranded DNA (ssDNA) from bacteriophage M13mp19. Sequences were uniform from sample to sample and plate to plate. Note that the samples from each plate were loaded simultaneously, but that the samples from the plate sequenced after 2.5 h were loaded later.

Sequencing with *Bst* DNA Polymerase

Sequenase did not prove to be an ideal DNA polymerase for the system, since it is labile even at low temperatures, and the manufacturer does not recommend storing the diluted enzyme for more than 60 min at 4°C (12). To achieve our goal of long periods of unattended operation, it was essential to use an enzyme that was stable in solution at 4°C at the working dilution. The DNA polymerase from *Bacillus stearothermophilus* (*Bst* DNA polymerase) was reported to be extremely stable (4,15) and effective for DNA sequencing even after 14 days of storage at room temperature. Therefore, we investigated its usefulness for the robot. Using *Bst* DNA polymerase, the robot generated sequences with excellent and uniform band intensity that were reproducible from plate to plate, even on plates sequenced 4 h apart (Figure 4).

DISCUSSION

Previously, protocols were developed to carry out DNA sequencing on pipetting workstations (1–3,5–7,13, 14,16,17). The protocols demonstrated the usefulness of automation in terms of consistent quality and improved throughput. The disadvantage of pipetting workstations was that relatively few reactions could be performed without human intervention or attention. A commercially available sequencer (CATALYST) was developed that was capable of unattended operation, but the use of a non-disposable pipet tip and reaction vessel requires long wash cycles to prevent cross-contamination (1). We developed a robot that is efficient, versatile and can perform DNA sequencing reactions completely unattended after initial setup. In its current configuration, it can perform 240 DNA sequencing reactions in 7 h. Addition of more storage space for microplates prior to DNA sequencing and for completed reactions will enable the robot to perform 480 DNA sequencing reactions in 14 h. Therefore, a total of 720 reactions could be performed per day by combining a 7-h daytime run with a 14-h nighttime run. The total operator time required for the 720 reactions would be about one hour, one-half hour per run. The throughput of the robot is about 0.6 reactions per minute or about 3 times slower than the most rapid implementation of automated sequencing reported in which a Biomek® 1000 workstation (Beckman Instruments) achieved a throughput of 48 reactions in 25 minutes (16). Speed was, in part, achieved by a one-step sequencing protocol, since fluorescently labeled primers were used, and it was not necessary to use the two-step protocol to obtain efficient incorporation of label. The CATALYST is limited to doing 24 reactions at a time before requiring operator intervention. A significant advantage of our robot is that the throughput is sustainable over long runs involving many samples and with little operator intervention.

Development of an algorithm that enabled the robot to "sense" the bottoms of the wells with the pipet tips, as well as tip wiping and inclusion of nonionic detergents, made it possible to pipet small volumes precisely and reproducibly. Consequently, reactions were reproducible from well to well, not only on consecutive plates, but also on plates that were sequenced 4 h (Figure 4) or more apart.

The robot has performed over 3500 reactions in our laboratory in about 30 separate runs of 96 to 192 templates per run without a robot-related problem. Failed reactions were traced to either human error or to poor template quality. The majority of the templates used in our DNA core facility were asymmetric PCR (8) products generated from cDNA or genomic DNA samples from patient material or from clones randomly selected from cDNA libraries. In

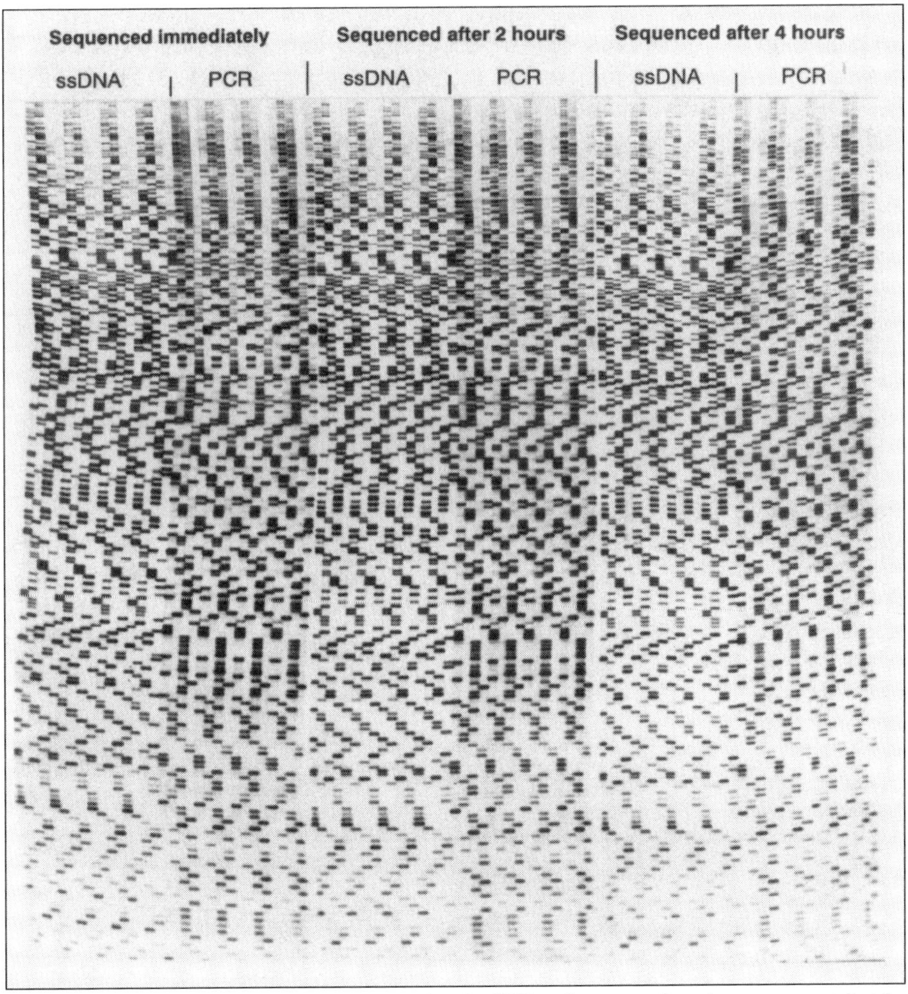

Figure 4. Sequences generated by the robot using *Bst* DNA polymerase. Reactions were started at the times indicated with four plates identical to those in Figure 3. Again, sequences were uniform from sample to sample and plate to plate.

addition, we also employed single-stranded bacteriophage M13 templates. The successful sequencing of a wide range of different templates by our robot suggests the system will be useful in many DNA core laboratories.

The capacity of the robot greatly exceeds the capacity of most single laboratories to electrophorese and analyze the reaction products using either conventional equipment or automated analyzers. Therefore, as a major component in a core DNA sequencing facility, the robot described here can serve the needs of a research center with diverse sequencing interests and demands. The robot, as part of a core facility, can increase throughput and reliability of DNA sequencing reactions for several small laboratories that can electrophorese and analyze the reaction products with existing equipment.

Another application of the robot is in diagnostic DNA sequencing for genetic mutations in patients where speed, quality and reproducibility are of paramount importance. Analysis of the same gene or same cDNA fragments from different patients makes it possible to rapidly produce large numbers of DNA sequencing templates, especially if the PCR (8) is employed. The large number of templates generated by such a protocol can be sequenced easily using the robot. The system described here was developed for DNA sequencing with radiolabeled nucleotides, but it can readily be employed for sequencing with nonradioactive labels such as the fluorescent labels currently employed by several commercial instruments for the electrophoresis and analysis of DNA sequencing reactions.

ACKNOWLEDGMENTS

The authors thank S. Madhatheri and A. Jokinen for expert technical assistance. This work was supported by grants from the National Institutes of Health (AR 38188 and HL 45996), the American Heart Association, Southeastern Pennsylvania Affiliate, the Lucille P. Markey Charitable Trust and the March of Dimes/Birth Defects Foundation.

REFERENCES

1. **Cathcart, R.** 1990. Advances in automated DNA sequencing. Nature *347*:310.
2. **Frank, R., A. Bosserhoff, C. Boulin, A. Epstein, H. Gausepohl and K. Ashman.** 1988. Automation of DNA sequencing reactions and related techniques: a workstation for micromanipulation of liquids. Bio/Technology *6*:1211-1213.
3. **Harrison, D., C. Baldwin and D.J. Prockop.** 1993. Use of an automated workstation to facilitate PCR amplification, loading agarose gels and sequencing of DNA templates. BioTechniques *14*:88-97.
4. **Lu, Y.-y., S.-y. Ye and G.-f. Hong.** 1991. Large fragment of DNA polymerase I from *Bacillus stearothermophilus* (*Bst* polymerase) is stable at ambient temperature. BioTechniques *11*:464-466.
5. **Mardis, E.R. and B.A. Roe.** 1988. Automated methods for Sanger dideoxynucleotide sequencing and single-stranded DNA isolation. FASEB J. *2*:A1125.
6. **Mardis, E.R. and B.A. Roe.** 1989. Automated methods for single-stranded DNA isolation and dideoxynucleotide DNA sequencing reactions on robotic workstation. BioTechniques 7:840-850.
7. **Martin, W.J., J.R. Washington, J.R. Galinsky, M. Gallagher, R.W. Davies, M.S. Beck and S.G. Oliver.** 1985. Automation of DNA sequencing: a system to perform the Sanger dideoxy sequencing re-

actions. Bio/Technology *3*:911-915.
8. **Saiki, R., S. Scharf, F. Faloona, K. Mullis, G. Horn, H.A. Erlich and N. Arnheim.** 1985. Enzymatic amplification of β-globin genomic sequences and restriction site analysis for diagnosis of sickle cell anemia. Science *230*:1350-1354.
9. **Sanger, F., S. Nicklen and A.R. Coulson.** 1977. DNA sequencing with chain terminating inhibitors. Proc. Natl. Acad. Sci. USA *74*:5463-5467.
10. **Tabor, S. and C.C. Richardson.** 1987. DNA sequence analysis with a modified bacteriophage T7 DNA polymerase. Proc. Natl. Acad. Sci. USA *84*:4767-4771.
11. **Tabor, S. and C.C. Richardson.** 1989. Selective inactivation of the exonuclease activity of bacteriophage T7 DNA polymerase by *in vitro* mutagenesis. J. Biol. Chem. *264*:6447-6458.
12. **United States Biochemical.** 1990. Step-by-step protocols for DNA sequencing with Sequenase Version 2.0, 5th Edition. Cleveland, OH.
13. **Wilson, R.K., S. Clark, A.S. Yuen and L.E. Hood.** 1988. Automation of dideoxynucleotide DNA sequencing reactions using a BIOMEK 1000 Robotic Workstation. FASEB J. *2*:A1125.
14. **Wilson, R.K., A.S. Yuen, S.M. Clark, C. Spence, P. Arakelian and L.E. Hood.** 1988. Automation of dideoxynucleotide DNA sequencing reactions using a robotic workstation. BioTechniques *6*:776-787.
15. **Ye, S.-Y. and G.-F. Hong.** 1987. Heat-stable DNA polymerase I large fragment resolves hairpin structure in DNA sequencing. Sci. Sin. *30*:503-506.
16. **Zimmerman, J., T. Dietrich, H. Voss, H. Erfte, C. Schwager, J. Stegemann, N. Hewitt and W. Ansorge.** 1992. Fully automated Sanger sequencing protocol for double-stranded DNA. Methods Mol. Cell. Biol. *3*:39-42.
17. **Zimmerman, J., H. Voss, C. Schwager, J. Stegemann and W. Ansorge.** 1988. Automated Sanger dideoxy sequencing reaction protocol. FEBS Lett. *233*:432-436.

Update to:

Robotic Automation of Dideoxyribonucleotide Sequencing Reactions

Helena Kuivaniemi and Gerard Tromp
Wayne State University School of Medicine, Detroit, MI, USA

We used dideoxyribonucleotide sequencing in our laboratory mainly for two purposes: *(i)* to identify unique cDNAs expressed in the human aorta, and *(ii)* to screen for mutations in candidate genes in patients with heritable disorders. Both of these projects required a large amount of sequencing, and so together with our colleagues James Earley and Darwin J. Prockop, we developed a robot to carry out standard Sanger dideoxyribonucleotide sequencing reactions efficiently and with minimal human intervention. To date, we used the robot to sequence 465 independent clones from a human aorta cDNA library, and generated about 86 500 nucleotides (nt) of unique sequences (manuscript submitted).

Our strategy for candidate gene analysis was to generate asymmetrically amplified cDNA products using the polymerase chain reaction (PCR). The asymmetric PCR products were sequenced using sequencing primers spaced about 200 to 300 nt apart to ensure high confidence that no mutations went undetected. The first candidate was the gene for the α1 chain of the homotrimeric type III procollagen (COL3A1). We determined the sequence of 3232 nt of COL3A1 cDNA from a total of 130 individuals with either familial aneurysms or the Ehlers Danlos syndrome type IV, a heritable disorder of the connective tissue. We analyzed about 420 000 nt of non-overlapping, non-redundant, sequences (1,2,4–7). In the case of patients with osteogenesis imperfecta, another heritable connective tissue disorder, we sequenced cDNAs for two independent genes, namely the α1 and α2 chains of the heterotrimeric type I procollagen to screen for mutations (8,9). For each of the 12 individuals, 4379 nt of α1 and 4200 nt of α2 cDNA sequences, and a total of about 103 000 nt were analyzed. Since in all cases there was a significant overlap between the sequence generated by consecutive primers, the true number of nucleotides sequenced in the mutation analysis is at least double the 523 000 nt. Since the templates for DNA sequencing were asymmetric PCR products, transcripts from both alleles were potentially present. In most cases, polymorphisms were present to confirm that both alleles of the patient were analyzed. Taking these factors into consideration, the number of nucleotides analyzed from the patient material was over 2 000 000 nt.

Ease of loading sequencing gels was greatly improved by the availability of combs that are compatible with the multichannel format (e.g., 4× comb from Owl Scientific, Woburn, MA, USA), and the development of loading tools such as multichannel micropipets (Proline 0, 5-10 multi-channel pipet, Biohit OY, Helsinki, Finland) and multichannel syringe pipet (8-channel syringe pipet; Hamilton company, Reno, Nevada). The interpretation of sequencing gels from patient samples, in which the same region is analyzed from several patients, is simplified by using a loading schema in which several reactions terminated with the same dideoxyribonucleotide are loaded in adjacent lanes (1,5,7,8, and Figure 1 of update). Typically, the samples from 6–12 patients were loaded. A significant advantage was that it was easy to identify the appearance of a new band where the other samples lacked the band. In addition, the blocks of adjacent lanes could be interpreted like a conventional sequencing ladder to identify the location.

The DNA sequencing reactions are nowadays carried out using ^{33}P-dATP in our laboratory. Comparison of different enzymes revealed that *Bst* DNA polymerase was the most stable and gave the most reproducible results (3).

DNA sequencing is a useful tool for mutation detection, since in many genes the spectrum of mutations is large. It is particularly useful for detecting small deletions, insertions and single base changes as it can identify both the site and the nature of the mutation. In addition, it is considered a very accurate method for mutation detection in that it is both sensitive and specific.

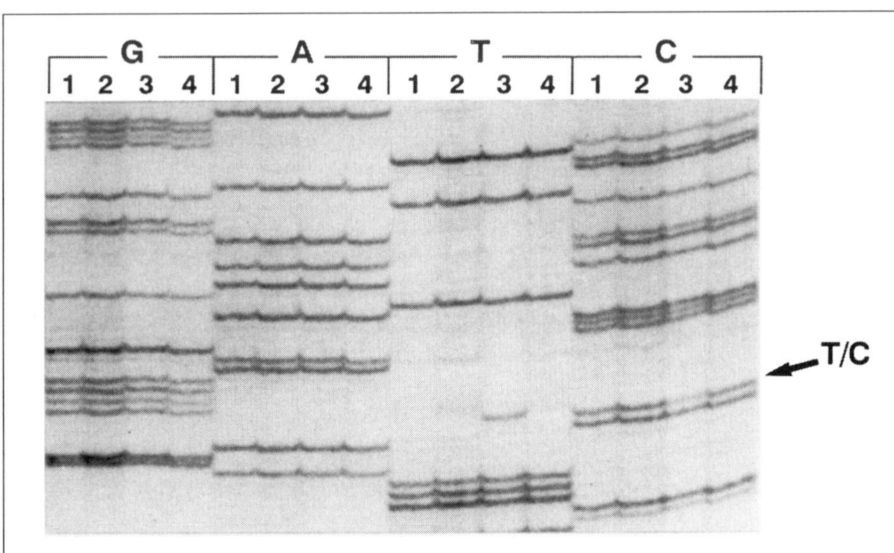

Figure 1. Example of block loading of sequencing reactions and ease with which a single base mutation can be detected. Asymmetric PCRs from four different individuals, numbered 1 through 4, were sequenced and were loaded such that all termination reactions with the same dideoxyribonucleotide were adjacent to each other in a block.

REFERENCES

1. **Anderson, D.W., T. Edwards, M.H. Ricketts, H. Kuivaniemi, G. Tromp, C.A. Stolle, S.B. Deak and C.D. Boyd.** 1996. Multiple defects in type III collagen synthesis are associated with the pathogenesis of abdominal aortic aneurysms. Ann. NY Acad. Sci. *800*:216-228.
2. **Anderson, D.W., S. Thakker-Varia, G. Tromp, H. Kuivaniemi and C.A. Stolle.** 1997. A glycine (415) to serine substitution results in impaired secretion and decreased thermal stability of type III procollagen in a patient with Ehlers-Danlos syndrome type IV. Hum. Mutat. *9*:62-63.
3. **Earley, J.J., H. Kuivaniemi, D.J. Prockop and G. Tromp.** 1993. Efficient DNA sequencing on microtiter plates using dried reagents and Bst DNA polymerase. DNA Seq. *4*:79-85.
4. **Kuivaniemi, H., D.J. Prockop, Y. Wu, S.L. Madhatheri, C. Kleinert, J.J. Earley, A. Jokinen, C. Stolle, K. Majamaa, V.V. Myllylä et al.** 1993. Exclusion of mutations in the gene for type III collagen (COL3A1) as a common cause of intracranial aneurysms or cervical artery dissections: results from sequence analysis of the coding sequences of type III collagen from 55 unrelated patients. Neurology *43*:2652-2658.
5. **Madhatheri, S.L., G. Tromp, K.-H. Gustavson and H. Kuivaniemi.** 1994. Substitution of glutamic acid for glycine 589 in the triple-helical domain of type III procollagen (COL3A1) in a family with variable phenotype of the Ehlers-Danlos syndrome type IV. Hum. Molec. Genet. *3*:511-512.
6. **Tromp, G., A. De Paepe, L. Nuytinck, S. Madhatheri and H. Kuivaniemi.** 1995. Substitution of valine for glycine 793 in the triple-helical domain of type III procollagen in a mother and a child with the Ehlers-Danlos syndrome type IV. Hum. Mutat. *5*:179-181.
7. **Tromp, G., Y. Wu, D.J. Prockop, S.L. Madhatheri, C. Kleinert, J.J. Earley, J. Zhuang, Ö. Norrgård, R.C. Darling, W.M. Abbott et al.** 1993. Sequencing of cDNA from 50 unrelated patients reveals that mutations in the triple-helical domain of type III procollagen are an infrequent cause of aortic aneurysms. J. Clin. Invest. *91*:2539-2545.
8. **Zhuang, J., G. Tromp, H. Kuivaniemi, S. Castells, M. Bugge and D.J. Prockop.** 1996. Direct sequencing of PCR products derived from cDNAs for the proα1 and proα2 chains of type I procollagen as a screening method to detect mutations in patients with osteogenesis imperfecta. Hum. Mutat. *7*:89-99.
9. **Zhuang, J., G. Tromp, H. Kuivaniemi, S. Castells and D.J. Prockop.** 1996. Substitution of arginine for glycine at position 154 of the α1 chain of type I collagen in a variant of osteogenesis imperfecta: comparison to previous cases with the same mutation. Am. J. Med. Genet. *61*:111-116.

Fully Automated, Nonradioactive Solid-Phase Sequencing of Genomic DNA Obtained from PCR

Arndt Rolfs and Ines Weber

Free University of Berlin, Klinikum Steglitz, Berlin, Germany

ABSTRACT

Nonradioactive sequencing in combination with solid-phase template purification is a powerful method for sequence analysis, especially of PCR-generated fragments. To perform DNA sequencing under optimal conditions, it is necessary to obtain a well-purified and single-stranded (ss) DNA template. The disadvantage of ssDNA sequencing is that additional steps are required to generate ssDNA templates before any sequencing reactions can be carried out and is thus more time-consuming. Here we describe an automated solid-phase system for direct sequencing of ssPCR products that incorporates magnetic beads coated with streptavidin as solid support and a computer-controlled device (PolySeq™) with heating, magnetic and mixing functions, all of which are integrated into a robotic workstation (Biomek™ 1000). This solid-phase method is extremely useful for rapid template purification and strand separation of DNA obtained from PCR as well as for high-quality dideoxyribonucleotide chain termination sequencing with fluorescently labeled primers. The system allows the complete automation of the sequencing procedure starting with PCR amplicons. DNA sequencing results obtained with this system were reproducible and gave an excellent length of readable sequences with low background and an analysis capacity of 30 000–40 000 bp per week.

INTRODUCTION

In the last few years, DNA sequencing has become increasingly important to molecular biology and biotechnology for DNA analysis. Currently there is an attempt to determine the complete genome sequences of various species (e.g., the Human Genome Project). Such research requires technically easy, automated procedures that produce reliable and reproducible results. For these large-scale sequencing projects to be successful, different systems and procedures have to be combined and integrated (4,16,17).

(Reprinted from BioTechniques 17:782-787, 1994)

To perform DNA sequencing of genomic DNA obtained from polymerase chain reaction (PCR) under optimal conditions, it is necessary to obtain a well-purified and single-stranded (ss) DNA template. Unincorporated primers and deoxyribonucleoside triphosphates (dNTPs) have to be removed from the amplified DNA because they will impair the sequencing reaction. Furthermore, a ssDNA template avoids interference with the sequencing reaction due to reassociation of the double-stranded (ds) PCR product and leads to increased length and accuracy of readable sequences (11); additionally, less DNA template is required than for dsDNA. A number of methods for generating ssDNA have been described, including: 1) generation of ssDNA by asymmetric PCR (3,15); 2) generation of ssDNA by blocking-primer PCR (2); 3) exonuclease digestion of the phosphorylated strand of the dsDNA (5); and 4) immobilization of amplified biotinylated dsDNA by avidin and strand-specific elution using alkali, by using either: a) affinity agarose gel (10,13) or b) capturing biotinylated PCR product using streptavidin-coated paramagnetic beads (14).

The disadvantage of ssDNA sequencing is that the generation of ssDNA is associated with additional steps prior to the sequencing reaction and is thus more time-consuming. Also, most procedures lead to a loss in the yield of the DNA template. Furthermore, the isolation of ssDNA from a strand-separating gel involves a higher risk of cross-contamination with other PCR products.

In this paper we present a protocol for DNA-sequencing using a solid-phase method with streptavidin-coated paramagnetic beads. The protocol has the advantage that it can be performed as a fully-automated process followed by the sequencing reaction. Several methods for fully-automated DNA sequencing procedures have already been described (8,9,15,16). The problem with these methods is that they disregarded a cost-effective and fast strand separation method. The use of a 96-well microplate is recommended for automated laboratory work because most robotic workstations in the routine laboratory are adapted to that format.

Any device used for the automated separation of DNA containing one biotinylated strand has to meet the following requirements: 1) the production of a strong and homogenous magnetic field in all 96 wells of the microplate; 2) agitation, to resuspend the paramagnetic beads following separation steps; 3) temperature control up to 96°C; and 4) easy adaptability to an existing pipetting robot. Furthermore, the microplate must be able to be placed directly onto the heating/magnetic block to ensure a satisfactory heat transfer between the heating block and the sample.

Here we describe an automated solid-phase protocol for direct sequencing of genomic PCR products. It uses magnetic beads coated with streptavidin as solid support and an instrument (PolySeq™) with heating, magnetic and mixing functions that is integrated in a robotic workstation (Biomek™ 1000). This solid-phase method is extremely useful for template purification and strand separation of DNA obtained from PCR. Compared with dsDNA sequencing, this method yields consistently better results with increased lengths of readable sequences and reduced backgrounds. For the sequencing reaction, we used a nonradioactive dideoxyribonucleotide chain termination procedure (12) with fluorescent dye-labeled primers (7). A similar protocol for the semi-automated application of bidirectional solid-phase sequencing of plasmid DNA was published by Hultman and coworkers (6). This group also used a Biomek 1000 robotic workstation equipped with a heating/cooling unit but without an instrument for the automated separation of the paramagnetic beads.

MATERIALS AND METHODS

Primer Design

The biotinylated PCR product is produced using a 5′-biotin-labeled amplification primer. The primers can be synthesized on a Cyclone™ DNA synthesizer (Millipore, Bedford, MA, USA); the biotinylation is done by coupling biotin-phosphoramidite at the 5′ end during the last synthesis reaction step. The forward primer (P1) contains the specific amplification sequence [for the 5SrRNA Legionellae gene 5′ ACT-ATA-GCG-ATT-TGG-AAC-CA 3′] and a 5′ sequence identical to that of the M13 universal fluorescent sequencing primer [5′ TAA-AAC-GAC-GGC-CAG-TGC-CA 3′]. The non-biotinylated reverse primer (P2) contains the specific sequence of the 3′ primer [for the 5SrRNA Legionellae gene 5′ GCG-ATG-ACC-TAC-TTT-CGC-AT 3′] and a 5′ sequence identical to that of the universal reverse M13 fluorescent sequencing primer [5′ CAG-GAA-ACA-GCT-ATG-ACC 3′]. For PCR, the primers are used in equimolar ratio. After PCR, 40 μL of the specific biotinylated amplification product is pipetted into a 96-well microplate. Figure 1 demonstrates the procedure for producing a biotinylated PCR amplicon.

Construction of the Automated Magnetic Station (PolySeq)

PolySeq was designed by us and constructed by Polygen (Frankfurt, Germany) and the robotic workstation (Biomek 1000), which carries out the pipetting steps, is from Beckman Instruments (Fullerton, CA, USA). The current version of PolySeq consists of a block with integrated temperature, magnet and agitation unit as well as a Biomek software-linked controller. The heating block can accomodate a 96-well microplate with a U-bottom (Falcon®; Becton Dickinson, Lincoln, NJ, USA) and has a regulation precision of ±0.2°C of temperatures between 30° to 98°C. The heating block has an overall temperature uniformity of ±0.4°C. The heating performance is 1°C/s. The cooling is done by a built-in ventilator and the cooling performance is about 0.8°C/s. The heating block can be agitated with a stroke of ±1 mm. Length and frequency of agitation can be regulated by the software. Agitation causes the uniform redistribution of paramagnetic beads after separation has taken place. The maximum sample volume, which can be agitated without spilling, is about 200 µL. PolySeq also has a platform with several neodymium iron-boron magnets (Neo-Delta™; IBS Magnet, Köln, Germany) in the form of a narrow strip. These magnets can fit between the wells of the microplate (Figure 4) and attract the paramagnetic beads to the sidewalls of the wells. The platform is elevated to separate the beads (which occurs within 5–15 s) and is then lowered so that the beads can be resuspended in the next reaction step. The problem of evaporation occuring during the reaction in the uncovered microplates, and particularly during the heating steps, is solved by adding deionized, distilled water (usually 2 µL) after the annealing step. Moreover, during the heating steps, we cover the microplate with a flexible lid, which can be moved by the Biomek Side Loader.

The Biomek 1000 automated workstation (1) consists of a holder for a tip rack, two tray positions for tube trays and one position for the PolySeq. For

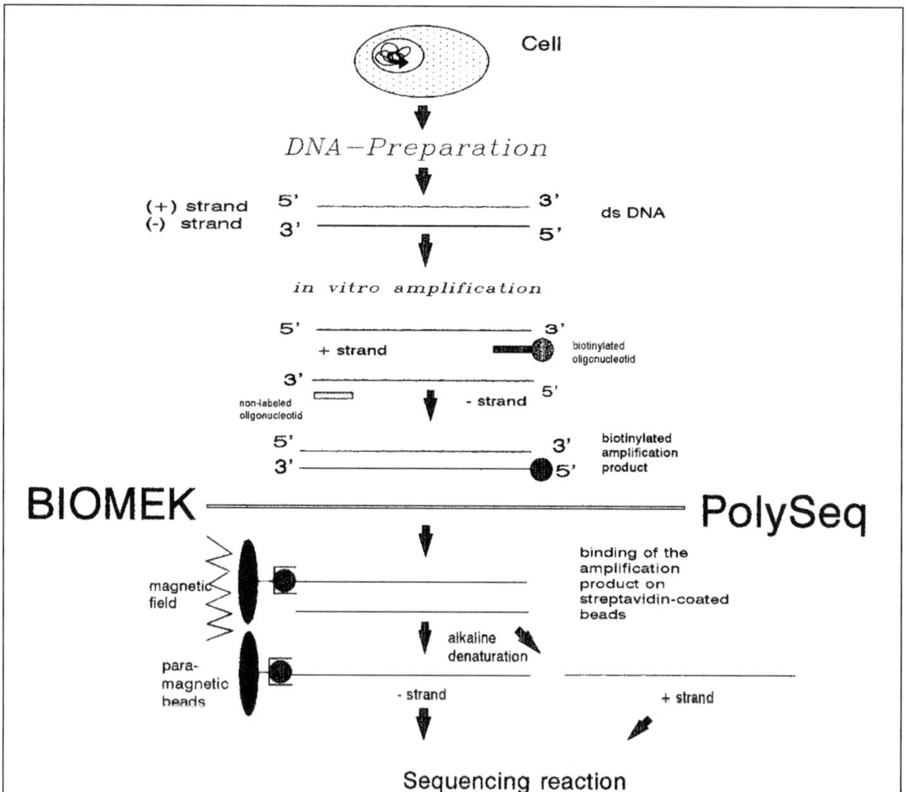

Figure 1. Schematic flow-sheet demonstrating the procedure for producing a biotinylated PCR product followed by the strand separation with streptavidin-coated paramagnetic beads using the PolySeq and Biomek 1000.

the pipetting procedures, four pipetting tools (from the Biomek 1000) are used: 1) P20, single-tip, 1–20 µL; 2) P200, single-tip, 20–200 µL; 3) MP20, eight-tip, 1–20 µL; and 4) MP200, eight-tip, 20–200 µL. The transfer operation can be completely automated with the addition of the Biomek Side Loader System, which consists of a robotic arm capable of transferring plates and tips between the Biomek working table and the storage area.

Automated Strand Separation with PolySeq

Forty microliters streptavidin-coated paramagnetic beads per well (M-280 Streptavidin Dynabeads®; Dynal AS, Oslo, Norway) are pipetted into a Falcon 96-well U-bottom flexible microplate. One milligram Dynabeads (ca. 100 µL, 6–7 × 10^8 beads per mL dissolved in phosphate-buffered saline [PBS]) bind about 25 pmol of a 300-bp dsPCR fragment. Forty microliters of binding buffer (10 mM Tris-HCl, pH 7.5, 100 mM NaCl, 1 mM EDTA, 0.15% Triton® X-100) are added for the equilibration of the beads. Using the magnets of the PolySeq platform, the beads were completely separated (usually about 40 s), and the supernatant removed and discarded. After the beads had been washed twice, 27 µL 20× SSPE (3 M NaCl, 2 mM NaH_2PO_4 × H_2O, 2 mM EDTA) plus 40 µL biotinylated PCR product are added and then well shaken by the agitation function of PolySeq. The reaction was incubated for 30 min at 37°C with gentle agitation steps every two min. During the incubation, the biotinylated PCR products are bound to the beads with streptavidin bridges. The beads can be separated again by the magnets. After removing and discarding the supernatant, 50 µL of denaturation buffer (0.1 M NaOH, 0.04 mM EDTA) are added and the beads resuspended by shaking and incubated for 3 min at room temperature. After this incubation, the beads are separated again, the supernatant removed, stored in a new reaction well and neutralized with 5 µL 1 M HCl. The supernatant contains the non-biotinylated strand and is stored for further sequencing. After the magnets have been removed, the beads are washed with 200 µL binding buffer and finally resuspended in 16 µL water and are now ready for the sequencing procedure.

Sequencing Reaction

Except for the M13 reverse primer, the sequencing reagents for the following solid-phase sequencing are from the Sequenase® Dye Primer Sequencing Kit (Catalog No. 401117; Applied Biosystem [ABI], Foster City, CA, USA). The kit contains 80 pmol of the "A"- and "C"-specific primers JOE and FAM, plus 160 pmol of the "G"- and "T"-specific primers TAMRA and ROX. The reaction procedures for the sequencing assay correspond to the manufacturer's recommendations. All pipetting steps required for the protocol are programmed with the Biomek 1000 software. Twenty-four sets of reactions were carried out in 96-well microplates.

The non-biotinylated minus-strand is precipitated by 95% ethanol, the pellet resuspended in 6 µL deionized distilled water and used for the *Taq* sequencing procedure (Taq Dye Primer Cycle Sequencing Kit, -21M13, Catalog No. #401119; ABI). The pipetting steps for the sequencing reaction are also done by the robotic workstation. We do not presently employ cycle sequencing when sequencing the non-biotinylated strand with *Taq*. The chain-termination reactions are performed at a temperature of 70°C for 15 min. The annealing step is similar to the procedure with Sequenase: 5 min at 65°C followed by slow cooling to room temperature within 10 min. Although the pipetting steps are not performed at 4°C and cooling after the execution of the annealing and termination steps is done only to room temperature, no significant difference in sequencing quality has been observed.

RESULTS

Quantification of PCR Products Necessary for Automated Pipetting and Sequencing Procedures

We have amplified a 144-bp DNA fragment (Figure 2) from the 5SrRNA gene from *Legionella pneumophila* (Oxford, Catalog No. 43110; ATCC, Rockville, MD, USA) as well as a 501-bp fragment from the envelope region of HIV-1 using specific oligonucleotides modified with the M13 and biotinylated reverse-M13 sequence as described in Materials and Methods. For the automated sequencing procedure with the PolySeq, different amounts of the PCR fragments (1000 ng, 500 ng, 250 ng, 60 ng and 30 ng) were used for testing the smallest amount of the amplicon that can be sequenced according our procedure. As shown in Figure 3, even amounts as little as 300 ng and 60 ng used in the automated sequencing procedures gave excellent quality in the

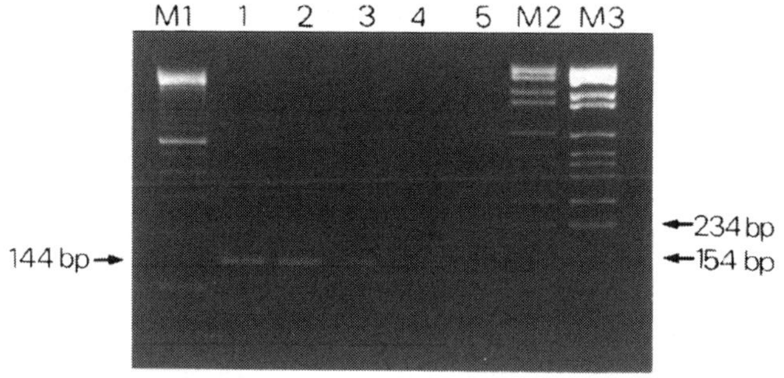

Figure 2. Agarose gel analysis of a 144-bp PCR fragment (5SrRNA) from *Legionella pneumophila* (ATCC 43110 Oxford). 10 µL, 7.5 µL, 5 µL, 2.5 µL and 1.25 µL (lanes 1–5, respectively) of a 100-µL PCR sample were separated in an ethidium bromide-stained 3% agarose/NuSieve® gel (FMC BioProducts, Rockland, ME, USA) in 1× TBE buffer (0.1 mM Tris-borate, 1 mM EDTA). Lane M1: 1 µL 100-bp ladder (Life Technologies, Gaithersburg, MD, USA), lanes M2 and M3: 1 µL and 2 µL of the Boehringer VI marker (Boehringer Mannheim, Mannheim, Germany), respectively. Semiquantification of PCR fragments reveals 60 ng product in lane 1 and about 8 ng in lane 5. Using the robotic workstation, 50–60 ng is sufficient for sequencing.

sequencing data. No differences in sequencing quality were found using PCR fragment amounts up to 60 ng. Below the cut-off point of about 60 ng, we were not able to obtain sufficient sequencing signals. Since the yield of an optimized 100-µL PCR assay (35 cycles) ranges mostly between about 500 ng and 1500 ng DNA, it will be possible to do at least five to ten sequencing reactions per PCR assay with our robot system.

Using the PolySeq for Increasing Sequencing Capacity

The automation process described in this paper includes automated strand separation of dsDNA followed by automated sequencing of the separated single strands. With this method one can determine the DNA sequence of twenty-four 350-bp PCR amplicons in 24 h using a totally computer-controlled system. Using this system one person can comfortably analyze 30 000 to 40 000 bp sequence data per week using the Model 373A DNA Sequencer (ABI) and have time left over for data analysis and other tasks.

Sequence Length and Error Analysis

Sequences for two PCR products (144-bp 5SrRNA *Legionella pneumophila* and 501-bp fragment from the env-HIV-1 gene) were determined as

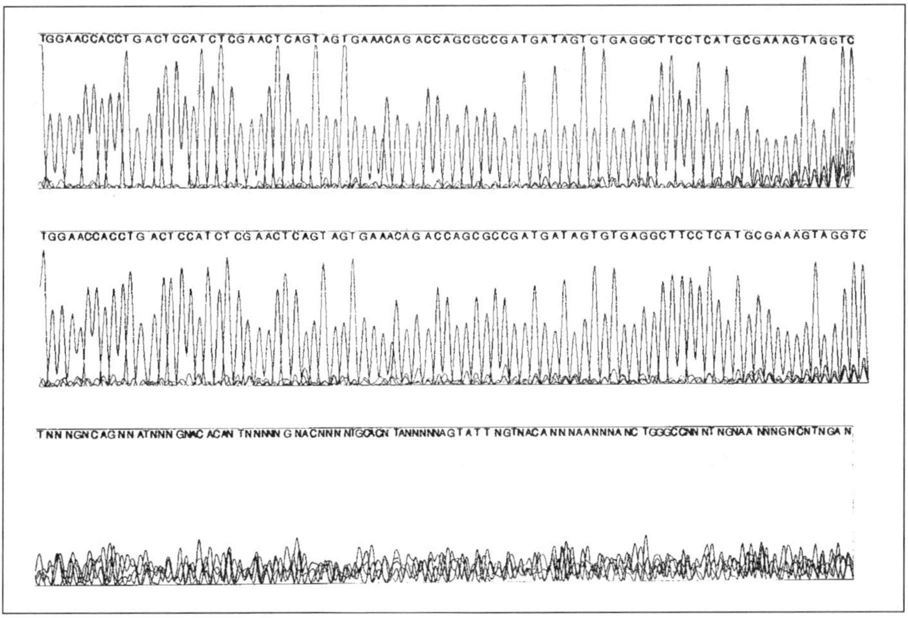

Figure 3. Demonstration of the Sequenase sequencing results of the single-stranded 144-bp 5SrRNA fragment of *Legionella pneumophila* using different amounts of DNA template for sequencing reaction. In the upper chromatogram about 300 ng template, in the middle chromatogram 60 ng and in the lower chromatogram 30 ng, were used respectively. There are no obvious differences in quality between the upper and the middle chromatograms. Below the cut-off of about 50–60 ng we were not able to get sufficient sequencing signals with the ABI 373A sequencer.

described above and were compared with manually determined reference sequences using the GeneWorks™ (Amsterdam, The Netherlands) computer program. Thirty-six 144-bp fragments and ninety-four 501-bp fragments were sequenced with Sequenase™ as well as with *Taq* DNA polymerase on the robotic workstation. Analysis of these fragments showed an error rate of about 1% for *Taq* DNA polymerase and about 0.3% for Sequenase (mismatch, insertion, deletion and ambiguity). After approximately the 300th base in a sequence, the error frequency appears to increase; therefore the quality of the 144-bp sequence was higher than that of the 501-bp sequence with both manual and automated sequencing. With nucleotides that exceed a length of 350 bp, the number of sequences available for comparison becomes too small and errors occur at a frequency of 6%–20%. The differences in the level of errors obtained from regions beyond 350 bp seem to depend mainly on the quality of the PCR assay.

DISCUSSION

The PolySeq is a new accessory for the Biomek 1000 that permits automated strand separation of dsPCR-generated DNA fragments as well as the sequencing reaction for both T7 polymerase and *Taq* DNA polymerase. In contrast to prior systems (6), our method allows the complete automatization process for single-stranded sequencing DNA obtained from PCR. The system

Figure 4. Prototype of the PolySeq with heating block, agitation unit and the magnets at the top of the pins (arrow)—1 per 4 wells—integrated in the Biomek 1000 robotic workstation.

configuration enables the automated sequencing reaction of about 30 000–40 000 bp per week using an ABI Model 373A DNA Sequencer and reduces the time necessary for the sequencing procedure by 50%–70%. Additionally, higher efficiency time is gained in which a technician can analyze data and pipet the next PCR assay while reactions are underway. DNA strand separation and sequencing done with the robotic workstation achieve an accuracy of about 99.7% using the T7 polymerase and 99.0% using the *Taq* DNA polymerase. These data agree with those from other groups doing their separation and sequencing procedures manually (8). Due to the uniform peak height, Sequenase seems to be ideal for detecting heterozygotic changes in the sequence. Nevertheless, we found no statistically significant differences in the ambiguity, rate of deletions, insertions and mismatches between the two enzymes used in the study.

The PolySeq achieves a highly uniform temperature distribution and has a regulation precision of ±0.2°C. Therefore, it is possible to design an automated assay including PCR, strand separation and sequencing. This means that an even greater increase in efficiency can be realized.

In summary, the fully-automated sequencing method described in this paper affords highly standardized sample handling, eliminates pipetting errors, and allows secure and efficient sample processing.

ACKNOWLEDGMENT

The authors would like to acknowledge both the expert technical assistance and advice from Mr. Roland Friedberger and the expert software skill of Andreas Kraft and Björn Heidrich. We thank Dr. Peter N. Robinson for his critical reading of the manuscript and Mr. Detlef Sasse (Beckman Instruments) for his patience and helpful suggestions.

REFERsENCES

1. **Frank, R., A. Bosserhoff, C. Boulin, A. Epstein, H. Gausepohl and K. Ashman.** 1988. Automation of DNA sequencing reactions and related techniques: a workstation for micromanipulation of liquids. Bio/Technology *6*:1211-1213.
2. **Gyllensten, U.B.** 1989. PCR and DNA sequencing. BioTechniques *7*:700-708.
3. **Gyllensten, U.B. and H.A. Erlich.** 1988. Generation of single-stranded DNA by the polymerase chain reaction and its application to direct sequencing of the HLA-DQA locus. Proc. Natl. Acad. Sci. USA *85*:7652-7656.
4. **Harrison, D., C. Baldwin and D.J. Prockop.** 1993. Use of an automated workstation to facilitate PCR amplification, loading agarose gels and sequencing of DNA templates. BioTechniques *14*:88-97.
5. **Higuchi, R.G. and H. Ochman.** 1989. Production of single-stranded DNA templates by exonuclease digestion following the polymerase chain reaction. Nucleic Acids Res. *17*:5865.
6. **Hultman T., St. Bergh, T. Moks and M. Uhlen.** 1991. Bidirectional solid-phase sequencing of in vitro-amplified plasmid DNA. BioTechniques *10*:84-93
7. **Knight, P.** 1988. Automated DNA sequencers. Bio/Technology *6*:1095-1096.
8. **Koop, B.F., L. Rowan, W.Q. Chen, P. Deshpande, H. Lee and L. Hood.** 1993. Sequence length and error analysis of Sequenase® and automated *Taq* cycle sequencing methods. BioTechniques *14*:442-447.

9. **Koop, B.F., R.K. Wilson, C. Chen, N. Halloran, R. Sciammis, J. Lindelin and L. Hood.** 1990. Sequencing reactions in microtiter plates. BioTechniques *9*:10-14.
10. **Mitchell, L.G. and C.R. Merril.** 1989. Affinity generation of single-stranded DNA for dideoxy sequencing following the polymerase chain reaction. Anal. Biochem. *178*:239-242.
11. **Rolfs, A., I. Schuller, U. Finckh and I. Weber-Rolfs.** 1992. PCR: Clinical Diagnostics and Research. Springer Verlag, New York, Heidelberg.
12. **Sanger, F., S. Nicklen and A.R. Coulson.** 1977. DNA sequencing with chain-termination inhibitors. Proc. Natl. Acad. Sci. USA *74*:5463-5467.
13. **Stahl, S., T. Hultman, A. Olsson, T. Moks and M. Uhlen.** 1988. Solid phase DNA sequencing using the biotin-avidin system. Nucleic Acids Res. *16*:3025-3038.
14. **Syvänen, A.C., T. Hultman, K. Aalto-Setälä, H. Söderlund and M. Uhlen.** 1991. Genetic analysis of the polymorphism of the human apolipoprotein E using automated solid-phase sequencing. Genet. Anal. Tech. Appl. *8*:117-123.
15. **Wilson, R.K., C. Chen and L. Hood.** 1990. Optimization of asymmetric polymerase chain reaction for rapid fluorescent DNA sequencing. BioTechniques *8*:184-189.
16. **Wilson, R.K., S. Clark, A.S. Yuen and L. Hood.** 1988. Automation of dideoxynucleotide DNA sequencing reactions using a robotic workstation. BioTechniques *6*:776-787.
17. **Wilson, R.K., S. Clark, A.S. Yuen and L. Hood.** 1988. Automation of dideoxynucleotide DNA sequencing reaction using a Biomek 1000 robotic workstation. FASEB J. *2*:A1125.

A More Stringent Choice of Primers Can Improve the Performance of Fluorescent Automated DNA Sequencers

G. De Bellis, M. Manoni, R. Pergolizzi, M.E. Redolfi and M. Luzzana

Consiglio Nazionale delle Richerche, Istituto di Tecnologie Biomediche Avanzate, Milano, Italy

ABSTRACT

Some primers frequently used in the double-stranded dideoxy DNA sequencing technique with radioactive markers are not suited for fluorescent detection. In fact oligonucleotides have a different annealing efficiency related to their base sequence, and this is reflected in nonequivalent results, particularly in fluorescent automated DNA sequencing. We present a method for the evaluation of primer performance in automated DNA sequencers and show its application to the search for a better set of primers for pBluescript vector.

INTRODUCTION

In recent years, the push towards automating DNA sequencing techniques has been necessitated by large-scale sequencing projects. The most outstanding improvement in this field has been the introduction of automated DNA sequencers based on real-time detection of electrophoresed DNA bands by means of laser-induced fluorescence (1,5,7). Although the performances of such systems are usually satisfactory, we have recently shown that the choice of primers in fluorescent sequencing is a relevant problem (3). Here we explain how this problem arises and how to solve it by using nearest-neighbor thermodynamic parameters to choose the right primer sequence. We have applied these concepts to the design of a new set of primers for pBluescript® clones, thus allowing the best performance from automated sequencers using this vector.

MATERIALS AND METHODS

Materials

All reagents for oligonucleotide synthesis were from Applied Biosystems

(Reprinted from BioTechniques 13:892-897, 1992)

Table 1. Thermodynamic Parameters for Test Primers

	Primer	Sequence	ΔG°	T_m
A)	KS	CGAGGTCGACGGTATCG	-29.9	82
B)	M13	GTAAAACGACGGCCAGT	-28.4	77
C)	T3	ATTAACCCTCACTAAAG	-23.0	63
D)	T7	AATACGACTCACTATAG	-20.0	58
E)	SK	TCTAGAACTAGTGGATC	-20.6	60
F)	RevM13	AACAGCTATGACCATGA	-23.6	69
G)	P13	GACCATGATTACGCCAA	-27.1	76
H)	P51	AAGGGAACAAAAGCTGG	-28.4	76
I)	P79	GGTGGCGGCCGCTCTAG	-34.4	88
L)	P112	TCGAATTCCTGCAGCCC	-31.2	83
M)	P172	TATAGGGCGAATTGGGT	-29.1	76

The listed free energy values of annealing are calculated according to (2) with an initiation value of 5 kcal/mol. ΔG° is in kcal/mol. T_m has been calculated with a primer concentration of 8×10^{-8} M and 1 M NaCl using the equation (4):

$$Tm = \frac{\Delta H^\circ}{\Delta S^\circ + R \times 1n(C/4)} - 273.15$$

where T_m = melting temperature of the duplex in °C, ΔH° = enthalphy and ΔS° = entropy of duplex formation (2), R = gas constant and C = concentration of the primer. Salt effects are very important in DNA-duplex stability, and the conditions used in the annealing step of the DNA sequencing procedure differ from those used here (1 M NaCl); thus the values listed (and plotted in Figure 2) can be used only as a measure of relative stability among primers.

(Foster City, CA, USA). Solvents for primer HPLC purification were from Merck (Darmstadt, Germany). Reagents were biochemical grade (Fluka, Buchs, Switzerland). pBluescript SK vector was from Stratagene (La Jolla, CA, USA).

Primer Synthesis and Labeling

Primers (for sequences see Table 1) were synthesized using standard phosphoramidite chemistry with Aminolink II (Applied Biosystems) as 5′ linker. After deprotection, they were ethanol precipitated to remove ammonium ions. The pellet, 25% of a 0.2-µmol synthesis, was dissolved in 20 µL $NaHCO_3$ 0.5 M, pH 9.0, and 6 µL of fluorescein-*N*-hydroxy-succinimide (Boehringer Mannheim, Mannheim, Germany) in dimethyl sulfoxide (80 mg/mL). After 4 h, the solution was loaded on a PD-10 column (Pharmacia Biotech, Uppsala, Sweden), eluted with 0.1 M triethylammonium acetate, pH 7.0, and the fast moving colored band purified by reverse phase HPLC. The primers were UV quantified and diluted to 3.0 µM in TE.

Sequencing Reactions

For each reaction we used 6 µg of midiprep purified Bluescript SK vector, 6 pmol of each primer and the Autoread T7 sequencing kit (Pharmacia Biotech). The reactions were loaded completely on the A.L.F.™ DNA Sequencer (Pharmacia Biotech) or the EMBL DNA sequencer (European Molecular Biology Laboratory, Heidelberg, Germany) (1) and run under standard conditions.

RESULTS

Although the performance of the EMBL DNA sequencer and Pharmacia Biotech A.L.F. are satisfactory when using primers with good annealing efficiencies, such as M13 and KS, we have already shown (3) that the results are much poorer with primers such as T3 and T7. These primers have lower efficiencies even though they are commonly used in dideoxy DNA sequencing with radioactive marker yielding good sequences. We also found that SK and Reverse M13 perform very poorly (data not shown).

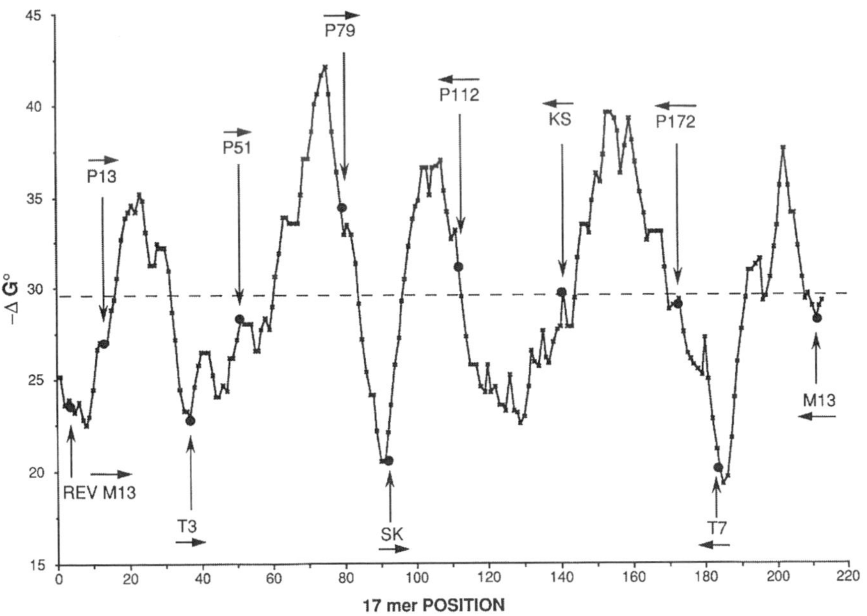

Figure 1. Plot of annealing free energies for all the possible 17-mer primers along the pBluescript SK plasmid polylinker. Indicated are ΔG° for all the tested primers. The dashed line shows the mean ΔG° value for 17-mers in this region. The horizontal arrows indicate the polymerization direction. All the primers could be used either on pBluescript SK or KS apart from p172, in which the last two 3′ bases do not match on pBluescript KS.

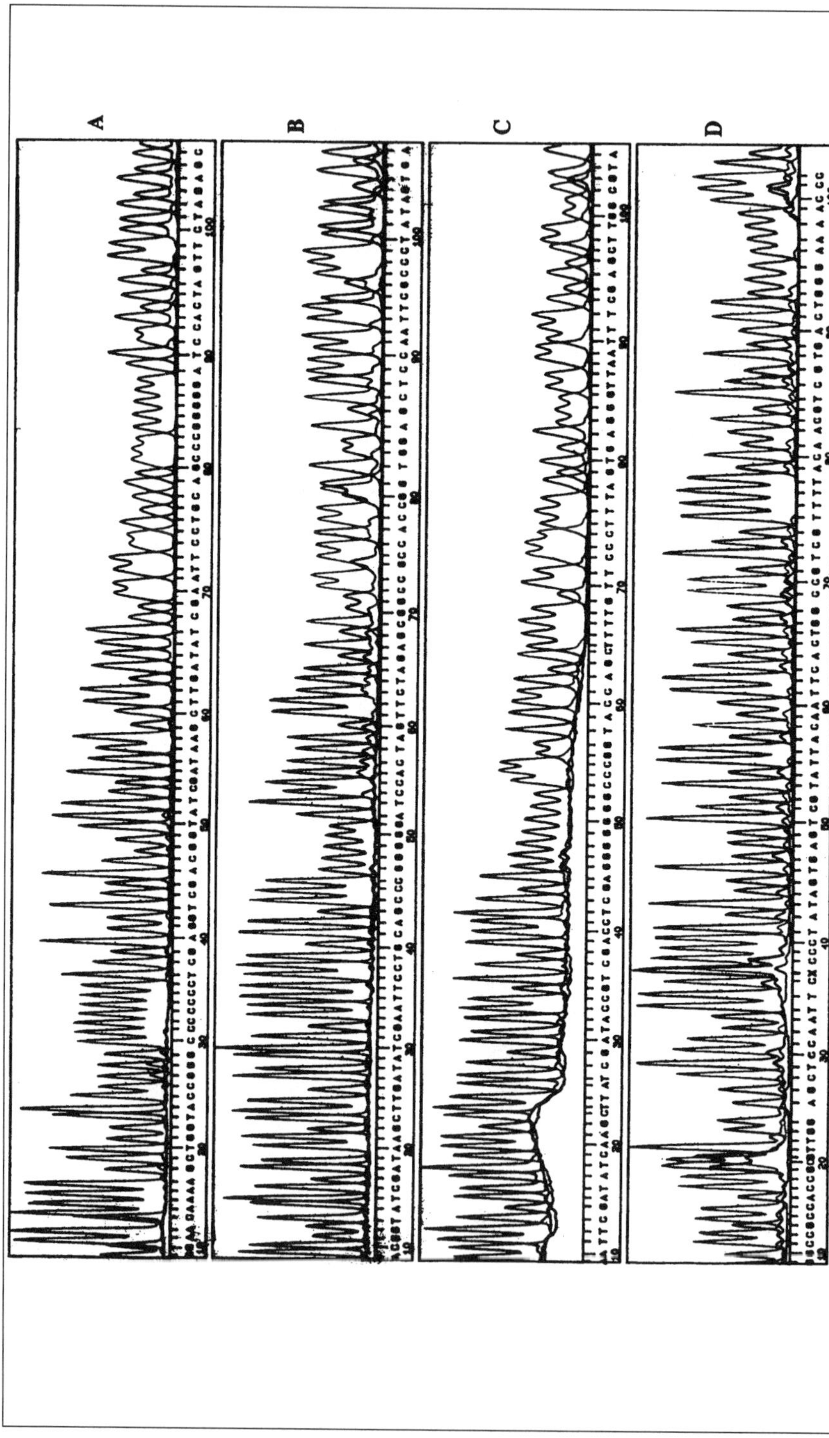

Figure 2. Plot of raw data coming from A.L.F. DNA Sequencer for primers P13 (A), P51 (B), P79 (C) and P112 (D), using 6 µg of midiprep purified Bluescript SK vector, 6 pmol of each primer and the Autoread T7 sequencing kit.

To understand how different performances arise among primers and to introduce a criterion for correct primer choice, we applied the nearest-neighbor thermodynamic model that gives a quantitative response to the analysis of primer-template noncovalent binding energies (2,4). Table 1 (A–F) shows $\Delta G°$ (free energy of annealing) and T_m (melting temperature) for the standard set of primers in the pBluescript vector calculated with this model. The listed values clearly point out the strong differences among primers, thus explaining the sequencing results reported (3) in terms of diverse annealing efficiency.

To validate these results and to find a better primer set for this vector, we analyzed the pBluescript SK polylinker (230 base pairs) by calculating the thermodynamic parameters for all possible 17-mers along its sequence using a program by W. Rychlik (6). Figure 1 graphically represents the $\Delta G°$ values plotted against the relative position of these oligonucleotides, which were examined to evalute their annealing efficiency, looking also at their base composition, autocomplementarity and homology to DNA sequences present in GenBank® (Mountainview, CA, USA). We tried to find the primers that could best replace the ones we found to be very inefficient in automated DNA sequencing (SK, T3, T7 and RevM13), locating their position near the previous set. In doing so, we designed, synthesized, and used for sequencing five new primers, listed in Table 1 (G–M) and positioned as shown in Figure 1.

Their performances were tested on several clones yielding consistently good sequences in terms of signal intensity and number of readable bases. As an example, during the sequencing of cosmid fragments cloned in pBluescript ranging in size from 0.3 to 5.0 kb, we sequenced an average of 430 bases using this new set of primers. Figure 2 shows the sequencing results using this new set of primers on pBluescript vector.

DISCUSSION

In dideoxy termination DNA sequencing, the annealing process consists of the incubation of a single-stranded (or denatured double-stranded) DNA molecule with an oligonucleotide (primer, which sequence is complementary to a particular region in the target DNA) at a certain temperature, allowing the two molecules to find the correct base matching. Then the primed DNA is divided into four aliquots; a polymerase, deoxyribonucleoside triphosphates (dNTPs) and a ddNTP are added to allow the chain extension and the base-specific termination. The yield of the annealing process directly affects the polymerase processing and therefore the signal intensity in the sequencing ladder on the gel. For this reason, the primer should be chosen with great attention to avoid poor sequencing results, searching for the best sequence using nearest-neighbor thermodynamic parameters to evaluate primer efficiency, as demonstrated with the new set of primers for pBluescript SK clones.

Several factors affect the yield of readable sequences in DNA sequencing. DNA quality and quantity are probably the most important and sometimes the least predictable, varying from sample to sample. Gel quality and electrophoretic parameters are very important and can be controlled more easily. The results we have presented also state the necessity for a careful choice of primers particularly in fluorescent automated DNA sequencing because of the relatively low sensitivity of this technique in comparison with the traditional radioactive system.

It is difficult to set a precise limiting value for the thermodynamic parameters describing the annealing reaction because these are not the only ones to be considered for consistent results. However, in our experience, primers having T_m higher (or $\Delta G°$ lower) than that for M13 are suited for fluorescent sequencing under a broad range of conditions.

In conclusion, we have shown that primer choice is critical for obtaining good results on automated DNA sequencers and that by using nearest-neighbor thermodynamic parameters, the best primer in the DNA region of interest can be chosen. This criterion has been applied to the design of a novel set of primers for pBluescript SK clones and should be of interest to anyone working on automated DNA sequencers.

ACKNOWLEDGMENTS

This work was funded by CNR, Target projects "Biotechnology and Bioinstrumentation" and "Genetic Engineering". We thank J. Hatton for carefully reading the manuscript and S. Debernardi for technical assistance. We wish also to thank Dr. W. Rychlik for his OLIGO software.

REFERENCES

1. **Ansorge, W., B. Sproat, J. Stegemann, C. Schwager and M. Zenke.** 1987. Automated DNA sequencing: ultrasensitive detection of fluorescent bands during electrophoresis. Nucleic Acids Res. *15*:4593-4602.
2. **Breslauer, K.J., R. Frank, H. Blocker and L.A. Marky.** 1986. Predicting DNA duplex stability from the base sequence. Proc. Natl. Acad. Sci. USA *83*:3746-3750.
3. **De Bellis, G., M. Manoni, R. Pergolizzi, P. Vezzoni and M. Luzzana.** 1990. Primer design in fluorescent DNA sequencing. Nucleic Acids Res. *18*:4951-4952.
4. **Freier, S.M., R. Kierzek, J.A. Jaeger, N. Sugimoto, M.H. Caruthers, T. Neilson and D.H. Turner.** 1986. Improved free energy parameters for predictions of RNA duplex stability. Proc. Natl. Acad. Sci. USA *83*:9373-9377.
5. **Prober, J.M., G. Trainor, R. Dam, F.W. Hobbs, C.W. Robertson, R.J. Zagursky, A.J. Cocuzza, M.A. Jensen and K. Baumeister.** 1987. A system for rapid DNA sequencing with fluorescent chain terminating dideoxynucleotides. Science *238*:336-341.
6. **Rychlik, W. and R.E. Rhoads.** 1989. A computer program for choosing optimal oligonucleotides for filter hybridization, sequencing and in vitro amplification of DNA. Nucleic Acids Res. *17*:8543-8551.
7. **Smith, L.M., J. Saners, R.J. Kaiser, P. Hughes, C. Dodd, C.R.Connell, S.B.H. Kent and L.E. Hood.** 1986. Fluorescent detection in automated DNA sequencing analysis. Nature *321*:674-679.

Update to:

A More Stringent Choice of Primers Can Improve the Performance of Fluorescent Automated DNA Sequencers

G. DeBellis, M. Manoni, R. Pergolizzi, M.E. Redolfi and M. Luzzana
Consiglio Nazionale delle Richerche, Instituto di Tecnologie Biomediche Avanzate, Milano, Italy

After the publication of this paper, the issue of primer choice in fluorescent automated DNA sequencing has raised some interest, although to a lesser extent to the huge amount of literature concerning primers choice in DNA amplification. In the years following this publication, we have had citations from independent groups (1–6). We applied the criteria for primer choice to completing a cosmid sequencing project and to the development of a new fluorescent labeling technique (7). The results we gained confirmed the importance of this issue to the final sequencing result.

Furthermore, we found that purification (by HPLC) of 5′-labeled primers was another key factor affecting the sequencing results. The sequencing signal strength directly reflects the labeled/unlabeled primer ratio. In our experience, the on-column 5′ labeling does not yield a synthesis product of high enough purity for optimal fluorescence of sequencing results, which renders HPLC purification mandatory.

Now we plan to apply the concept of primer choice in fluorescent DNA sequencing to PCR templates.

REFERENCES

1. **Steffens, D.L., S.L. Sutter and S.C. Roemer.** 1993. An alternate universal forward primer for improved automated DNA sequencing of M13. BioTechniques *15*:580-582.
2. **Scott, S.W. and X. Ge.** 1995. The nucleotide-sequence of citrus leaf rugose virus RNA-1. J. Gen. Virol. *76*:3233-3238.
3. **Iraqi, F. and E.J. Smith.** 1995. Organization of the sex-linked late-feathering haplotype in chickens. Anim. Genet. *26*:141-146.
4. **Skory, C.D. and S.N. Freer.** 1995. Cloning and characterization of a gene encoding a cell-bound, extracellular beta-glucosidase in the yeast Candida wickerhamii. Appl. Environ. Microbiol. *61*:518-525.
5. **Scott, S.W. and X. Ge.** 1995. The complete nucleotide-sequence of RNA-3 of citrus leaf rugose and citrus variegation ilarviruses. J. Gen.Virol. *76*:957-963.
6. **Zeng, L.W. and M. Kreitman.** 1996. Simple strategy for sequencing cDNA clones. BioTechniques *21*:446-451.
7. **DeBellis, G., R. Pergolizzi, S. Debernardi, L. Invernizzi and M. Luzzana.** 1995. Fluorescent automated DNA sequencing by limited primer labeling. BioTechniques *19*:66-70.

Cloning and Direct Sequencing of Plant Promoters Using Primer-Adapter Mediated PCR on DNA Coupled to a Magnetic Solid Phase

Mari Espelund and Kjetill S. Jakobsen

University of Oslo, Oslo, Norway

ABSTRACT

*A method that allows amplification and direct sequencing or cloning of an unknown DNA segment flanked by a known sequence is described using barley genomic DNA. The method avoids the step of circularization necessary for inverse PCR by ligation of primer-adapters to restricted genomic DNA. Specificity is achieved in the first amplification step; linear PCR with a biotinylated primer complementary to the known flanking sequence (primer 1-B) produces a single-stranded product that is purified employing streptavidin-coated magnetic beads. After this step, which removes genomic DNA, two rounds of exponential PCR are performed, first with the adapter-primer and primer 1 and second with primer 1 substituted by a nested primer 2. If the second primer is biotinylated, the product can be sequenced directly using solid-phase sequencing. We have employed this method to sequence directly and to clone the promoters of two late embryogenesis-abundant (*Lea*) genes (B19.4 and B19.3) from barley.* Lea *B19.4 and B19.3 encode putative desiccation-protective proteins that act in the final stages of embryogenesis and have previously been cloned as cDNAs. We demonstrate here that their proximal promoter regions are very similar (80% identity) and that both contain putative abscisic acid-responsive elements.*

INTRODUCTION

Cloning of genes from large eukaryotic genomes requires extensive protocols including construction of gene libraries, screening and characterization of positive clones. In plants, construction of genomic libraries of high complexity has been especially difficult because of the high content of 5-methyl-cytosine. The use of *mcr⁻* host cells has to a certain extent circumvented this problem (12). For the cloning of single genes from, for instance, gene-tagged lines, the traditional method of screening gene libraries is laborious and

(Reprinted from BioTechniques 13:74, 1992)

needs to be supplemented by alternative methods.

Direct cloning of an "unknown" DNA region located outside a known sequence is possible using a technique termed inverse polymerase chain reaction (PCR) (19,26,27). This procedure involves restriction of genomic DNA and ligating the free ends under conditions that favor circularization of the DNA. The final step is then amplification using oligonucleotides complementary to the 5′ and 3′ ends of the known sequences. However, the circularization step is unreliable and often concatamers are observed.

Recently a method based on "ligation-mediated single-sided PCR," originally intended for in vivo footprinting (18), has been developed that allows direct cloning of the shark Po promoter (8). This method uses linker ligation and subsequent amplifications with a linker-primer and multiple specific primers ("nested" primers) to obtain specificity. The ligation-mediated single-sided PCR protocol involves multiple PCRs and subsequent purifications on agarose gels.

In order to avoid the purification steps required because of unspecificity in the PCR, an additional step was introduced (21) that involves linear amplification of the target sequence with a biotinylated primer and separation of the product with the aid of streptavidin-coated magnetic beads (15). This strategy was employed in combination with ligation of oligo-cassettes to restricted DNA to directly amplify and sequence unknown regions of a yeast artificial chromosome (YAC) clone within the background of total yeast DNA (21). The method has recently been used for the extension of a human-derived microclone (22). We have extended the application of this strategy to involve direct amplification, sequencing and cloning of promoter sequences from a large plant genome, namely, barley. The barley genome is approximately twice the size of the human genome (1,2), and contains, in addition, like most cereals, a high proportion of 5-methylcytosine (25).

We were interested in cloning the promoters of the barley B19 genes, which are expressed in developing embryos and belong to the diverse group of late embryogenesis-abundant (*Lea*) genes (7). Some *Lea* genes are responsive to osmotic and desiccation stress and to the plant hormone abscisic acid (ABA), and some may be involved in protection against desiccation damage during the final stages of plant embryogenesis or during early germination (10). The B19 genes constitute a small gene family showing high sequence homology (7). For the direct cloning strategy, we chose primers complementary to a region conserved between the members of the gene family. This allowed us to directly sequence and clone two B19 promoters using the same set of primers.

MATERIALS AND METHODS

Isolation of Genomic DNA, Preparation of Adapters and Primers

Genomic DNA was isolated from freeze-dried leaves of barley (*Hordeum*

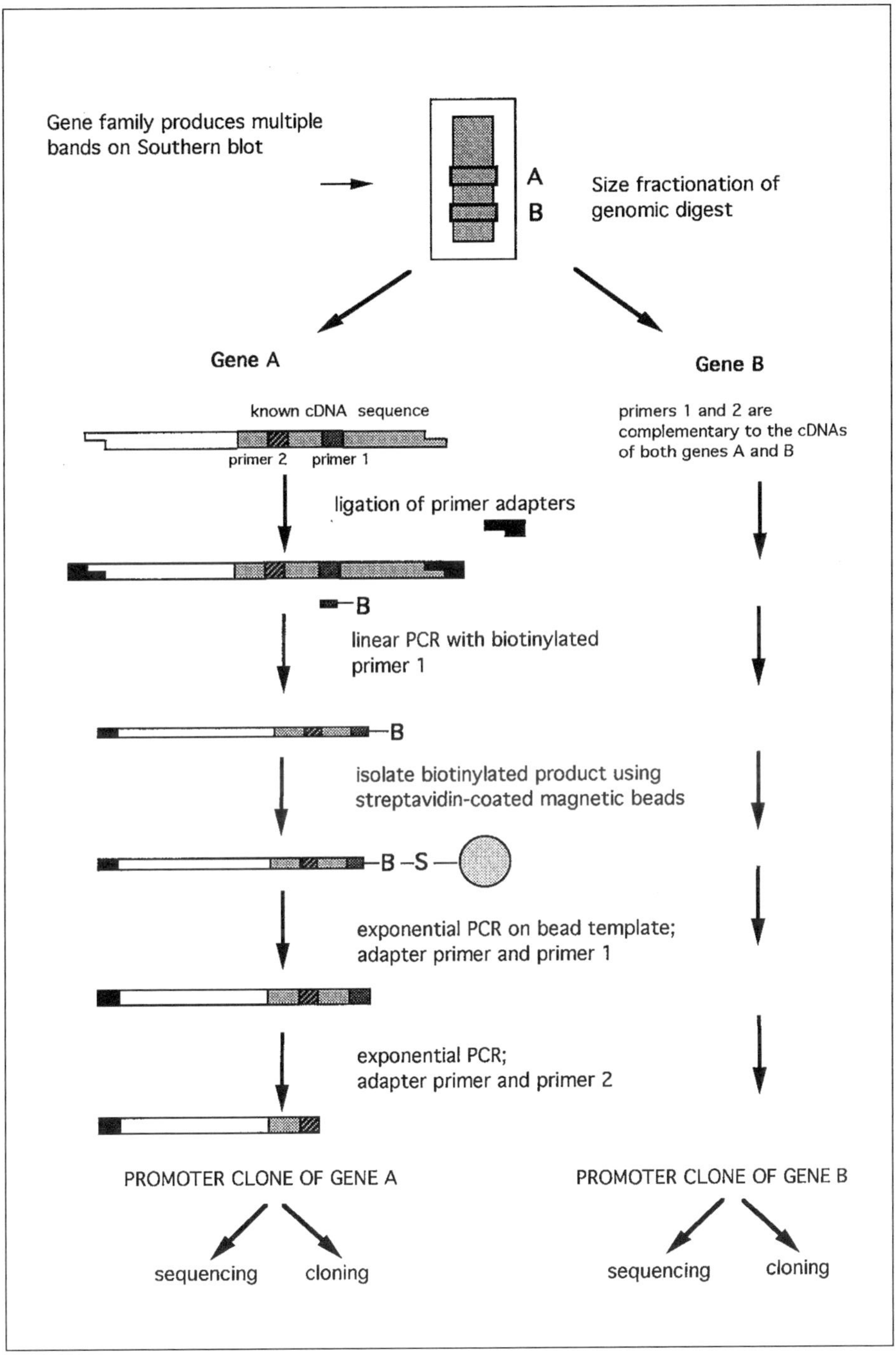

Figure 1. Schematic overview of the cloning strategy involving primer-adapter mediated PCR coupled to the use of magnetic beads.

vulgare L. var. distichum cv. Bomi) using proteinase K and phenol/chloroform extraction (3). Synthesis of oligonucleotides was performed on an Applied Biosystems Model 381A DNA synthesizer (Applied Biosystems, Foster City, CA, USA). The oligos were biotinylated at the NH_2-modified 5′ end of the oligo using a biotin-x-NHS-ester according to the manufacturer's recommendations (Clontech Laboratories, Palo Alto, CA, USA). The oligo sequences used are as follows: adapter-oligo, 5′-GATCCGAAGCTTGGGGTGAATT CGC-3′; adapter-primer, 5′-GCGAA TTCACCCCAAGCTTCG-3′; primer 1, 5′-GCCTTCGGCAAGGTGCTCC TGCG-3′ and primer 2, 5′-ACCTCTC CTGCTGACCGGAAGCCAT-3′. Kinasing of the adapter-oligo and annealing of the adapter to generate an intact *Sau*3AI site was performed according to Reference 29. Prior to experiments with genomic DNA, the quality of the adapters was tested by ligation to a *Sau*3AI-digested fragment and subsequent PCR amplification using the adapter-primer.

PCR Cloning Procedure

*Sau*3AI-digested genomic DNA (50 µg) was separated on a 1% agarose gel, the size fractions of approximately 500 and 900 bp were excised from the gel and purified using a GeneClean® kit (Bio 101, La Jolla, CA, USA). Ligation of adapters was performed in a total volume of 20 µL of adapter-ligation buffer (66 mM Tris-HCl, pH 7.6, 10 mM $MgCl_2$, 10 mM dithiothreitol [DTT], 0.3 mM ATP, 1 mM spermidine-HCl, 200 µg/mL bovine serum albumin [BSA]) with 200 ng adapters, 2–5 µg genomic DNA and 1 U T4 DNA ligase (GIBCO BRL/Life Technologies, Gaithersburg, MD, USA). The ligation reaction was incubated at 16°C overnight, and the non-ligated adapters were removed by spin-column purification (Sephacryl® S-300, Pharmacia LKB Biotechnology AB, Uppsala, Sweden). The columns were equilibrated with PCR buffer (50 mM KCl, 10 mM Tris-HCl, pH 8.3, 1.5 mM $MgCl_2$) and eluted in 50 µL volume.

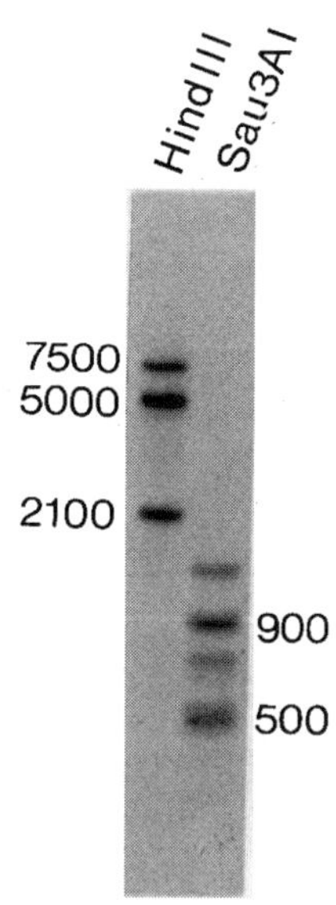

Figure 2. Genomic Southern blot of barley DNA. The blot was hybridized to a 230-bp subclone of B19.4 cDNA, which corresponds to the first part of the gene. 10 µg of digested genomic DNA were loaded to each lane, the restriction enzyme used and the approximate size of the hybridizing bands are indicated in the figure.

The cDNA complementary

primer 1 (see Figure 1) was first used for a linear amplification of the target sequence. The reaction mixture of 100 µL contained 200 µM deoxynucleoside triphosphates (dNTP), 1× PCR buffer (Promega, Madison, WI, USA), 20 pmol biotinylated primer 1, 30 µL of adapter-ligated genomic DNA template and 1 U *Taq* DNA polymerase (Promega). The temperature program was a two-step program of 50 cycles, 95.5°C/30 s and 70°C/2 min 30 s on the PCR machine PHC-2 (Techne, Cambridge, UK). The linear biotinylated products were bound to Dynabeads® M-280 Streptavidin (Dynal A.S., Oslo, Norway); 0.25 mg beads were washed prior to binding (twice with 1 M NaCl in TE buffer (10 mM Tris-HCl, pH 8.0, 1 mM EDTA) and once with 1× PCR buffer) and then incubated for 5 min at room temperature (RT) with the amplified product. After binding, the supernatant containing the genomic DNA was removed by fixing the beads with the magnet MPC-E (Dynal) and discarding the supernatant. The beads were washed 3 times with 1 M NaCl in TE buffer, 3 times with TE buffer and once with PCR buffer.

Exponential PCR was performed on the single-stranded template bound to the beads, and the 100-µL reaction mixture contained 200 µM dNTP, 1× PCR buffer, 50 pmol of each of the adapter-primer and the non-biotinylated primer 1 and 1 U *Taq* DNA polymerase. The temperature program was 35 cycles, 95.5°C/30 s, 62°C/1 min and 72°C/1 min. A second exponential PCR of 35 cycles was performed under the same conditions as above using 1 µL of the obtained PCR product in 100 µL reaction mixture and replacing primer 1 with the nested primer 2 (see Figure 1).

The PCR products were subjected to Southern blot analysis and sequenced directly using the solid-phase-sequencing method for single-stranded templates (15). After confirmation of the identity, the PCR products were "blunted" using Klenow polymerase (4) and subcloned into the *Sma*I site of pBluescript® (Stratagene, La Jolla, CA, USA). The inserts were then completely sequenced on both strands using the sequencing method above.

Southern Blots and Hybridizations

Southern gels, blotting and hybridizations were as described (9,16). Hybridization probes were ^{32}P-labeled, single-stranded DNA generated using biotinylated single-stranded DNA templates bound to Dynabeads M-280 Streptavidin (6).

RESULTS

Gene Structure of the B19 Family

From previous experiments involving cloning and sequencing of different B19 cDNAs, it was shown that the B19 gene family in barley consists of at least three closely related members (7). The cDNA clones, which are very similar both at the nucleotide and protein level, have been designated B19.1,

B19.3 and B19.4 due to the number of an internal 20-amino acid motif (present 1, 3 and 4 times in the genes, respectively) (7). A genomic Southern blot hybridized to the 5′ end of B19.4 (230 bp *Xho*I/*Eco*RI fragment) is shown in Figure 2. This probe cross-hybridizes to all B19 genes, producing three bands with *Hin*dIII-digested DNA. Digestion with *Sau*3AI gives two prominent bands and two weaker bands. Taking the cDNA sequence data of B19.4 into account, only one of the stronger bands could arise from B19.4. More extensive restriction mapping has shown that the three genes are tightly clustered (data not shown).

Basic Amplification Concept and Strategy

The basic concept of the method is to employ an "internal" primer complementary to a known sequence in the cDNA in combination with an "external" adapter-primer (Figure 1). First, primer adapters are ligated onto the genomic DNA digested with a suitable enzyme (here *Sau*3AI is used), then a linear PCR is performed with the internal primer (primer 1-B), which is biotinylated. Since the linear PCR product is biotinylated, it can then be purified from the rest of the genomic DNA with the aid of streptavidin-coated magnetic beads. After the magnetic purification, an exponential PCR is carried out using the internal primer in combination with the adapter-primer. An extra round of PCR with a nested internal primer (primer 2) and the adapter-primer can be performed to achieve increased specificity. Finally, the product can be either sequenced directly using the magnetic sequencing approach of Hultman et al. (15) or cloned.

An obvious strategy for amplification of the barley B19 promoters would be to use gene-specific internal primers. This approach requires no size fractionation of DNA prior to amplification. Because of high similarity in the first part and the coding region of the genes and because the middle part of two of the genes contains multiple repeats (7), which is unfavorable for PCR, we decided to use internal primers complementary to a conserved region of the genes. Since it was not clear which one of the *Sau*3AI fragments represented B19.4, we chose to separate the *Sau*3AI-cut genomic template DNA into two size fractions

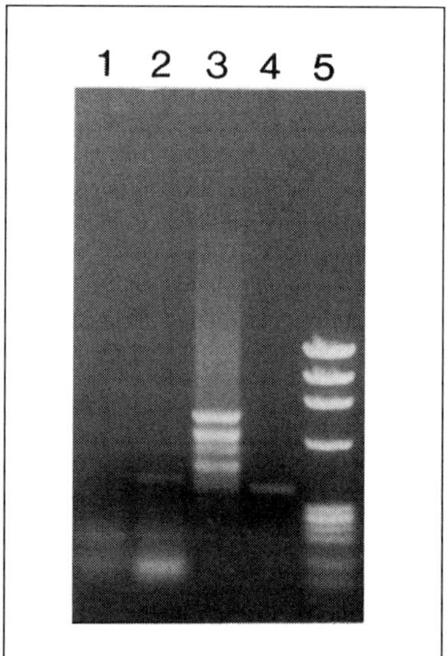

Figure 3. Ethidium bromide-stained agarose gel showing the amplified genomic fragments. Lane 1, first exponential PCR on the 900-bp fraction of *Sau*3AI-cut genomic DNA; lane 2, second exponential PCR using the product of lane 1 as template; lane 3, first exponential PCR on the 500-bp fraction of *Sau*3AI-cut genomic DNA; lane 4, second exponential PCR using the product of lane 3 as template; lane 5, ϕX174/*Hae*III.

before cloning. The expected result should be a B19.4 promoter clone from one of the two fractions and another B19 promoter from the other fraction.

Amplification of the B19 Promoters

The amplification procedure, which is outlined in Figure 1, involves as a first step restriction with *Sau*3AI and ligation of primer adapters to the different DNA size fractions. Then, 50 cycles of linear amplification are performed using an internal biotinylated primer (primer 1-B). The primer used complements perfectly to B19.4 and has one and three mismatches, respectively, to B19.3 and B19.1. The biotinylated linear PCR product was purified from the rest of the genomic DNA with streptavidin-coated magnetic beads and subjected to exponential PCR using the adapter-primer and primer 1 as described in Materials and Methods. The result of this first round of exponential PCR is visualized on the agarose gel in Figure 3 (lanes 1 and 3) and on the Southern blot (hybridized to B19.4/230-bp fragment) in Figure 4 (lanes 2 and 5). Figure 3 clearly indicates unspecific amplification, and Figure 4 (lanes 2 and 5) shows weak hybridization to fragments of expected sizes, whereas most of the produced DNA did not hybridize. By using the primer 1 alone, it was evident that this primer produced non-hybridizing products (not shown). Therefore, a second round of exponential amplification was performed using the adapter-primer and a nested primer (primer 2). The result was increased specificity and amount of product (Figure 3, lanes 2 and 4; Figure 4, lanes 3 and 6). Figure 4 (lanes 4 and 7) shows that PCR with primer 2 alone clearly decreases the amount of hybridizing product, demonstrating that the adapter-primer is necessary for exponential amplification.

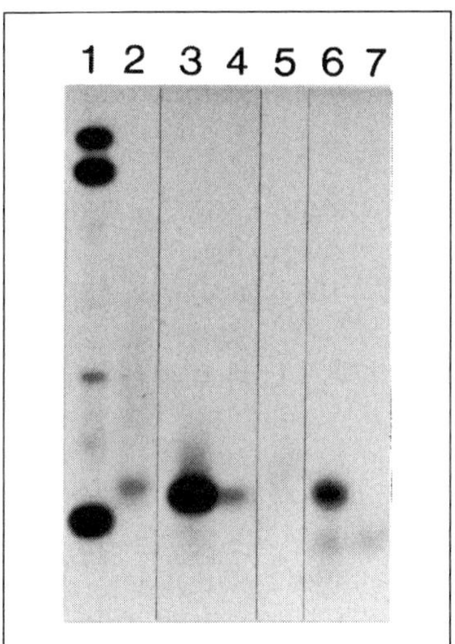

Figure 4. Southern blot of the PCR products. The blot was hybridized to the same probe (B19.4/230 bp) as in Figure 2. Due to varying amounts of products, the figure had to be combined from different exposures. Lane 1, marker bands of 3859 bp, 3196 bp, 970 bp, 730 bp and 306 bp; lane 2, first exponential PCR on the 500-bp fraction of *Sau*3AI-cut genomic DNA; lane 3, second exponential PCR using the product of lane 1 as template; lane 4, as lane 3 but without the adapter-primer; lane 5, first exponential PCR on the 900-bp fraction of *Sau*3AI-cut genomic DNA; lane 6, second exponential PCR using the product of lane 5 as template; lane 7, as lane 6 but without adapter-primer.

Confirmation of the PCR Products by Sequence Analysis

The PCR products were reamplified with a biotinylated primer 2 for direct use in magnetic solid-phase sequencing (14,15). After cloning of the non-biotinylated products into the *Sma*I site of pBluescript, the complete sequence was determined. Both clones were flanked by one adapter-primer and one primer 2. The clone generated from the 500-bp fraction overlapped with 118 bp to the B19.4 cDNA, whereas the other clone most likely was the B19.3 promoter, since it did not overlap with B19.1. However, our data could not confirm this, because the B19.3 cDNA was incomplete. Thus, in a PCR with a primer complementary to the "probable B19.3 promoter" and a gene-specific primer to the 3′-UTR, we amplified a product that obviously was the genomic B19.3 sequence, and it was overlapping the promoter clone with 117 bp [Figure 5 and (7)]. When sequencing the cloned B19.4 and B19.3 inserts, it was evident that a few differences due to errors during the PCRs were present in some clones (a total of 3 misincorporations in 4 clones sequenced). The errors correspond to an average mutation rate of 1.7×10^{-5} errors per nucleotide incorporated per cycle, which is in the range of previous estimations on the fidelity of the *Taq* DNA polymerase (11,23). Fewer rounds of PCR, which can be achieved by omitting the first round of exponential PCR with primer 1 and using the nested primer (primer 2) directly, should, of course, result in less errors. However, this strategy will reduce the yield significantly and has not been employed here.

Sequences of the B19.4 and B19.3 Promoters are Highly Similar

The sequences of the two PCR clones are presented in Figure 5. Shown is the B19.3 sequence, and the B19.4 sequence is presented where it differs from B19.3. Primer sequences are indicated. Both upstream regions contain consensus TATA boxes. The 5′-UTRs are 88% identical, while somewhat less similarity is seen in the 5′ upstream region. Still, the identity is around 78% in this region. Both B19.4 and B19.3 promoter regions contain multiple ACGT motifs, which has been shown to be the core recognition sequence of several DNA-binding proteins (28). Some of the ACGT-motifs in the B19 promoter sequences resemble the putative abscisic acid-response elements (ABRE) in the promoter of the wheat Em gene (13). However, they are oriented in the opposite direction of the Em ABRE elements. Notably, three regions of 7 to 10 bp in length present in B19.3 are not found in B19.4. This is interesting in light of the differential response of the two genes on salt stress (7).

DISCUSSION

Employing a strategy that involves ligation of primer-adapters, linear PCR, purification of biotinylated product and finally exponential PCR with nested primers, we have shown that direct sequencing and cloning of unknown promoter sequences can be accomplished from large genomes. Once having decided on the restriction sites to use for primer-adapter ligation, the cloning and sequencing can be performed within one week. We have also shown that combined with size fractionation before adding primer adapters, our method can be used to clone related genes or DNA sequences with one set of primers.

The B19 genes are members of a gene family of which we have previously cloned three cDNAs. Using the same set of primers and the protocol described here, we cloned two of the promoters. We also unsuccessfully attempted to clone a promoter from a third size fraction containing an 800-bp fragment hybridizing weakly to B19.4 (Figure 2). This hybridizing fragment is either B19.1 or a fourth member of the gene family. Between primer 1 in the experiment and the corresponding sequence of B19.1, there are 3 mismatches, which would explain why we did not obtain a PCR clone from that template.

The limitations of this method are mainly the same as for PCR in general; for instance, there are problems in obtaining long products. The primer choice is, of course, the crucial point of the procedure. The presence of structural features like GC-rich regions and inverted repeats may also cause difficulties. Recently we have amplified a 1200-bp, GC-rich sequence from barley using primer-adapters with *Hin*dIII as cloning site (K. Hollung and K.S. Jakobsen, unpublished results). In this case a "touchdown" PCR (5) was performed, since the usual PCR procedures failed to give a product.

The strategy we have employed here involves, as a first step, some restriction mapping using the known cDNA sequence as a probe. On the basis of

```
B19.3   GATCCTGGCTACGTGTCCACTCGTGGACAGCACAAGACAACGCACGAGTC   50
B19.4   ******A**********AC*AG*AC*********G***G**T********

        GCTAGCATTGGCAGCTCAATATCCGCTTCCTCGGACAACCGTGCACACT-  100
        ******************---***----------*A****-------**A

        CAT-TCGGTGCCCATGCAAGTTACCACGTGTCGTCCAACCAGAGGCTTGC  150
        ***G****************C******A****A***C*************

        ATGCACGACACGTCTTAGACTCTTGATACAGCAGGAGCGCGGAGAGGTAC  200
        *********G****--------********A*****A*AT**********

        ACGCGCCGGCTCCTAGCGCC--AAGCTCCTCCTTGTCGGCTATATAAA  250
        *****************-**T*CG***********C*C***********T

        CGCGACGTCGCTAGCTCCTCCGCCATCAGCATCAGGCAATCACAGAGCAC  300
        *T*************C************C*********************

        CAAGAGCAGTCGCATACCAAAACCCTAAGTCGGCTTACGGTTTCCGT--A  350
        **C*******A******-*********CA**CA*****T********TT*

        GCTAGGTCCCAGGCAGCGTGCAATGGCTTCCGGTCAGCAGGAGAGGT     397
        ******************C****************************
```

Figure 5. Sequence comparison of the two genomic PCR clones for the promoters of B19.3 and B19.4. Asterisks represent identical base pairs between B19.3 and B19.4, and dashed lines in both sequences show deleted base pairs. The primer used for the second exponential PCR of the cloning procedure is indicated by a line above the complementary sequence, and the primer used for generation of the internal B19.3 clone (in combination with a gene-specific primer in the 3′-UTR) is underlined. The TATA box is shown in bold, and the putative ABRE elements are boxed. The cDNA start of B19.4 is indicated by a vertical arrow, and putative translation starts by a horizontal arrow.

this information, the primer-adapter oligos are synthesized. An alternative strategy, which avoids restriction mapping, would be to use a limited number of restriction enzymes and corresponding primer-adapters blindly. This approach will also have a potential for genomic walking. If lack of a suitable restriction enzyme site is a problem, we suggest using partial digestion with a "frequent cutter" and subsequent size fractionation on an agarose gel.

This method is not only suitable for cloning promoters but also a simple way to compare promoters from related genes (gene families). It can be used to extend incomplete cDNAs and to clone regions that are unclonable by means of conventional methods. Furthermore, the strategy employing linear PCR with a biotinylated primer can be used to improve the "ligation-mediated single-sided PCR" method, which is currently being used in genomic methylation studies (20). Here, purification of the linear PCR products after ligation of linker by means of streptavidin-coated magnetic beads, as well as similar purification of the exponential PCR products, should result in increased specificity and higher yield and possibly eliminate the need for blotting and hybridization of the sequencing gel. Finally, this method has potential as a rapid way to obtain sequence information and to clone the flanking regions of a sequence-tagged site. In plants, this would be of particular interest since the use of transposable elements such as Ac and Mu to generate mutations in metabolic, developmental or hormonal pathways and the use of T-DNA insertional mutagenesis in the dissection of plant development have recently proven to be very powerful techniques (17,24). For the analysis of tagged sequences, one set of primers directed to a known region, for instance, a transposon sequence, can be used to clone the sequence adjacent to the integration site in a number of individual plants.

The advantages of magnetic beads as a solid phase in manipulation of nuleic acids opens up perspectives for "automatic gene cloning." Magnetic beads are already being used in robotic sequencing strategies (14), and employing current technology, a simple robot in combination with a magnetic workstation and a thermocycler should be able to handle most of the steps described here.

Note added in proof: The sequence data for the B19.3 and B19.4 gene fractions have been deposited with EMBL. The EMBL data library accession numbers are X65951 and X65952, respectively, for sequences B19.3 and B19.4.

ACKNOWLEDGMENTS

This work was supported in part by a grant from the Norwegian Research Council for Science and Humanities (NAVF) to K.S.J. We are grateful to Erik Hornes and Elisabeth S. Jensen for providing oligonucleotides and for helpful discussions. Thanks are also due to Ole Martin Gulliksen and Randi Amundsen for help in the initial stages of this project. Robin A.P. Stacy is gratefully acknowledged for technical assistance in the lab, as well as for help on computer graphics and for reading this manuscript. We also thank Kristin Hollung for assisting with computer analysis and for helpful comments on the manuscript.

REFERENCES

1. **Arumuganathan, K. and E.D. Earle.** 1991. Nuclear DNA content of some important plant species. Plant Mol. Biol. Rep. *9*:208-218.
2. **Bennett, M.D. and J.B. Smith.** 1976. Nuclear DNA amounts in angiosperms. Phil. Trans. R. Soc. Lond. B *274*:227-274.
3. **Blin, N. and D.F. Stafford.** 1976. A general method for isolation of high molecular weight DNA from eucaryotes. Nucleic Acids Res. *3*:2303-2308.
4. **Davis, L.G., M.D. Dibner and J.F. Battey.** 1986. Basic Methods in Molecular Biology, Elsevier Science Publishing, New York.
5. **Don, R.H., P.T. Cox, B.J. Wainwright, K. Baker and J.S. Mattick.** 1991. 'Touchdown' PCR to circumvent spurious priming during gene amplification. Nucleic Acids Res. *19*:4008.
6. **Espelund, M., R.A. Prentice Stacy and K.S. Jakobsen.** 1990. A simple method for generating single-stranded DNA probes labeled to high activities. Nucleic Acids Res. *18*:6157-6158.
7. **Espelund, M., S. Sæbøe-Larssen, D.W. Hughes, G.A. Galau, F. Larsen and K.S. Jakobsen.** 1992. Late embryogenesis-abundant genes encoding proteins with different numbers of hydrophilic repeats are regulated differentially by abscisic acid and osmotic stress. Plant J. *2*:241-252.
8. **Fors, L., R.A. Saavedra and L. Hood.** 1990. Cloning of the shark Po promoter using a genomic walking technique based on the polymerase chain reaction. Nucleic Acids Res. *18*:2793-2799.
9. **Galau, G.A., D.W. Hughes and L. Dure.** 1986. Abscisic acid induction of cloned cotton late embryogenesis-abundant (*Lea*) mRNAs. Plant Mol. Biol. *7*:155-170.
10. **Galau, G.A., K.S. Jakobsen and D.W. Hughes.** 1991. The controls of late dicot embryogenesis and early germination. Physiol. Plantarum *81*:280-288.
11. **Goodenow, M., T. Huet, W. Saurin, S. Kwok, J. Sninsky and S. Wain-Hobson.** 1989. HIV-1 isolates are rapidly evolving quasispecies: evidence for viral mixtures and preferred nucleotide substitutions. J. Acquir. Immune Defic. Syndr. *2*:344-352.
12. **Graham, M.W., J.P. Doherty and D.M. Woodcock.** 1990. Efficient construction of plant genomic libraries requires the use of mc^r-host strains and packaging mixes. Plant Mol. Biol. Rep. *8*:18-27.
13. **Guiltinan, M.J., W.R. Marcotte and R.S. Quatrano.** 1990. A plant leucine zipper protein that recognizes an abscisic acid response element. Science *250*:267-271.
14. **Hultman, T., S. Bergh, T. Moks and M. Uhlén.** 1991. Bidirectional solid-phase sequencing of *in vitro*-amplified plasmid DNA. BioTechniques *10*:84-93.
15. **Hultman, T., S. Ståhl, E. Hornes and M. Uhlén.** 1989. Direct solid phase sequencing of genomic and plasmid DNA using magnetic beads as solid support. Nucleic Acids Res. *17*:4937-4946.
16. **Maniatis, T., E.F. Fritsch and J. Sambrook.** 1982. Molecular Cloning: A Laboratory Manual. Cold Spring Harbor Laboratory Press, Cold Spring Harbor, NY.
17. **Meinke, D.W.** 1991. Perspectives on genetic analysis of plant embryogenesis. Plant Cell *3*:857-866.
18. **Mueller, P.R. and B. Wold.** 1989. *In vivo* footprinting of a muscle specific enhancer by ligation mediated PCR. Science *246*:780-786.
19. **Ochman, H., A.S. Gerber and D.L. Hartl.** 1988. Genetic applications of an inverse polymer chain reaction. Genetics *120*:621-625.
20. **Pfeifer, G.P., S.D. Steigerwald, P.R. Mueller, B. Wold and A.D. Riggs.** 1989. Genomic sequencing and methylation analysis of ligation mediated PCR. Science *246*:810-813.
21. **Rosenthal, A. and D.S.C. Jones.** 1990. Genomic walking and sequencing by oligocassette mediated polymerase chain reaction. Nucleic Acids Res. *18*:3095-3096.
22. **Rosenthal, A., R.N. MacKinnon and D.S.C. Jones.** 1991. PCR walking from microdissection clone M54 identifies three exons from the human gene for the neural cell adhesion molecule L1 (CAM-L1). Nucleic Acids Res. *19*:5395-5401.
23. **Saiki, R.K., D.H. Gelfand, S. Stoffel, S.J. Scharf, R. Higuchi, G.T. Horn, K.B. Mullis and H.A. Erlich.** 1988. Primer-directed enzymatic amplification of DNA with a thermostable DNA polymerase. Science *239*:487-491.
24. **Schmidt, R.J., F.A. Burr and B. Burr.** 1987. Transposon tagging and molecular analysis of the maize

regulatory locus *opaque-2*. Science *238*:960-963.
25.**Shapiro, H.S.** 1976. Handbook of Biochemistry and Molecular Biology. CRC Press, Cleveland, p. 313-318.
26.**Silver, J. and V. Keerikatte.** 1989. J. Cell Biochem. (Suppl.) *13E*:306.
27.**Triglia, T., M.G. Peterson and D.J. Kemp.** 1988. A procedure for *in vitro* amplification of DNA segments that lie outside the boundaries of known sequences. Nucleic Acids Res. *16*:8186.
28.**Weisshaar, B., G.A. Armstrong, A. Block, O.C. Silva and K. Hahlbrock.** 1991. Light inducible and constitutively expressed DNA-binding proteins recognizing a plant promoter element with functional relevance in light responsiveness. EMBO J. *10*:1777-1786.
29.**Wu, R., T. Wu and A. Ray.** 1987. Adaptors, linkers and methylation. Methods Enzymol. *152*:343-348.

Update to:

Cloning and Direct Sequencing of Plant Promoters Using Primer-Adapter Mediated PCR on DNA Coupled to a Magnetic Solid Phase

Mari Espelund and Kjetill S. Jakobsen
University of Oslo, Oslo, Norway

The method described in our original paper has not been modified. In our lab, a different adapter PCR strategy in connection with magnetic bead capture has recently been designed for rapid cloning and sequencing of microsatellites (U.H. Refseth, B.M. Fangan and K.S. Jakobsen, submitted). In contrast to the original article, microsatellites are captured directly by hybridization to a microsatellite repeat-containing oligonucleotide attached to a magnetic bead. Following hybridization capture, the hybridizing single-stranded microsatellites are amplified and simultaneously made double-stranded by adapter-PCR. The isolated microsatellites can be TA-cloned for sequencing and further analysis. The microsatellite isolation protocol is very rapid compared to conventional strategies. Of the recombinant clones obtained, 70% or more contain microsatellites, corresponding to at least a 40-fold enrichment for repeats like AC/GT. In principle, this strategy could be optimized for the isolation of any high- to medium-abundant repeated sequence (e.g., proviruses, transposons, LINEs, SINEs and minisatellites).

In contrast to the protocol described by Refseth et al., the advantage of the original strategy is the linear PCR amplification step that allows single copy genes/promoters to be isolated. In particular, the method is very useful for obtaining promoter sequences of the individual members of a gene family. This can be done either by enrichment of each (or a few) member(s) by agarose gel purification prior to PCR, or selective linear PCR amplification using gene-specific primers. The first alternative has been demonstrated by Holk et al. (2) for cloning of the 5′ ends and the promoters of five closely related genes (DC 2.15) associated with the onset of somatic embryogenesis in carrot.

The principle of the method, which is to amplifiy a specific sequence using a biotinylated primer and subsequently isolate this sequence from the bulk of DNA by binding to a magnetic solid phase, has been used for several other related applications. One example is the extension of incomplete cDNAs (1). Here, cloning of both 5′-ends and 3′-ends was demonstrated, and it is shown that the method is more specific than the conventional RACE methods. In addition, a similar strategy has been developed for chromosome walking (see Reference 3).

For ligation-mediated PCR (LM-PCR), a protocol based on our method for magnetic solid-phase purification of the sequence of interest has significantly improved the quality of genomic sequencing ladders as well as footprinting patterns (4). In this case, a particular gene or sequence, is enriched by an extension product-capture protocol involving linear PCR with a specific biotinylated primer and then ligation of an adapter in the 3′ end. Following ligation, the product is captured on streptavidin beads, the minus strand eluted, and PCR is carried out with a specific primer and the adapter primer similarly as described in our paper.

ACKNOWLEDGMENTS

We thank Unn Hilde Refseth for discussions regarding the microsatellite cloning approach and Kristin Hollung for comments on the manuscript. The work described here, and in the original article, was supported by grants from the Norwegian Research Council (NFR) to ME and KSJ (Grant No. 100368/410) and to KSJ (Grant No. 107622/420).

REFERENCES

1. **Charnock-Jones, D.S., M. Platzer and A. Rosenthal.** 1994. Extension of cDNAs (ESTs) by biotin/streptavidin-mediated walking using the polymerase chain reaction. J. Biotechnol. *35*:205-215.
2. **Holk, A., R. Kaldenhoff and G. Richter.** 1996. Regulation of an embryogenic carrot gene (DC 2.15) and identification of its active promoter sites. Plant Mol. Biol. *31*:1153-1161.
3. **Rosenthal, A.** 1992. PCR amplification techniques for chromosome walking. Trends Biotechnol. *10*:44-48.
4. **Törmänen, V.T., P.M. Swiderski, B.E. Kaplan, G.P. Pfeifer and A.D. Riggs.** 1992. Extension product capture improved genomic sequencing and DNase I footprinting by ligation-mediated PCR. Nucleic Acids Res. *20*:5487-5488.

A General Approach for PCR-Amplification and Sequencing of Chloroplast DNA from Crude Vascular Plant and Algal Tissue

B.M. Fangan, B. Stedje, O.E. Stabbetorp[1], E.S. Jensen[2] and K.S. Jakobsen

Biological Institute, University of Oslo; [1]Botanical Garden and Museum; [2]Dynal AS., Oslo, Norway

ABSTRACT

A universal and reliable sequencing approach suitable for evolutionary and phylogenetic studies of most plant species has been developed. The initial step in the approach is PCR with primers from conserved regions in the chloroplast genome. Prior to PCR a simple DNA extraction is carried out where lysed tissue is embedded in agarose and dialyzed. Due to the use of one biotinylated primer in the PCR, the PCR products can be bound to magnetic streptavidin-coated beads and then sequenced directly by the solid-phase approach. The regions used for sequencing and subsequent analyses are the intron of the trnL *gene and the intergenic spacer between the* trnL *and the* trnF *genes in the chloroplast DNA. Comparisons of the sequences obtained from closely and distantly related species show that the different chloroplast DNA regions harbor variations suitable for phylogenetic inference from the subspecific to the interfamiliar level. The strategy can be used both for manual and automated sequencing.*

INTRODUCTION

The utilization of DNA sequence data for phylogenetic studies of plants has become increasingly important during the last few years. The main limitations for a more extensive use of DNA sequence data so far have been the high cost and time-consuming methods. The appearance of the polymerase chain reaction (PCR) (32,36,37) has been a major step forward in making a rapid pathway from plant tissue to DNA sequence. Several approaches including asymmetrical PCR (20), direct sequencing of denatured DNA fragments (31), "cycle sequencing" (7) and solid-phase sequencing (23,24,26) allow

(Reprinted from BioTechniques 16:484-494)

direct sequencing of PCR products. Of these methods, direct solid-phase sequencing using magnetic beads is among the simplest and the most reliable (22). In solid-phase sequencing, an asymmetrically biotinylated DNA fragment is obtained through PCR, and the DNA is captured, purified and denatured by the use of streptavidin-coated magnetic beads. The solid-phase sequencing approach is advantageous for automated generation of templates and sequencing (16,23,24).

Still there are many cumbersome steps in obtaining DNA sequences from plants. DNA isolation procedures have mostly involved phenol extractions and/or CsCl ultracentrifugation. (e.g., References 4 and 30). Several miniprep methods have been suggested. The numerous variants of the cetyltrimethylammonium bromide (CTAB) method of Saghai-Maroof et al. (35) are all dependent on extractions with organic solvents and ethanol precipitation (e.g., Reference 13). More simple methods of DNA extraction for PCR, avoiding the extraction, centrifugation or precipitation steps, have been reported recently (6,28).

The chloroplast-encoded large subunit of ribulose-1,5-biphosphate carboxylase (*RbcL*) has been the DNA sequence most extensively used for plant phylogenics (reviewed in Reference 8). A segment of the nuclear gene alcohol dehydrogenase 2 (19) and a segment of the 28S ribosomal RNA gene (10) are examples of other DNA regions that have been proposed for phylogenetic studies. All these slow-evolving coding sequences have proved to be most suitable for phylogenetic studies at family level and above. DNA sequence data have had a more limited application for phylogenetic and systematic studies at and below the generic level. Both untranscribed nuclear intergenic spacer (IGS) sequences and internal transcribed spacer (ITS) sequences in the ribosomal RNA genes have, however, been used in phylogenetic inference (e.g., References 2,3,27).

Taberlet et al. (41) designed primers for three non-coding regions between three chloroplast tRNA genes (*trnL*, *trnF* and *trnT*), which may be a useful tool for phylogenetic studies. Using these primers, they were able to amplify DNA from mosses, ferns and seed plants. However, the utility of these sequences for phylogenetic studies has not been thoroughly tested. The nucleotide sequences of the tRNA genes are generally more conserved throughout the plant kingdom than protein-coding genes and the large ribosomal RNA genes. Accordingly, the design of universal primers for PCR based on the tRNA genes should be easier than for most other genes.

The objective of this work has been to develop a universal and reliable sequencing approach suitable for complete automation for evolutionary and phylogenetic studies of most plant species. In this paper we describe a PCR sequencing approach based on sets of nested or hemi-nested primers derived from the highly conserved chloroplast *trnL* and *trnF* genes. The use of biotinylated primers and magnetic beads allows direct sequencing of unpurified PCR products. Since high-throughput sequencing of a large number of species is dependent on simple DNA extraction procedures, we have used lysed plant tissue embedded in agarose (for dialysis against sterile water) as a source for PCR. We have sequenced PCR products obtained from the *trnL* (UAA) intron and the intergenic spacer IGS between the *trnL* and the *trnF* genes (tRNA-IGS) from a variety of species. The results show that these regions are useful in phylogenetic studies at different taxonomic levels.

MATERIALS AND METHODS

This study is based on species listed in Table 1. Information on localities and voucher specimens can be obtained from the authors on request. Agarose plugs of lysed cells from all the sequenced species (except *Picea abies*) were prepared from fresh tissue as described below.

Sample Preparation for PCR Amplification

Plant tissue was homogenized, incubated with proteinase K, mixed with low melting point agarose (No. A-9419; Sigma Chemical, St. Louis, MO, USA), dialyzed against sterile water and stored as agarose plugs at 4°C according to Brunel (6). Total genomic DNA from *P. abies* was isolated from callus according to Dellaporta et al. (11). Total genomic DNA from the other species listed in Table 1 was prepared either by using the Green Gene™ Plant DNA Isolation Kit as indicated by the manufacturer (Clontech, Palo Alto, CA, USA) or by the CTAB method of Lichtenstein and Draper (29).

Primer Design

In addition to using biotinylated and non-biotinylated versions of the primers described by Taberlet et al. (41), two new sets of primers were designed. The primer sequences are presented in Figure 1A. By comparing the tRNA gene sequences from *Nicotiana tabacum* (38), *Oryza sativa* (21) and *Marchantia polymorpha* (33), we designed a set of nested primers with 3′ ends complementary to the -21 M13 universal primer. These primers were designed to lay within the tRNA genes. The C-nested primer was allowed, however, to extend 13 bp into the intron. The tRNA-IGS 5′ primer (EM13, Figure 1) is identical with the previously published primer except for the added -21 M13 sequence. The OLIGO® (Version 3.3) primer analysis software (National BioScience, Plymouth, MN, USA) (34) was used to optimize the oligonucleotide primers for the PCRs.

The oligonucleotides were biotinylated at the NH_2-modified 5′ end using a biotin-x-NHS-ester according to the manufacturer's recommendations (Cambridge Research Biochemicals, Northwich, UK). Oligonucleotide syntheses were performed on Applied Biosystems 381A and 391 DNA Synthesizers (Applied Biosystems, Foster City, CA, USA), and on a Beckman System 200A DNA Synthesizer (Beckman Instruments, Fullerton, CA, USA).

The primers supplied with the *Taq* Dye-Primer Cycle Sequencing Kit

Table 1. Survey of the Species Investigated, Including PCR Results

Taxon	PCR Product Intron	PCR Product Spacer
Ceramiaceae (Red algae):		
Ceramium nodosum	+	-
Ceramium strictum	+	+
Gelidium pulchrum	+	-
Psilotaceae (Whisk fern):		
*Psilotum triquetrum****	na	++
Marsileaceae (Fern):		
Marsilea quadrifolia	+	++
Pinaceae (Gymnosperm):		
* *Picea abies**	++	++
Brassicaeaea (Dicotyledon):		
* *Arabidopsis thaliana*	++	++
Caprifoliaceae (Dicotyledon):		
*Leycesteria formosa****	+	++
Davidiaceae (Dicotyledon):		
*Davidia involucrata****	na	++
Dipsaceaceae (Dicotyledon):		
*Scabiosa columbaria****	na	++
Sambucaceae (Dicotyledon):		
*Viburnum carlesii****	na	++
Solanaceae (Dicotyledon):		
Nicotiana tabacum	++	++
* *Petunia hybrida*	++	++
Hypoxidaceae (Monocotyledon):		
*Hypoxis rigidula****	na	+
* *H. villosa*	++	++
Hyacinthaceae (Monocotyledon):		
*Ornithogalum tenuifolium***	++	+
*O. sessiliflorum***	++	+
Drimia altissima	++	++
D. elata	++	++
Drimiopsis botryoides	++	++
* *Scilla autumnalis*	++	++
* *S. camerooniana*	++	++
* *S. hyacinthina (from India and Zimbabwe)*	+	++
* *S. obtusifolia*	+	++
* *S. rigidifolia*	++	+
Amaryllidaceae (Monocotyledon):		
Crinum abyssinicum	++	+
C. asiaticum	++	++
C. crassicaule	++	++
* *C. jagus*	++	++
C. latifolium	++	++
C. lugardiae	-	+
C. macowanii	++	+
C. variabile	-	++
C. zeylanicum	++	++
Asphodelaceae (Monocotyledon):		
Aloe exceisa	++	++
Poaceae (Monocotyledon):		
Hordeum vulgare	++	++
Zea mays	+	++
*Z. luxurians****	na	+

Table 1. (Continued)

+: Weak PCR Product, ++: Strong PCR product, -: No PCR product, na: PCR amplification not performed.

*: Species from which sequences are published here. PCR amplifications were performed on agarose plugs except where otherwise noted: Total, cellular DNA isolation by *Green Gene Kit, **according to Dellaporta et al. (11), ***the CTAB method of Lichtenstein and Draper (29).

(Applied Biosystems, Part No. 901482) were utilized in the automated DNA sequencing.

PCR Amplification

PCR amplifications were performed directly on agarose plugs. The plugs were melted at 68°C and 4 µL added to the following reaction mixture: 50 mM KCl, 10 mM Tris-HCl pH 8.3, 2 mM $MgCl_2$, 1 µM of each primer, 200 µM of each deoxyribonucleoside triphosphate (dNTP) and 0.5–1 U of *Taq* DNA polymerase (Promega, Madison, WI, USA). The total reaction volume was 50 µL.

A two-step system of amplifications was designed to generate PCR products for sequencing (Figure 1B). In the first PCR, we used a program for "touchdown" PCR (12) with non-modified primers (the primer pairs C/D and E/F for the intron and tRNA-IGS, respectively) starting at an annealing temperature of 65°C decreasing to 55°C over eight cycles and with 27 cycles at the target temperature. The other temperature cycling parameters were denaturation at 95°C for 0.5–1 min, annealing for 0.5–1 min and polymerization for 1–2 min at 72°C, depending on the species amplified. For some of the species tested (the lilioid monocots), a regular program (95°C for 30 s, 50°C for 30 s and 72°C for 60 s) also gave good results.

In the second PCR, we used two sets of modified primers necessary for sequencing both cpDNA regions in both directions (C-Biotin/D-Nested-M13 or D-Biotin/C-Nested-M13 for the intron region, and E-biotin/F-Nested-M13 or F-Biotin/E-M13 for the tRNA-IGS; Figure 1B). In this reaction, 0.5–1.0 µL PCR products from the first PCR were added to the same reaction mixture as described above and run 10–20 cycles in the second PCR depending on the yield of PCR products in the first-round amplification. The temperature cycling parameters were 95°C for 20 s, 50°–55°C for 20 s and 72°C for 50 s.

The reactions were run on a Hybaid thermal reactor Model No. HBTR1 (Hook and Tucker Instruments Ltd., Croyden, England) and a Biometra TRIO-Thermoblock (Biometra Biomedizinische Analytic GmbH, Göttingen, Germany) for the first- and second-round amplifications, respectively.

Solid-Phase Sequencing

Sequencing was performed using a modified version of the solid-phase strategy of Hultman et al. (23). Twenty-five microliters of biotinylated PCR products were immobilized to Dynabeads® M-280 Streptavidin (Dynal AS.,

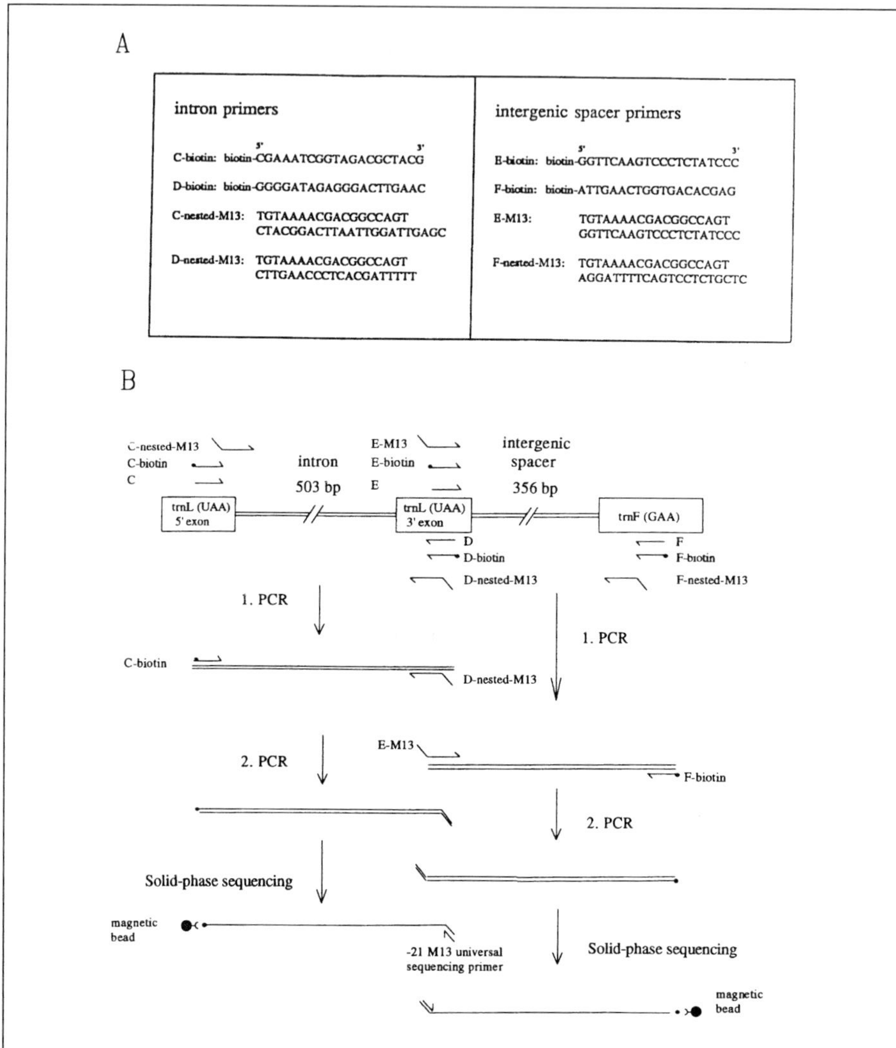

Figure 1. A. Base sequence (5′ to 3′) of the primers used in this study to amplify the two non-coding chloroplast DNA regions: The *trnL* intron and the tRNA-IGS. B. A two-step strategy for PCR amplification of two non-coding cpDNA sequences for direct solid-phase sequencing. The primers used are represented as arrows relative to the *Nicotiana* chloroplast genome. The positions of the 3′ end of the primers relative to the *Nicotiana* chloroplast genome are C/C-biotin, 49317; C-nested-M13, 49335; D/D-biotin, 49855; D-nested-M13, 49842; E/E-biotin/D-M13, 49873; F/F-biotin, 50272; F-nested-M13, 50253. In the first PCR, non-modified primers were used to amplify the desired cpDNA region (primers C and D for the intron; primers E and F for the tRNA-IGS). In the second PCR, each region was amplified with two different primer combinations (one biotinylated, and one containing a sequence complementary to the -21 M13 universal primer) depending on the desired direction of sequencing. For each cpDNA region, only one of the combinations is shown. When generating PCR products for sequencing the intron, the two primer combinations C-biotin/D-nested-M13 and D-biotin/C-nested-M13 were used for sequencing from the 3′ end and 5′ end, respectively. Similarly, for the tRNA-IGS, PCR products were generated using the primer combination E-Biotin/F-Nested-M13 for sequencing the 3′ end, and F-Biotin/E-M13 for sequencing the 5′ end.

Oslo, Norway). The Dynabeads (30 µL) were washed with 6× standard saline citrate (SSC) prior to binding and incubated for 5 min at room temperature with the amplified product (made 6× SSC by the addition of 20× SSC and H_2O). After binding, the supernatant containing excess primer and genomic DNA was removed by fixing the Dynabeads with a magnet (MPC-E; Dynal AS.) and discarding the supernatant. The template was made single-stranded by denaturation with alkali and magnetic separation (15). Standard dideoxy sequencing reactions were performed on the single-stranded immobilized template using the Sequenase® Kit (Version 2.0; United States Biochemical, Cleveland, OH, USA).

The sequencing reactions were manually run on a Model S2 Sequencing Gel Electrophoresis Apparatus (GIBCO BRL/Life Technologies, Gaithersburg, MD, USA).

Automated sequencing was performed using an Applied Biosystems 373A sequencer. The templates were prepared for sequencing in the same fashion as above, but the sequencing reactions were performed with a *Taq* Dye-Primer Cycle Sequencing Kit. Sequences were analyzed using the SecEd™ (Applied Biosystems) software.

Alignment and Statistical Analyses

The sequences were aligned both manually and by utilizing the PILEUP procedure of the GCG package (18). Similarity between sequences was calculated as the proportion of the sites being equal after alignment (common gaps excluded). Statistical calculations were performed according to Conover (9).

RESULTS

The Amplification and Sequencing Strategy

The basic idea of the sequencing strategy is to design PCR primers in highly conserved regions in the chloroplast genome (tRNA genes) and to amplify and sequence the regions between the tRNA genes using solid-phase sequencing (Figure 1B). First, the specific region is amplified using non-modified primers. A second PCR is then normally performed with one biotinylated and one nested primer (hemi-nested amplification). Applying a nested primer enhances the specificity of the amplification, and the resulting biotinylated PCR product can be purified using streptavidin-coated magnetic Dynabeads. In the case of the E-primer, however, it was not possible to design a good nested primer. Finally, the PCR product can be sequenced directly using the magnetic sequencing approach of Hultman et al. (24).

PCR Amplification of Chloroplast DNA from Diverse Species Directly from Lysed Tissue Embedded in Agarose Plugs

To optimize the procedure, we have employed lysed tissue embedded in agarose plugs as substrate for the PCRs. Performing PCR amplification directly on these samples proved to work satisfactorily for most of the tested species. In preliminary experiments, we also tested PCR on plant tissue squeezed onto a nylon membrane (28). This approach gave less reproducible results than the agarose plug method. We have amplified the *trnL*(UAA) intron and the tRNA-IGS between the *trnL* and *trnF* genes from representatives of the major lineages of vascular plants and from some algal species. The species used in this investigation and a survey of other PCR amplifications performed on several other species are presented in Table 1.

PCR amplification of the intron and the tRNA-IGS gave products of approximately 550 and 400 bp, respectively. For most of the species investigated, we found that in the second PCR less than 1 µL of first-round PCR product needed to be amplified to get enough template for sequencing. No further purification steps were necessary to obtain readable sequences. An example of an ethidium bromide-stained agarose gel showing second-round PCR products from algae, monocots and dicots is presented in Figure 2. Only small size differences are observed between the species in the two regions.

Manual and Automated Solid-Phase Sequencing

From both manual and automated sequencing of the amplified PCR products, we were able to read sequences of between 250 and 300 bp with few ambiguities. In the cases where both manual and automated sequencing were applied, the two independent strategies gave identical results. We were, however, able to increase the length of interpretable sequence by 50–60 bp when running automated sequencing. By sequencing in both directions, we got a

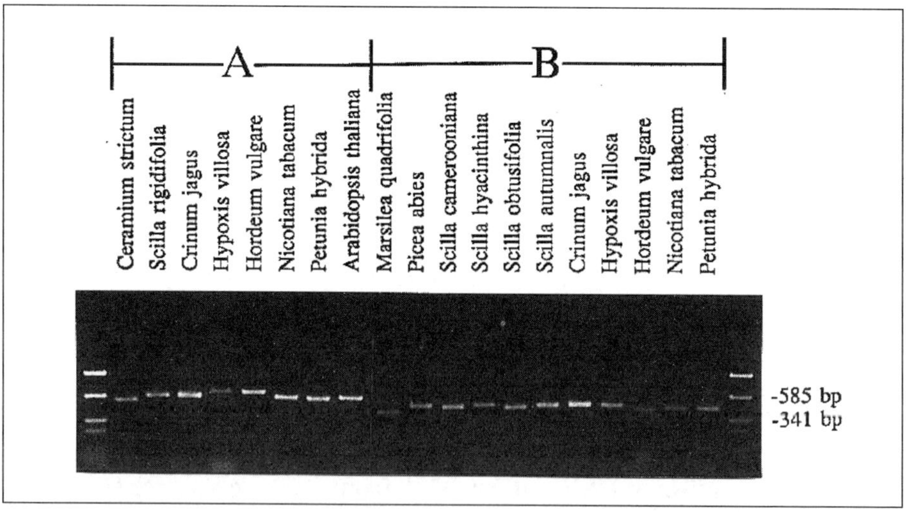

Figure 2. Results of second-round PCR amplification of non-coding chloroplast DNA regions. Panel A shows PCR products from amplification of the *trnL* (UAA) intron. Panel B shows PCR products from amplification of the IGS between *trnL* (UAA) 3′ exon and the *trnF* (GAA) gene.

considerable overlap. Due to several terminations 25–30 bp from the 5′ end of the tRNA-IGS, we were not able to get interpretable results from this region. This problem was only partly resolved by substituting dGTP with dITP and by running automated cycle sequencing.

In this study we present *trnL*(UAA) intron sequences from 9 species and tRNA-IGS sequences from 11 species. Our sequencing results from *Nicotiana* and *Zea mays* are identical to the published sequences of these species (38,39). We have selected species that represent the major lineages of vascular plants, closely related genera and closely and distantly related species within one genus (*Scilla*). In Figures 3 and 4, the results from the intron and the tRNA-IGS regions are aligned with some previously published sequences. Sequences from closely related species were easily aligned manually, whereas the more distantly related species (e.g., *Nicotiana* and *Picea*) differ radically, especially in the tRNA-IGS.

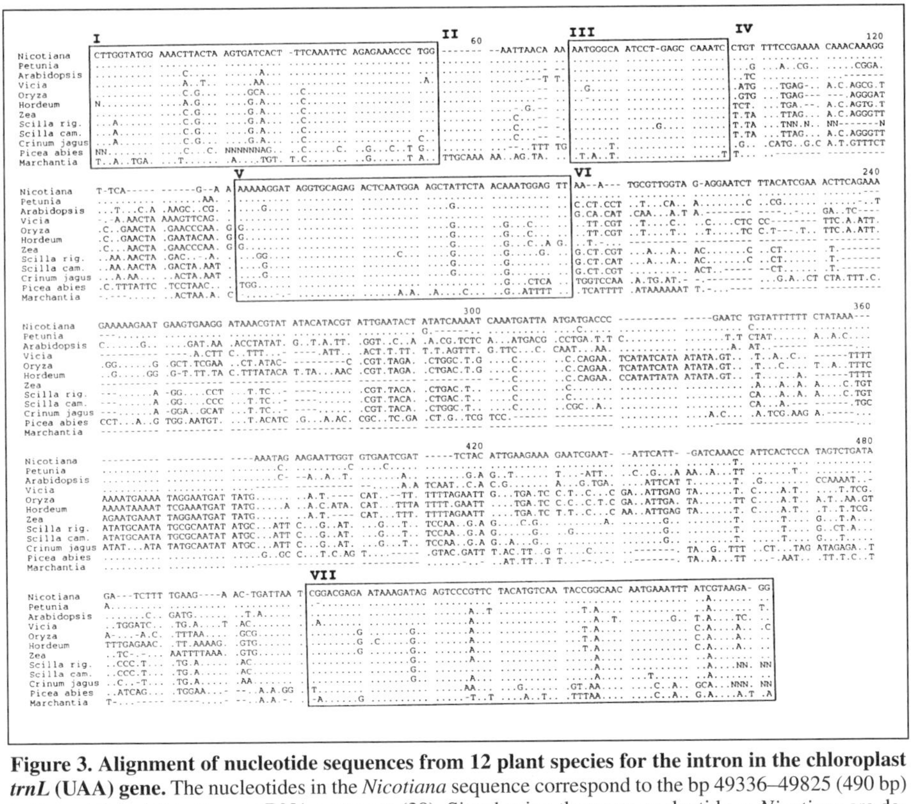

Figure 3. Alignment of nucleotide sequences from 12 plant species for the intron in the chloroplast *trnL* (UAA) gene. The nucleotides in the *Nicotiana* sequence correspond to the bp 49336–49825 (490 bp) in the published *Nicotiana* cpDNA sequence (38). Sites having the same nucleotide as *Nicotiana* are denoted by dots. Gaps are represented by (-) and uninterpretable sites by (N). The *Marchantia*, *Vicia* and *Oryza* sequences are taken from (33), (5) and (21), respectively. All the species were sequenced manually. *Hordeum*, *Zea*, *Scilla camerooniana*, *Crinum*, *Nicotiana*, *Petunia* and *Picea* were also automatically sequenced. The four boxed regions (labeled I, III, V and VII) represent conserved parts of the intron. The other regions (labeled II, IV and VI) are more variable.

Comparison of the Intron Sequences

The intron can be divided into four conserved regions (region I, III, V and VII in Figure 3) and three variable regions (region II, IV and VI). The conserved regions are similar to the regions pointed out by Bonnard et al. (5) to be essential for the formation of secondary structure.

In the conserved regions, the similarities range from 0.70 to 0.99 in pairwise comparisons between species (Table 2). In the variable regions, the sequence similarities are reduced to 0.12–0.94, the lower figures being obtained in pairs of distantly related species. This implicates that a proper alignment cannot be found in these cases. For the three more closely related groups of species, *Nicotiana/Petunia, Oryza/Hordeum/Zea* and two *Scilla* species, the sequence similarities for the conserved regions are 0.99, 0.98 and 0.98, respectively. The corresponding similarities for the variable regions are 0.93, 0.55–0.76, and 0.94. There is a high correlation between the similarities in the conserved and the variable region (Spearman's r = 0.877, *P*<.0001) (9).

```
                                                                         60                                                                120
Nicotiana           AAAAGACTAT TTCACTCCCC AACT------ --ATTTATCC GACCCCCTTT CCTTAGC--- -GGTTCCAAA T-----TCCT TATCTTTCTC ATTC-ACTCT ATTCTTTTAG AAATGG----
Petunia             NNNNNNNNNN G....CAGTT NNNNNNNNN. ..-....N.. ..-...T... .......... ....C..... .......... .......... ..C....... .......... ..........
Hypoxis             .....C.C.. ..T...T..N ....A..... ..-....... TCNTTTT..C .T.AGAT... .....-.... CAAAAT..AA .......... .......... ...T...C.C ..........
Crinum jagus        NNNNNNNNNN NNNNNNNNNN ....ATTTAT CT.......T TCTTTTT... .A.A...... .....-.... GAAAAT..AA .......... ......T... .C.....C.C ..........
Scilla autumnalis   NNNNNNNNNN NNNNNNNNNN NNNNNN.... ..-N...-.T TC.TTTT.-. .A.A...... .....-.... G.AAAT..AA .......... ......T... .C.....CGC ....A.....
S. obtusifolia      NNNNNNNNNN NNNNN.T.TT G...A..... ..-....... TC.TTTT--. .G.A..T... .....-.... G.AAAT..AA .......... ......T... .C.....CGC ....A.....
S. camerooniana     NNNNNNNNNN NNNNNNNTT- -..NA..... ..-....C-. TC...TT... .G.A...... .....-N... ..AAAT..AA .......... ......T.A. .C.....CGC ....A.....
S. hyacinthina,I    NNNNNNNNNN NNNNNNNNNN N...AT.... ..-.NN-... TC...TT... .G.A...... .....-N... ..AAAT..AA .......... ...T..T.A. .C.....CGC ....A.....
S. hyacinthina,Z    NNNNNNNNNN NNNNNNNNNN N...A..... ..-....... CC...TT... .G.A..G... .....-.... ..AAAT..AA .......... ......T.A. .C.....CGC ....A.....

Hordeum             ---------- ---------- ---------- -----AATCC TCTTTTTT-C TTTTATCAAT GGGTTTAAGA T-----TGAT TAGCTTTCTC ATTCTACTCT TTTC----AC AAA-GGAAAG
Zea                 ..AACCCTCT TTTATTCCCT AACCATAGT. ..TGTT.... .-........ .......... .......... .......C.C .......... .......... ..-....... ......----
Oryza               ..AACCCTCT TTTATTCC.T AACTCTAGT. ..ATTT.... .G.......- ---...T... A......... ......---- ---------- ---------- ---....-- ---.------

                                                                        180                                                                240
Nicotiana           --ATTTGAGC G----TAAAT GGCTTTCTCT TAT--CACAA GTC------- TTGTGATATA TATG------ -----ATAC- ----ACATAG AAATGAACGT C-TTTGAGCA AGG-AATCCC
Petunia             .......... .......... .......... .......... .......... .......... .......... .......... .......... ........A. .......... ..........
Hypoxis             ....CC..A. AGAAA.CTT. T.A----.T. ....C..A.T T--....... -.AG---... G..A...... ...TG....C CGC....--- -........A. A..A..G... ........G.
Crinum jagus        ...CCC..A. ATAAA.CTT. T.A----... ...AC..ATT T--....GGT ...A---... G..A...... ...TG....C CGT....--- -........A. A..A..GT.. .......T..
Scilla autumnalis   ....CC.GA. AGAAA.CTT. ..A----... ....C.TT.T T--...AGGT ...A---... G..A...... ...CG....C TGT....--- -........A. A..A..GT.. .......T..
S. obtusifolia      ...CCC.G.. AGAAA.CCT. ..A----... ....C.CT.T TCTATTAGGT ...A---... G..A...... ...TG....C TGT....--- -........A. A..A..GTT. .......T..
S. camerooniana     ....CC.GA. AGAAA.CCT. ---------. ....C.CT.T T--...AGGT ...A---... G..A...... ...CG....C TGT....--- -........A. A..A..GT.. .......T..
S. hyacinthina,I    ....CC.GA. AGAAA.CCT. ..A----... ....C.CT.T T--...AGGT ...A---... G..A...... ...CG..N.C TGT....--- -........A. A..A..GT.. .......T..
S. hyacinthina,Z    ....CC.GA. AGAAA.CCT. ..A----... ....C.CT.T T--...AGGT ...A---... G..A...... ...CG....C TGT....--- -........A. A..A..GT.. .......T..

Hordeum             GAATGCGAAG AGAACTCAAT GGA----TCT TAT-CC-TAT TC-A------ TTGA---ATA G--------- -----ATTTC TTTTTTAT-- -------TAG AGTATCGGCA AGA-AATCTT
Zea                 -.G.....CA .......... .-.A...... ....G..... .......... ..A....... .ATGATTTCT TTTTT..... ........TT ATTAGAG... .......... ..G......C
Oryza               ---------- -----..... ...A...--- ---.--.--- --.-...... ----...-.. C......... .......... .......... .........T .......... ..G....G.C

                                                                        300                                                                360
Nicotiana           TA--GTTGAA TGATTCCCTA TCAATATCAT TACTCATAC- -------TGA AACTTACAAA GT----CATC TTTTTG-AAG ATCGAAGAAA TTCCCC---- -GGCTTTGAG AAAATT----
Petunia             .......... C.......G. .......... .......... .......... .......... .......... .......... .......... .......... ...A.....T ....--....
Hypoxis             ..TTA..... .C....A.A. .AC....... ..TCTT...G TTTACAAA.. T..AA.G..T .......T.. .....TTT.T .GTT...... ...AGGAACT A..---.C-- ...-......
Crinum jagus        C.TTA..... .C....A.AG ..C....... ..T..T...A TT.....--- C..AA.G... .......T.. .........A ...T...... ...GGGGACT A..---.C-- .....-....
Scilla autumnalis   C.TTA..... .T....A.AG C.C....... ..TCTT...A TT.....--- C..AA.A... .......T.. .......... ...T...... ...GGGGACT A..---.C-- ....--....
S. obtusifolia      C.TTA..... .T...----- ---------- ---------. .......--- ---------- --....---- --........ ..TT...... ...GGGGACT A..---.C-- ..........
S. camerooniana     C.TTA..... .T....A.A. C.C....... ..TC.T...A TT.....--- C..AA.A... ......TT.. .....A.... ...T...... ...GGGGACT A..---.C-- ..........
S. hyacinthina,I    C.TTA..... .T....A.AG C.C....... ..TC.T...A TT.....--- C..AA.A... ......TT.. .....A.... ...T...... ...GGGGACT A..---.C-- ..........
S. hyacinthina,Z    C.TTA..... .T....A.AG C.C....... ..TC.T...A TT.....--- C.TAA.A... ......TT.. .......... ..AT...... ...GGGGACT A..---.C-- ..........

Hordeum             GGTTATTCAC T--------- -------C-- TATTTTTA-A GT-------- -ATTATTTAA GTAAACCAT- --GC-ACAAT GCATAGGG-C TACCCCC-CC CC--TTTC-- AAA-TTTTGA
Zea                 .A.....A.T .......... ........GT .T........ .......... .....-.... ....G..... ..C....... .......A.. ......-T.. ..A.....CT ..-.......
Oryza               .A.....A.. .......... ........GA .....--... A......... .....-.... A..GG.TT.C TTTGT..... .......A.. .G.....T.. ..A.....C. ......----

                                                                        420
Nicotiana           -TTT--AAT- -CTAC----- ---TTTT-G- ----TCCTTG TAATTGACA- ---------- TAGACCCCAG TTCTCTAATA AAATGAGGAT ACTACATTGG GAA-TA
Petunia             A...G..... .......... .......... .........T .......... .......... .......... .......... .......... ....G..... ......
Hypoxis             ....G....A .----..... .TT....A.. .....A.CCT .......... .......... ....TA.A.. .CA...GC.- -----.A... GA.GT.C-.. T.GA.G
Crinum jagus        ....G....A ...T-..... .TT....A.T TTAG..TA.T .....T.... .......... ....TA.A.. .AT....C.- -----..... GA.G.GC-.. ...A.C
Scilla autumnalis   ....GG...A G----..... .TT....A.. ......TA.T .....T...T ACATATTACA ....TA.A.. .A.....C.- -----.A... GA.G..C-.. ...A.G
S. obtusifolia      .........A G----..... .TT....A.. ......TA.T .....T...T ACATATTACA ....TA.A.. .A.....C.- -----.A... GA.G..C-.. ...A.G
S. camerooniana     .........A G----..... ..T....A.. ......TA.T .....T...T ACATATTACG ....TA.A.. .A.....C.- -----..... GA.G..C-.. ...A.G
S. hyacinthina,I    .........A G----..... ..T....A.. ......TA.T .....T...T ACATATTACG ....TA.A.. .A.....CC- -----..... GA.G..C-.. ...A.G
S. hyacinthina,Z    .........A G----..... ..T....A.. ......TA.T .....T...T ACATATTACG ....TA.A.. .A.....C.- -----..... GA.G..C-.. ...A.G

Hordeum             ATTTGAAATA -CTTTAATTG ATTTTTTAG- ----TCCTTT TAATTGACA- ---------- TAGATACAAA TACTCTACT- -----AGGAT GATGCACAAG AAAA-G
Zea                 ..--.G.... .....-.... .......... .......C.. .......... .......... .....G.... ....T..... ......A... .......... ...G..
Oryza               -....G-..- .--------- ---------. ...------- --........ .......... .......... .......... .......... .......... ......
```

Figure 4. Alignments of nucleotide sequences from 12 plants for the tRNA-IGS. The nucleotides in the *Nicotiana* sequence correspond to the base pairs 49876–50231 (356 bp) in the published *Nicotiana* cpDNA sequence (38). The dicots and the lilioid moncots are aligned together, whereas the grasses are aligned separately. The two alignments are then aligned tentatively to each other. All the species were sequenced manually. *Nicotiana, Petunia, Hypoxis villosa, Crinum jagus, S. obtusifolia, S. camerooniana, S. hyacinthina* and *Zea* were also automatically sequenced. The notation is as in Figure 3, except that the three lower sequences are compared to *Hordeum* and not to *Nicotiana*.

Table 2. Similarity Matrices Based on Alignment of the *trnL* Intron

	Oryz	Hord	Zeam	Srig	Scam	Crin	Vici	Nico	Petu	Arab	Pice	Marc
Oryz		0.76	0.65	0.47	0.46	0.47	0.27	0.34	0.34	0.33	0.27	0.12
Hord	0.98		0.55	0.46	0.46	0.43	0.28	0.37	0.36	0.36	0.29	0.12
Zeam	0.98	0.98		0.36	0.34	0.35	0.24	0.27	0.30	0.22	0.20	0.15
Srig	0.93	0.93	0.94		0.94	0.84	0.42	0.55	0.55	0.45	0.32	0.14
Scam	0.94	0.94	0.94	0.98		0.83	0.43	0.55	0.56	0.46	0.32	0.15
Crin	0.93	0.93	0.93	0.97	0.98		0.40	0.51	0.52	0.43	0.31	0.14
Vici	0.89	0.89	0.89	0.90	0.91	0.90		0.36	0.36	0.31	0.25	0.15
Nico	0.90	0.90	0.90	0.92	0.94	0.92	0.93		0.93	0.60	0.33	0.14
Petu	0.90	0.90	0.90	0.92	0.94	0.92	0.94	0.99		0.58	0.31	0.15
Arab	0.92	0.92	0.92	0.94	0.95	0.94	0.93	0.95	0.95		0.33	0.16
Pice	0.83	0.82	0.83	0.82	0.82	0.82	0.82	0.82	0.82	0.83		0.17
Marc	0.71	0.72	0.73	0.71	0.72	0.71	0.70	0.72	0.72	0.73	0.76	

Similarities based on the conserved part of the intron are shown below the diagonal and based on the variable parts of the intron above the diagonal. Oryz: *Oryza sativa* (21), Hord: *Hordeum vulgare*, Zeam: *Zea mays*, Srig: *Scilla rigidifolia*, Scam: *S. camerooniana*, Crin: *Crinum jagus*, Vici: *Vicia faba* (5), Nico: *Nicotiana tabacum*, Petu: *Petunia hybrida*, Arab: *Arabidopsis thaliana*, Pice: *Picea abies*, Marc: *Marchantia polymorpha*.

Comparison of the tRNA-IGS Sequences

The tRNA-IGS cannot be divided into conserved and variable regions (Figure 4). *Marchantia*, *Marsilea* and *Picea* differed too much from each other and from the angiosperms to be included in a multiple alignment, whereas the grasses could only be aligned to the angiosperms in parts of the sequence. We have therefore aligned the dicots, *Nicotiana* and *Petunia*, together with the lilioid monocots, and aligned the grasses separately.

Between the dicots and the lilioid monocots, we found sequence similarities ranging from 0.52 to 0.62 (Table 3). Within the grasses, the similarities are between 0.53 and 0.76. The lilioid monocots show higher similarities (0.67–0.98). The *Scilla* species expose sequence similarities ranging from 0.80 to 0.98, while *Nicotiana* and *Petunia* are 94% identical. The total range of variation in sequence similarity is from 0.52 to 0.98, i.e., in about the same range as found for the angiosperms in the variable parts of the intron.

DISCUSSION

By applying a strategy that involves preparing crude plant lysates, running two rounds of PCR, and direct solid-phase sequencing, we have demonstrated that chloroplast DNA sequences can readily be obtained from a wide

Table 3. Similarity Matrices Based on the tRNA-IGS Region after Alignment

	Oryz	Hord	Zeam	Scam	ShyI	Sobt	Saut	ShyZ	Crin	Hypo	Nico	Petu
Oryz												
Hord	0.53											
Zeam	0.56	0.76										
Scam												
ShyI				0.97								
Sobt				0.80	0.82							
Saut				0.93	0.94	0.83						
ShyZ				0.96	0.98	0.83	0.94					
Crin				0.82	0.83	0.73	0.86	0.83				
Hypo				0.75	0.76	0.67	0.78	0.76	0.79			
Nico				0.59	0.61	0.52	0.60	0.61	0.61	0.59		
Petu				0.60	0.60	0.52	0.60	0.61	0.62	0.59	0.94	

Oryz: *Oryza sativa* (21), Hord: *Hordeum vulgare*, Zeam: *Zea mays*, Scam: *Scilla camerooniana*, ShyI: *S. hyacinthina*, India, Sobt: *S. obtusifolia*, Saut: *S. autumnalis*, ShyZ: *S. hyacinthina*, Zimbabwe, Crin: *Crinum jagus*, Hypo: *Hypoxis villosa*, Nico: *Nicotiana tabacum*, Petu: *Petunia hybrida*.

variety of vascular plants utilizing a single approach and the same sets of primers. No purification of plant DNA is required. Lysed tissue can be used for PCR amplification, and the PCR products can be sequenced directly using magnetic beads. The use of nested primers in generating PCR products for sequencing have been shown to enhance the quality of the sequencing reactions (22,43). This is also the case in this work, but it should be noted that the sequence interpretability obtained from amplification/sequencing with primer E was in some cases almost as good as with the nested primers.

The Intron and Spacer Sequences Can Be Used to Infer Phylogenies at Different Systematic Levels

To evaluate the variation in "the tRNA region," we have compared the chloroplast gene *rbcL* from *Z. mays* (17), *O. sativa* (21), *Petunia hybrida* (1) and *Picea pungens* (EMBL Acc. No. X58136) with the *trnL* intron sequence for the same species (except *P. abies* substituting for *P. pungens*). The comparison indicates parallel patterns of variation in the two regions. *P. pungens* have a similarity of 0.86 to the three analyzed angiosperms in the *rbcL* sequence, while *P. abies* show a similarity of 0.82–0.83 to the same angiosperms in the conserved part of the intron. *Oryza* and *Zea* are highly similar in both regions (0.96 in *rbcL*, 0.98 in the conserved intron part), while *Petunia* shows the same degree of similarity to the two cereals in the two regions (0.86–0.87 in *rbcL* vs. 0.90 in the conserved part of the intron).

The conserved regions of the *trnL* intron have a rate of base substitution similar to *rbcL*. The intron regions are, however, not subjected to codon bias, as is the case with *rbcL*, and should therefore be a good choice for further analyses among distantly related species. In the variable parts of the intron, however, the similarities between species of the same family range from 0.55 to 0.94, suggesting that this region is an excellent tool for phylogenetic analyses at or below the family level.

In the tRNA-IGS, the degree of similarity between the *Scilla* species varies within a range comparable to what has been found in the IGS of nuclear rDNA in Cucurbitaceae (27) and in the ITS of nuclear rDNA from the subtribe Madiinae in Compositae (2,3). We also observe intraspecific variation, at least when the species is sampled from different geographical regions (cf. *Scilla hyacinthina*, Figure 4). Intraspecific variation has previously been documented in the ribosomal ITS (40). The tRNA-IGS thus harbor a variation most suitable for phylogenetic inference at lower systematic levels.

Automation

All steps in this approach, including DNA preparation, can be handled by a robotic workstation, for instance a Biomek®-1000 robot (Beckman Instruments, Fullerton, CA, USA) equipped with a magnetic workstation and a thermal cycler (42) or a Catalyst™ (Perkin-Elmer/Applied Biosystems). It has been previously demonstrated using a robotic workstation that template preparation (16) as well as the sequencing reactions (24,42) with regards to both plasmid DNA and genomic DNA sequencing can be performed automatically using Dynabeads. The fact that we found no discrepancies between the results from manual and automated sequencing supports the view that an automated system would give reliable results.

Applicability to Other Sequences

The basic concept of this approach could easily be adapted to other regions of the chloroplast genome or to mitochondrial and nuclear genes, provided that sequences suitable for designing efficient primers are known. In some cases it may be desirable to compare sequences adjacent to this chloroplast tRNA region for evolutionary studies. One approach will then be to use information from other published sequences to design primers complementary to other tRNA genes such as the conserved *trnT* sequences adjacent to the *trnL*(UAA) 5′ exon (41). Another approach, independent of known sequences outside the *trnL* and *trnF* region, involves ligation of primer adaptors, linear PCR with a biotinylated primer inside the *trnL*/*trnF* region, purification of the PCR product by magnetic beads and subsequent PCR and direct solid-phase sequencing. Such an approach can be used to extend any known sequence in any direction provided that a suitable restriction site outside the known region exists (14).

Perspectives

The non-coding cpDNA regions studied in this work have great potential for resolving phylogenetic relationships. In particular, the *trnL* intron, which harbors large variations in rates of base substitution, should be an excellent tool in phylogenetic inference at different taxonomic levels. By designing primers inside the conserved intron regions for groups of closely related species, short and highly variable sequences can be amplified. Most suitable for this is the large variable intron region (region IV, Figure 3). A simplified, automated sequencing strategy that involves running only one or two of the four sequencing reactions and treating the sequencing ladder as a binary code analogous to the system for DNA fingerprinting suggested by Jeffreys et al. (25) may generate sufficient variation for phylogenetic inference.

Note added in proof: The sequence data have been deposited in the EMBL data library under the following accession numbers: X74572 (*Petunia hybrida* tRNA intron); X74573 (*Arabidopsis thaliana* tRNA intron); X74574 (*Hordeum vulgare* tRNA intron); X74575 (*Scilla rigidifolia* tRNA intron); X74576 (*Scilla camerooniana* tRNA intron); X74577 (*Picea abies* tRNA intron); X74729 (*Crinum jagus* tRNA intron); X74578 (*Petunia hybrida* tRNA IGS); X75479 (*Hypoxis villosa* tRNA IGS); X74581 (*Scilla autumnalis* tRNA IGS); X74582 (*Scilla obtusifolia* tRNA IGS); X74583 (*Scilla camerooniana* tRNA IGS); X74584 (*Scilla hyacinthina* I tRNA IGS); X74585 (*Scilla hyacinthina* Z tRNA IGS); X74586 (*Hordeum vulgare* tRNA IGS).

ACKNOWLEDGMENTS

This work has been supported by grants from The Research Council of Norway (NFR) and The Fridtjof Nansen Foundation for the Advancement of Science. We want to express our gratitude to Inger Nordal for initiating the project and for invaluable support in all phases of the work. We also want to thank Erik Hornes (Dynal AS., Oslo, Norway) for helpful discussions. Thanks are also due to John Doebley (University of Minnesota) for providing seeds from different subspecies of *Zea mays*, to Jan Rueness for providing the algal samples, to Praveen Sharma (NLH-Ås, Norway) for providing total cellular DNA from *Picea abies* and to John Stacy for helping with the computer graphics and sequence analysis. Astrid Rostrup is acknowledged for reading the manuscript.

REFERENCES

1. **Aldrich J., B. Cherney, E. Merlin and J. Palmer.** 1986. Sequence of the *rbcL* gene for the large subunit of ribulose biphosphate carboxylase-oxygenase from petunia. Nucleic Acids Res. *14*:9534.
2. **Baldwin, B.G.** 1992. Phylogenetic utility of the internal transcribed spacers of nuclear ribosomal DNA in plants: An example from the Compositae. Mol. Phylogenet. Evol. *1*:3-16.

3. **Baldwin, B.G.** 1993. Molecular phylogenetics of Calycadenia (Compositae) based on its sequences of nuclear ribosomal DNA: Chromosomal and morphological evolution reexamined. Am. J. Bot. *80*:222-238.
4. **Blin, N. and D.F. Stafford.** 1976. A general method for isolation of high molecular weight DNA from eukaryotes. Nucleic Acids Res. *3*:2303-2308.
5. **Bonnard, G., F. Michel, J.H. Weil and A. Steinmetz.** 1984. Nucleotide sequence of the split $tRNA^{Leu}$(UAA) gene from *Vicia faba* chloroplasts: evidence for structural homologies of the chloroplast $tRNA^{Leu}$ intron with the intron from the autospliceable *Tetrahymena* ribosomal RNA precursor. Mol. Gen. Genet. *194*:330-336.
6. **Brunel, D.** 1992. An alternative, rapid method of plant DNA extraction for PCR analyses. Nucleic Acids Res. *20*:4676.
7. **Carothers, A.M., G. Urlaub, J. Mucha, D. Grunberger and L.A. Chasin.** 1989. Point mutation analysis in a mammalian gene: Rapid preparation of total RNA, PCR amplification of cDNA and *Taq* sequencing by a novel method. BioTechniques *7*:494-499.
8. **Clegg, M.T.** 1993. Chloroplast gene sequences and the study of plant evolution. Proc. Natl. Acad. Sci. USA *90*:363-367.
9. **Conover, W.J.** 1980. Practical Non-Parametric Statistics, 2nd Edition. John Wiley & Sons, New York.
10. **Cullings, K.W.** 1992. Design and testing of a plant-specific PCR primer for ecological and evolutionary studies. Mol. Ecol. *1*:233-240.
11. **Dellaporta, S.L., J. Woods and J.B. Hicks.** 1983. A plant DNA minipreparation: Version II. Plant Mol. Biol. Rep. *1*:19-21.
12. **Don, R.H., P.T. Cox, B.J. Wainwright, K. Baker and J.S. Mattick.** 1991. 'Touchdown' PCR to circumvent spurious priming during gene amplification. Nucleic Acids Res. *19*:4008.
13. **Doyle, J.J. and J.L. Doyle.** 1987. A rapid DNA isolation procedure for small quantities of fresh leaf tissue. Phytochem. Bull. *19*:11-15.
14. **Espelund M. and K.S. Jakobsen.** 1992. Cloning and direct sequencing of plant promoters using primer-adapter mediated PCR on DNA coupled to a magnetic solid phase. BioTechniques *13*:74-81.
15. **Espelund, M., R.A. Prentice Stacy and K.S. Jakobsen.** 1990. A simple method for generating single-stranded DNA probes labeled to high activities. Nucleic Acids Res. *18*:6157-6158.
16. **Fry, G., E. Lachenmeier, E. Mayrand, B. Giusti, J. Fischer, L. Johnston-Dow, R. Cathcart, E. Finne and L. Kilaas.** 1992. A new approach to template purification for sequencing applications using paramagnetic particles. BioTechniques *13*:124-131.
17. **Gaut, B.S., S.V. Muse, W.D. Clark and M.T. Clegg.** 1992. Relative rates of nucleotide substitution at the *rbcL* locus of monocotyledonous plants. J. Mol. Evol. *35*:292-303.
18. **Genetics Computer Group.** 1991. Program manual for the GCG Package, Version 7. Genetics Computer Group, Madison, WI.
19. **Goloubinoff, P., S. Pääbo and A.C. Wilson.** 1993. Evolution of maize inferred from sequence diversity of an *Adh2* gene segment from archaeological specimens. Proc. Natl. Acad. Sci. USA *90*:1997-2001.
20. **Gyllensten, U.B. and H.A. Erlich.** 1988. Generation of single-stranded DNA by the polymerase chain reaction and its application to direct sequencing of the HLA-DGA locus. Proc. Natl. Acad. Sci. USA *85*:7652-7656.
21. **Hiratsuka, J., H. Shimada, R. Whittier, T. Ishibashi, M. Sakamoto, M. Mori, C. Kondo, Y. Honji, C.-R. Sun, B.-Y. Meng, Y.-Q. Li, A. Kanno, Y. Nishizawa, A. Hirai, K. Shinozaki and M. Sugiura.** 1989. The complete sequence of the rice (*Oryza sativa*) chloroplast genome: Intermolecular recombination between distinct tRNA genes accounts for a major plastid DNA inversion during the evolution of the cereals. Mol. Gen. Genet. *217*:185-194.
22. **Hopgood, R., K.M. Sullivan and P. Gill.** 1992. Strategies for automated sequencing of human mitochondrial DNA directly from PCR products. BioTechniques *13*:82-92.
23. **Hultman, T., S. Ståhl, E. Hornes and M. Uhlén.** 1989. Direct solid phase sequencing of genomic and plastid DNA using magnetic beads as solid support. Nucleic Acids Res. *17*:4937-4946.
24. **Hultman, T., S. Bergh, T. Moks and M. Uhlén.** 1991. Bidirectional solid-phase sequencing of *in vitro*-amplified plasmid DNA. BioTechniques *10*:84-93.
25. **Jeffreys, A.J., A. MacLeod, K. Tamaki, D.L. Neil and D.G. Monckton.** 1991. Minisatellite repeat coding as a digital approach to DNA typing. Nature *354*:204-209.
26. **Kaneoka, H., D.R. Lee, K.-C. Hsu, G.C. Sharp and R.W. Hoffman.** 1991. Solid-phase direct DNA sequencing of allele-specific polymerase chain reaction-amplified HLA-DR genes. BioTechniques *10*:30-34.
27. **King, K., R.A. Torres, U. Zentgraf and V. Hemleben.** 1993. Molecular evolution of the intergenic spacer in the nuclear ribosomal RNA genes of Cucurbitaceae. J. Mol. Evol. *36*:144-152.

28. **Langridge, U., M. Schwall and P. Langridge.** 1991. Squashes of plant tissue as substrate for PCR. Nucleic Acids Res. *19*:6954.
29. **Lichtenstein, C.P. and J. Draper.** 1985. Genetic engineering of plants, p. 67-119. *In* D.M. Glover (Ed.), DNA Cloning, Vol. II. IRL Press, Oxford.
30. **Maniatis, T., E.F. Fritsch and J. Sambrook.** 1982. Molecular Cloning: A Laboratory Manual. Cold Spring Harbor Laboratory Press, Cold Spring Harbor, NY.
31. **Mitchell, L.G. and C.R. Merril.** 1989. Affinity generation of single stranded DNA for dideoxy sequencing following the polymerase chain reaction. Anal. Biochem. *178*:239-242.
32. **Mullis, K. and F. Faloona.** 1987. Specific synthesis of DNA *in vitro* via a polymerase catalysed chain reaction. Methods Enzymol. *55*:335-350.
33. **Ohyama, K., H. Fukuzawa, T. Kohchi, H. Shirai, T. Sano, S. Sano, K. Umesono, Y. Shiki, M. Takeuchi, Z. Chang, S.-i. Aota, H. Inokuchi and H. Ozeki.** 1986. Complete nucleotide sequence of liverworth *Marchantia polymorpha* chloroplast DNA. Plant Mol. Biol. Rep. *4*:148-175.
34. **Rychlik, W. and R.E. Rhoads.** 1989. A computer program for choosing optimal oligonucleotides for filter hybridization, sequencing and *in vitro* amplification of DNA. Nucleic Acids Res. *17*:8543-8551.
35. **Saghai-Maroof, M.A., K.M. Soliman, R.A. Jorgensen and R.W. Allard.** 1984. Ribosomal DNA spacer-length polymorphisms in barley: Mendelian inheritance, chromosomal location, and population dynamics. Proc. Natl. Acad. Sci. USA *81*:8014-8018.
36. **Saiki, R.K., D.H. Gelfand, S. Stoffel, S.J. Scharf, R. Higuchi, G.T. Horn, K.B. Mullis and H.A. Erlich.** 1988. Primer-directed enzymatic amplification of DNA with a thermostable DNA polymerase. Science *239*:487-491.
37. **Saiki, R.K., S. Scharf, F. Faloona, K.B. Mullis, G.T. Horn, H.A. Erlich and N. Arnheim.** 1985. Enzymatic amplification of β-globin genomic sequences and restriction site analysis for diagnosis of sickle cell anemia. Science *230*:1350-1354.
38. **Shinozaki, K., M. Ohme, M. Tanaka, T. Wakasugi, N. Hayshida, T. Matsubayasha, N. Zaita, J. Chunwongse, J. Obokata, K. Yamaguchi-Shinozaki, C. Ohto, K. Torazawa, B.Y. Meng, M. Sugita, H. Deno, T. Kamogashira, K. Yamada, J. Kusuda, F. Takaiwa, A. Kata, N. Tohdoh, H. Shimada and M. Sugiura.** 1986. The complete nucleotide sequence of the tobacco chloroplast genome. Plant Mol. Biol. Rep. *4*:110-147.
39. **Steinmetz, A., E.J. Gubbins and L. Bogorad.** 1982. The anticodon of the maize chloroplast gene for $tRNA^{Leu}$(UAA) is split by a large intron. Nucleic Acids Res. *10*:3027-3037.
40. **Sytsma, K.J. and B.A. Schaal.** 1990. Ribosomal DNA variation within and among individuals of *Lisianthius* (Gentianaceae) populations. Plant Syst. Evol. *170*:97-106.
41. **Taberlet, P., L. Gielly, G. Pautou and J. Bouvet.** 1991. Universal primers for amplification of three non-coding regions of chloroplast DNA. Plant Mol. Biol. *17*:1105-1109.
42. **Wahlberg, J.A., A. Holmberg, S. Bergh, T. Hultmann and M. Uhlén.** 1992. Automated magnetic separation of DNA templates for solid phase sequencing. Electrophoresis *13*:547-551.
43. **Wahlberg, J.A., J. Lundeberg, S. Cox, B. Wahren and M. Uhlén.** 1992. Dynamic changes in HIV-1 quasispecies from azidothymidine(AZT)-treated patients. The FASEB J. *6*:2843-2847.

Update to:

A General Approach for PCR-Amplification and Sequencing of Chloroplast DNA from Crude Vascular Plant and Algal Tissue

Kjetill S. Jakobsen and Brita Stedje[1]
University of Oslo and [1]*Agricultural University of Norway, Oslo, Norway*

A UNIVERSAL PCR SEQUENCING SYSTEM FOR CHLOROPLAST DNA

The original work was based on a report that described a set of chloroplast tRNA PCR primers that were shown to amplify regions of chloroplast DNA from diverse species (9). We restricted our system to two regions, the *trnL* (UAA) intron and the spacer region between *trnL* 3′ exon and *trnF* (*trnL-F* IGS), and developed new sets of nested universal primers for both regions. The nested PCR systems were found to work for all types of vascular plants, including ferns and higher plants, as well as some algal species tested. In addition to developing a universal chloroplast PCR system, we also designed possibilities for dye-primer automatic sequencing by adding M13 universal primer tails to some of the primers. By including a simple procedure for obtaining PCR-ready DNA (i.e., lysed plant tissue embedded in low melting point agarose plugs; termed agarose plug method), we had designed a total system for sequence analysis of chloroplast DNA. Since the PCR primers are restricted to two regions, our PCR-sequencing system is suitable for diagnostic, phylogenetic and evolutionary problems.

NEW DEVELOPMENTS AND CHANGES IN THE ORIGINAL PROTOCOL

Since publication of this paper, we have utilized the two chloroplast regions in several projects. In particular, we used these regions for evolutionary and systematic studies of higher plants (8; O. Stabbetorp, B.M. Fangan and K.S. Jakobsen, submitted; B.M. Fangan, I. Nordal and K.S. Jakobsen, manuscript in preparation). Thus, some parameters and primers have been optimized since the original article was published. A new nested E primer (5′-AAAAATCGTGAGGGTTCAAG-3′) for the *trnL-F* IGS region has been developed (8). This primer works well for most dicots and monocots. The temperature program for first-round PCR of the *trnL-F* IGS has been changed to 95°C for 30 s, 50°C for 30 s, 72°C for 60 s (35 cycles). In addition, it has been found for many species that reducing the amount of material for PCR in connection with the agarose plug method from 4 µL to 0.5 µL per PCR, improves yield and PCR specificity.

We have amplified the *trnL* intron from several species of cyanobacteria and some algae belonging to the groups of Prymnesiophyceae, Chlorophyceae and Rodophyceae (K. Rudi and K.S. Jakobsen, manuscript in preparation; O.J. Dahlberg, D. Klaveness and K.S. Jakobsen, manuscript in preparation). For such purposes, it has been necessary to develop C and D primer sets and PCR programs more optimal for cyanobacteria and algae. These primer sequences are available upon request.

Sequencing of the *trnL* intron and *trnL-F* IGS can be performed both manually by the magnetic solid-phase approach described in the original paper, and automatically by either dye primer sequencing or by chain terminators. Both VISTRA™ (Amersham) and Perkin-Elmer/Applied Biosystems 373A sequencers have been used. In our experience, the use of thermostable polymerases (ThermoSequenase and AmpliTaq® DNA polymerase, FS) in cyclic sequencing has significantly improved the sequence quality compared to the conventional (non-cyclic) solid-phase sequencing originally described. In particular, we observe that the multiple-chain terminations frequently seen in conventional sequence ladders are virtually absent in automatic cyclic sequencing with the new thermostable polymerases.

A new, simple and more reliable protocol for PCR-ready DNA has also been designed (6). This protocol employs magnetic beads that bind nonspecifically to DNA in the presence of high salt concentrations and an anionic detergent. We have tested and optimized bead-DNA isolation protocols for dicots, monocots, gymnosperms, algae and bacteria (6). For most PCRs involving chloroplast DNA in our lab, this is now the preferred method. In contrast to the agarose plug method, the bead approach is simple to adapt for automation.

USE OF *trnL* INTRON AND *trnL-F* IGS IN EVOLUTIONARY STUDIES

In this paper we showed that the *trnL* intron and the *trnL-F* IGS are suitable for phylogenetic and evolutionary sequencing despite being relatively short DNA regions (on average 500 bp and 350 bp, respectively). The *trnL* intron has a slower evolutionary rate than the *trnL-F* IGS and is therefore well suited for studies at the intergeneric level in dicots and monocots (4,5; O. Stabbetorp, B.M. Fangan and K.S. Jakobsen, submitted). The *trnL-F IGS* can be used to resolve evolutionary relationships both between and within genera (1,8). However, for closely related species within certain genera, the *trnL-F* IGS does not harbor sufficient variation to unambiguously resolve the taxonomic relationships (5; B.M. Fangan, I. Nordal and K.S. Jakobsen, manuscript in preparation). For intergeneric studies it may often be useful to combine the two regions (4,8). If only one of the regions can be used, it should be combined with data from other sequences, e.g., *rbcL* or ITS (ribosomal transcribed spacers), to get enough characters for a reliable phylogenetic analysis. The *trnL* intron and the *trnL-F* IGS regions are generally monomorphic or show little variation at the subspecies or population level, although the *trnL* intron has been used to track postglacial migration routes of oak into Britain due to a 13-bp duplication within the P8 stemloop present in some British populations, but not in mainland Europe populations (3).

Both regions have been found to have a higher substitution rate than the *rbcL* gene (4; O. Stabbetorp, B.M. Fangan and K.S. Jakobsen, submitted). The *trnL* intron also has a higher substitution rate than another chloroplast intron used in evolutionary studies, the *rpoC1* intron, and it therefore gives a higher resolution among closely related taxa (2). Compared to the nuclear ITS 1 and 2, the *trnL* intron shows, on average, two times less sequence divergence within ten species of the genus *Genetiana* (5). However, the phylogenetic trees produced were highly concordant.

In addition to being a group 1 self-splicing intron present in both bacteria and higher plants, the *trnL* intron also has other intriguing properties from an evolutionary standpoint. The P8 stemloop region, which varies in length from 150 to 300 bp among higher plants, has 2.8 and 4.3 times higher substitution rates for dicots and monocots, respectively, compared to the intron core (O. Stabbetorp, B.M. Fangan and K.S. Jakobsen, submitted). The CpG dinucleotide bias, which is normal compared to the rest of the chloroplast genome in the core intron, is on average 3 and 11 times higher in P8 in dicots and monocots, respectively. Although no open reading frames can be found in P8, this region may have a different evolutionary origin than the rest of the intron (O. Stabbetorp, B.M. Fangan and K.S. Jakobsen, submitted).

Both the *trnL* intron and the *trnL-F* IGS have proven useful in studies of intrageneric relationships among *Echium* species (1) and among the related monocot genera (Poceae) *Hordeum*, *Triticum* and *Aegilopsis* (4). In a recent work (8) using both the *trnL* intron and the *trnL-F* IGS the phylogeny and systematics of some closely related genera belonging to the family Hyacinthaceae is addressed. The sequence data support the splitting off of the genus *Ledebouria* from *Scilla* and also demonstrate a surprisingly close genetic similarity between two genera (*Ledebouria* and *Drimiopsis*) of strongly diverging floral morphology.

Most the studies utilizing these two regions have so far been based on sequencing of PCR products. A simple and less expensive approach is RFLP analysis of the same PCR products, provided that suitable sites are present. Notably, this has been done in a study of the different genera of the Papaveraceae using restriction sites present in the *trnL* intron (7). For studies utilizing the *trnL* intron or *trnL-F* IGS for more diagnostic purposes such as evolutionary and phylogenetic investigations of closely related species or simply detection of a species (i.e., algae, cyanobacteria) or subspecies/strain (i.e., plants), diagnostic primers and other specialized PCR-based detection systems will most certainly be developed in the near future.

ACKNOWLEDGMENTS

We would like to thank professor Inger Nordal for sharing unpublished results and for helpful discussions. We would also like to acknowledge the unpublished work mentioned in this update carried out by colleagues in our lab. Supported by grants from the Norwegian Research Council (NFR) to KSJ (Grant No. 107622/420) and BS (Grant No. 610.93/038 and 101998/720).

REFERENCES

1. **Böhle, U-R., H. Hilger, R. Cerff and W.F. Martin.** 1994. Non-coding chloroplast DNA for plant molecular systematics at the infrageneric level, p. 391-403. *In* B. Scierwater, B. Streit, G.P. Wagner and R. DeSalle (Eds.), Molecular Ecology and Evolution: Approaches and Applications. Birkhäuser Verlag, Basel.
2. **Downie, S.S., D.S. Katz-Downie and K-J Cho.** 1996. Phylogenetic analysis of Apiaceae subfamily Apioideae using nucleotide sequences from the chloroplast *rpoC1* intron. Mol. Phylogenet. Evol. *6*:1-18.
3. **Ferris, C., R.P. Oliver, A.J. Davy and G.M. Hewitt.** 1995. Using chloroplast DNA to trace postglacial migration routes of oaks into Britain. Mol. Ecol. *4*:731-738.
4. **Gielly, L. and P. Taberlet.** 1994. The use of chloroplast DNA to resolve plant phylogenies: noncoding versus *rbcL* sequences. Mol. Biol. Evol. *11*:769-791.
5. **Gielly, L., Y-M. Yuan, P. Kupfer and P. Taberlet.** 1996. Phylogenetic use of noncoding regions in the genus *Genetiana* L.: chloroplast *trnL* (UAA) intron versus nuclear ribosomal internal transcribed spacer sequences. Mol. Phylogenet. Evol. *5*:460-466.
6. **Rudi, K., M. Kroken, O.J. Dahlberg, A. Deggerdal, K.S. Jakobsen and F. Larsen.** Rapid, universal method to isolate PCR-ready DNA using magnetic beads. BioTechniques (In press).
7. **Schwarzbach, A.E. and J.W. Kadereit.** 1995. Rapid radiation of North American desert genera of the Papaveraceae: evidence from restriction site mapping of PCR-amplified chloroplast DNA fragments. Plant System. Evol. (Suppl.) *9*: 159-170.
8. **Stedje, B.** Phylogenetic relationships and generic delimitation of *Scilla* L. (Hyacinthaceae) from morphological and DNA sequence data. Plant System. Evol. (In press).
9. **Taberlet, P., L. Gielly, G. Pautou and J. Bouvet.** 1991. Universal primers for amplification of three non-coding regions of chloroplast DNA. Plant Mol. Biol. *17*:1105-1109.

Alu-Vector PCR with Biotinylated Primers to Isolate YAC Ends Ready for Sequencing

Ricardo Fujita and Anand Swaroop

W.K. Kellogg Eye Center, University of Michigan, Ann Arbor, MI, USA

Yeast artificial chromosomes (YACs) are widely used for cloning large human genomic DNA fragments (1). End sequences from YAC clones provide sequence-tagged sites (STSs), which are useful for physical mapping studies (4). A number of methods may be employed for isolating the YAC end fragments; these include end rescue, inverse polymerase chain reaction (PCR), vectorette and Alu-vector PCR (reviewed in Reference 7). Most of these techniques are time-consuming and require purification of end fragments by preparative gel electrophoresis for obtaining high-quality sequence information. Alu-vector PCR is a comparatively simple method that involves DNA amplification using one primer derived from the Sup4 sequences in the YAC vector and another from the Alu-repeat sequence present in the human DNA insert (6). A major problem with this method is that many nonspecific products, including those of endogenous *Saccharomyces cerevisiae*'s Sup4 gene, are also obtained during PCR amplification (3). Here we report a modified Alu-vector PCR procedure for rapid isolation and sequencing of the human insert DNA from the ends of YAC clones. The improvements in our method are *(i)* the use of vector primers located in the pBR322 plasmid instead of the Sup4 region, facilitating higher specificity of amplification products, and *(ii)* the use of biotinylated vector primers, allowing a rapid isolation of the specific Alu-vector PCR product with the streptavidin-conjugated magnetic beads (5). Two strands of the selected amplified product are then separated and directly subjected to sequencing without further gel purification. The specific YAC end sequence is obtained using either a vector-primer or an Alu-primer.

DNA (0.2–2 ng) from the YAC clones spanning the DXS140 and OTC loci at the Xp21.1 chromosomal region (2) was used for obtaining the end sequences. Amplification was performed in a volume of 50 µL, which contained 1× PCR buffer (10 mM Tris-HCl pH 8.2, 50 mM KCl), 1 µM biotinylated vector-primer (b-inv14 or b-inv20), 0.4 µM Alu-primer (see Table 1 for primer

(Reprinted from BioTechniques 18:796-800, 1995)

sequences), 1.5 mM $MgCl_2$, 200 μM each dNTP and 2 units of AmpliTaq® (Perkin-Elmer, Norwalk, CT, USA). The thermal cycling profile was 10 cycles of 94°C for 45 s, 60°C for 45 s and 72°C for 2 min, followed by 25 cycles of 94°C for 45 s, 45°C for 45 s and 72°C for 2 min. To capture biotinylated Alu-vector PCR product, 40 μL of the PCR reaction were mixed with 20 μL of streptavidin-conjugated magnetic beads (Dynabeads® M-280; Dynal,

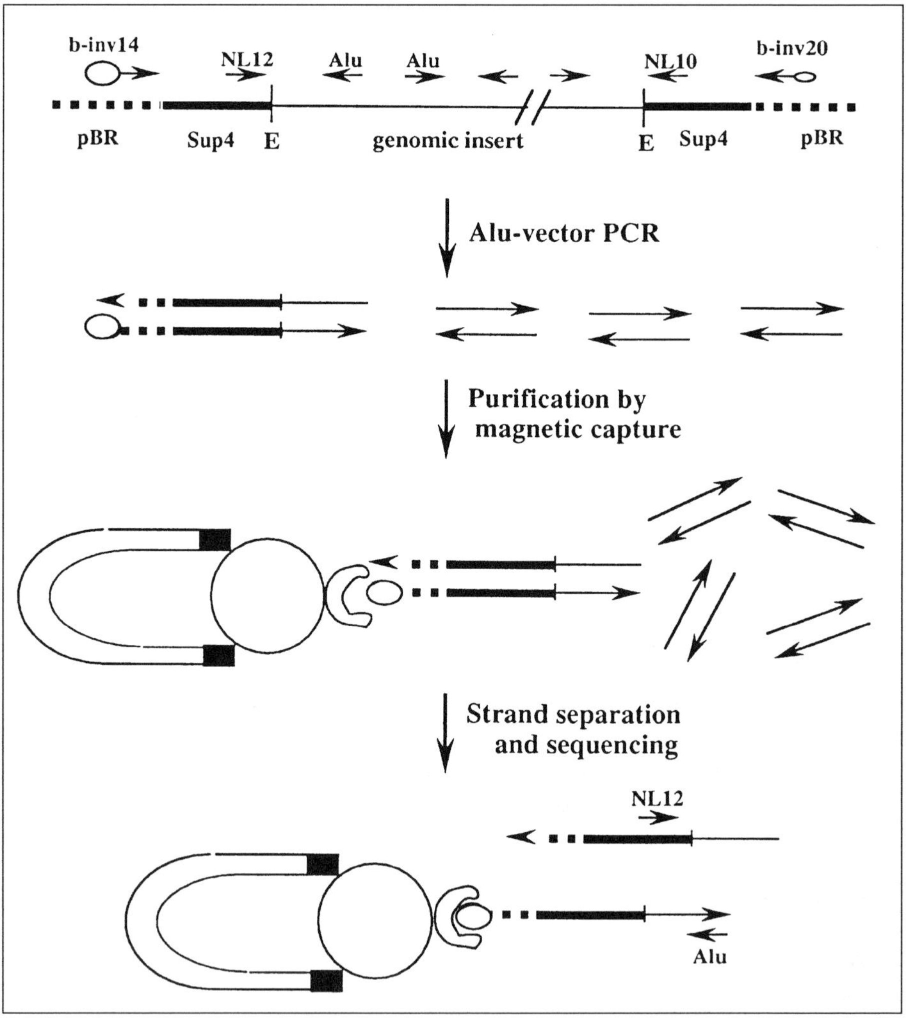

Figure 1. Schematic representation of the modified Alu-vector PCR method for isolating YAC ends. Position of the primers is shown on top by arrows (see Table 1 for the description of primers). The pBR322 (pBR), Sup4 and genomic insert DNA are indicated by dashed, thick and thin lines, respectively. E is the *Eco*RI cloning site. Arrows and lines with oval ends represent biotinylated primers and DNA. After PCR amplification, DNA is depicted as antiparallel arrows. The biotinylated strand is captured by streptavidin-conjugated magnetic beads attached to the magnet. In the lower section, NL12 and Alu indicate the primers used for sequencing.

Table 1. Oligonucleotide Primers Used for Alu-Vector PCR

Primer	Sequence	Origin (Reference)
b-inv20	5′ biotin - ATG CCG GCC ACG ATG CGT CCG GCG 3′	pBR - right arm
b-inv14	5′ biotin - ATG CGC ACC CGT TCT CGG AGC 3′	pBR - left arm
NL10	5′ CTC CCG GGG GCG AGT CGA ACG CCC 3′	Sup4 - right arm (3)
NL12	5′ CAA TTA AAT ACT CTC GGT AGC CAA G 3′	Sup4 - left arm (3)
NH34	5′ AAG TCG CGG CCG CTT GCA GTG AGC CGA GAT 3′	Alu der TC65 (6)
NA71	5′ CCT CGG CCT CCC AAA GTG CTG GGA TTA CAG 3′	Alu der 278 (3,6)

Great Neck, NY, USA), which were washed and resuspended in 2× binding and washing buffer (B&W, 1× = 10 mM Tris-HCl pH 7.5, 1 mM EDTA, 2 M NaCl) prior to use. After 20 min incubation at 37°C with gentle shaking, the Dynabeads were washed twice in 2× B&W. The beads were then resuspended in 8 µL of 0.1 *N* NaOH and incubated at room temperature for 10 min to release the non-biotinylated strand in solution. The non-biotinylated DNA was transferred to a fresh microcentrifuge tube and neutralized with 4 µL of 0.2 *N* HCl and 1 µL of 1 M Tris-HCl pH 7.5, prior to sequencing. The Dynabeads with the captured biotinylated strand were then washed once with 50 µL of 0.1 *N* NaOH, twice with 50 µL of B&W and once with 50 µL of TE (10 mM Tris-HCl pH 8.0, 1 mM EDTA). The beads were finally resuspended in 20 µL of water and directly used for obtaining the sequence of the biotinylated strand of the Alu-vector PCR product. Sequencing was performed with the Sequenase® Version 2.0 kit (United States Biochemical, Cleveland, OH, USA) using 7 µL of DNA and 0.5 pmol of an appropriate oligonucleotide (an Alu primer for the biotinylated strand and a nested vector primer—NL10 or NL12—for the non-biotinylated strand). Sequencing reactions were loaded on a 6% polyacrylamide-urea gel (Sequagel™; National Diagnostics, Atlanta, GA, USA).

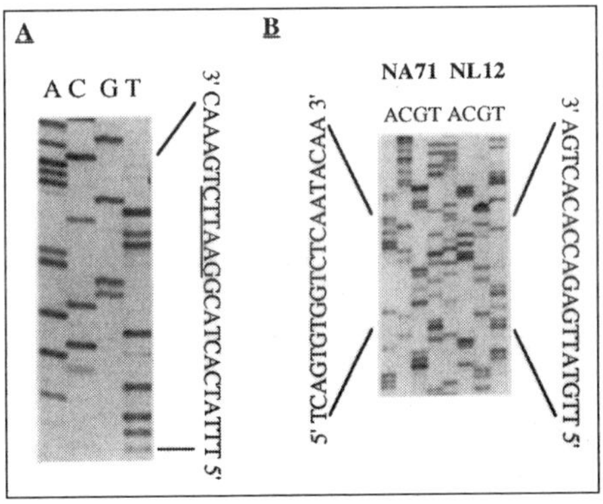

Figure 2. Sequence of the selected Alu-vector PCR product. DNA from the OTC-C YAC was amplified with the Alu primer NA71 and the biotinylated vector primer b-inv14 corresponding to the left arm. The amplified Alu-vector PCR product was purified by the method shown in Figure 1 and used for sequencing. (A) Sequence at the left arm-insert boundary was obtained from the non-biotinylated strand using the vector-primer NL12. *Eco*RI cloning site is underlined. Sup4 and insert sequences are at the 5′ and 3′ end of the *Eco*RI site, respectively. (B) Specificity of the Alu-vector PCR product was confirmed by sequence complementarity of the two strands. The sequence of the biotinylated strand, obtained with the Alu-primer NA71, is shown on the left, whereas the sequence of the non-biotinylated strand, obtained with the NL12 vector-primer, is shown on the right.

The use of vector primers derived from the pBR322 region of the YAC vector eliminates nonspecific amplification from the yeast DNA. The biotinylated vector primers allow specific capture of the target sequence from a mixture of Alu-PCR products by the streptavidin-conjugated magnetic beads. DNA from the two complementary strands is obtained without a gel extraction or additional purification step. The quality of generated sequence is equally good with both the nested vector primer (for the non-biotinylated strand) and the Alu primer (for the biotinylated strand).

The sequence at the vector-insert boundary of an Alu-vector PCR product demonstrates the specificity of the method (see Figure 2A, for the sequence of a product with the b-inv14 vector primer obtained by using the NL12 primer). The complementarity of sequence from the two strands confirms the specificity and uniqueness of the Alu-vector product (Figure 2B). A specific sequence can be obtained even in the presence of nonspecific products generated during the PCR (data not shown). The results, presented here, were also confirmed by sequencing the cloned YAC end fragments.

Other interspersed sequences or random short primers may be used to overcome the lack of an Alu sequence in a correct orientation within the YAC end segment. The use of glycerol, formamide or *Taq* Extender™ PCR additive (Stratagene, La Jolla, CA, USA) in the reaction or the use of other thermostable polymerases (e.g., r*Tth* DNA polymerase from Perkin-Elmer) should permit the amplification of long Alu-vector PCR products, even if the Alu sequence is several kilobases downstream of the vector primer. It should also be possible to extend this method to other species by using appropriate primers derived from the repetitive sequences. For a laboratory engaged in constructing YAC contigs, this protocol should lower costs in material and time, and can be easily automated.

ACKNOWLEDGMENTS

We thank Dr. A. Hanauer (Strasbourg, France) for the b-inv20, b-inv14 and NH34 primer sequences. This work was supported by a postdoctoral fellowship from Fight for Sight (R.F.), by NIH Grants EY07961 (A.S.) and HG00209 (Michigan Human Genome Center) and by the Retinitis Pigmentosa Foundation, Baltimore (A.S.).

REFERENCES

1. **Burke, D.T., G.F. Carle and M.V. Olson.** 1987. Cloning of large segments of exogeneous DNA into yeast by means of artificial chromosome vectors. Science *236*:806-812.
2. **Fujita, R., D. Ayon, L. Gieser, C. Skolnick, T. Yang-Feng, P. Sieving and A. Swaroop.** 1993. Genetic and physical analysis of the genomic region spanning X-linked retinitis pigmentosa locus RP3. Am. J. Hum. Genet. *53* (Suppl.) *A*:1290.
3. **Fujita, R., G. Sirugo, F. Duclos, H. Abderrahim, D. Le Paslier, D. Cohen, B.H. Brownstein, D. Schlessinger, J.L. Mandel and M. Koenig.** 1992. A 530 kb YAC contig tightly linked to Friedreich ataxia locus contains 5 CpG clusters and a new highly polymorphic microsatellite. Hum. Genet. *89*:531-538.

4. **Green, E.D. and M.V. Olson.** 1990. Chromosomal region of cystic fibrosis gene in yeast artificial chromosomes: a model for human genome mapping. Science *250*:94-98.
5. **Hultman, T.S., S. Bergh, T. Moks and M. Uhlén.** 1991. Bidirectional solid phase sequencing of *in vitro* amplified plasmid DNA. BioTechniques *10*:84-93.
6. **Nelson, D.L., S.A. Ledbetter, L. Corbo, M.F. Victoria, R. Ramirez-Solis, T.D. Webster, D.H. Ledbetter and C.T. Caskey.** 1989. Alu polymerase chain reaction: a method for rapid isolation of human-specific sequences from complex DNA sources. Proc. Natl. Acad. Sci. USA *86*:6686-6690.
7. **Silverman, G.A.** 1993. Isolating vector-insert junctions from yeast artificial chromosomes. PCR Methods Appl. *3*:141-150.

Update to:

Alu-Vector PCR with Biotinylated Primers to Isolate YAC Ends Ready for Sequencing

Ricardo Fujita and Anand Swaroop
University of Michigan, Ann Arbor, MI, USA

Sequences obtained from the YAC ends in Xp21 have allowed us to screen a YAC library and build a contiguous array of clones spanning more than 2 megabases. Additional clones are being analyzed with this procedure (Alu-vector PCR with biotinylated primers) to generate new STSs in the region and to verify the integrity of individual YAC clones (Fujita et al., in preparation).

In addition to NA71 and NH34, another Alu primer that yields good results is oligonucleotide NG76: 5′ CGA CCT CGA GAT CTC GGC TCA CTG CAA 3′ (1). Typically, a biotinylated vector primer is used in three different PCR reactions to enhance the probability of relevant amplification. For a new YAC clone, we prepare a pre-mixture for six reactions containing all the reagents (including DNA) except primers. This pre-mixture is added to tubes containing a particular pair of biotinylated vector primers (left or right) with one of the three Alu primers.

Before biotin selection, it is advisable to verify the quality and quantity of the PCR product on a 1.5% agarose gel. The absence of bands eliminates a sample from further analysis. However, the presence of multiple bands does not preclude the use of the sample. A safer procedure is to prepare a Southern blot of the agarose gel and hybridize it with the internal primers (NL10 for right arm and NL12 for left arm).

REFERENCE

1. **Fujita, R. and A. Swaroop.** 1996. Rapid YAC end sequencing by Alu-vector PCR and biotinylated primers, p. 227-237. *In* T. Meier and F. Fahrenholz (Eds.), BioMethods, Vol. 7: A Laboratory Guide to Biotin-Labelling in Biomolecule Analysis. Birkhauser Verlag AG, Basel.

A Method for Obtaining High-Quality Sequences from the Non-Biotinylated, Free ssDNA Remaining after Solid-Phase Sequencing

Youn-Ho Lee and Victor D. Vacquier

Marine Biology Research Division, Scripps Institution of Oceanography, University of California-San Diego, La Jolla, CA, USA

The sequencing of PCR-generated DNA products using streptavidin-coated magnetic beads as the solid support is an important technical advance. Briefly stated, one strand of the PCR product double-stranded (ds) DNA contains a biotinylated primer that binds the DNA to the magnetic beads as an avidin-biotin complex (2). After the reactants are washed away, the dsDNA is denatured with NaOH; this leaves the biotinylated single-stranded (ss) DNA bound to the bead and ready for dideoxy chain termination sequencing (4) using the Sequenase® Version 2.0 protocol (United States Biochemical, Cleveland, OH, USA). Compared with dsDNA sequencing, this method yields consistently better results with increased lengths of readable sequences and reduced backgrounds. However, sequencing the non-biotinylated, free, complementary DNA strand does not always yield such clean results. Although a bidirectional solid-phase sequencing method has been presented (1), obtaining consistently high-quality results remains problematical. Difficulties in this method (1) may arise from the pH, since neutralization of the NaOH following separation of the non-biotinylated strand requires delivery of an extremely accurate equivalent of acid. Here we describe a simple protocol for reproducibly obtaining excellent quality, 400–500 base readable length sequences of the non-biotinylated, free ssDNA remaining in the NaOH denaturing solution (Figure 1). Thus, both strands of the same PCR product can be sequenced simultaneously with considerable savings in time and expense.

Protocol for Sequencing the Non-Biotinylated, Free, Complementary DNA Strand

1) The 1–10 μg dsDNA PCR product is immobilized on Dynabeads® (Dynal, Great Neck, NY, USA) in a 1.5-mL microcentrifuge tube and washed

(Reprinted from BioTechniques 14:191-192, 1993)

with binding and washing buffer according to the manufacturer's instructions.

2) The Dynabeads are then resuspended in 30 μL of 0.1 *N* NaOH (room temperature, 10 min).

3) The beads are collected (1 min) on the side of the tube using the magnetic stand. The liquid phase containing the free, non-biotinylated, complementary DNA strand is transferred to a Centricon®-30 microconcentrator (Amicon, Beverly, MA, USA) filled with 2 mL of TE buffer (10 mM Tris-HCl, pH 7.6, 1 mM EDTA).

4) The beads, magnetically held to the tube wall, are washed once with 50 μL of 0.1 *N* NaOH, and the supernatant is combined with the TE in the Centricon. The Dynabeads are washed once with 100 μL of binding and washing buffer and twice with pure water. The immobilized, biotinylated ssDNA on the beads is ready for sequencing using the Sequenase Version 2.0 protocol.

5) The free, complementary ssDNA in TE buffer in the Centricon-30 is reduced to 100 μL. Two milliliters of pure water are added to this retentate and the volume again reduced to 100 μL. This washing step is repeated twice, and the final volume of 100 μL is transferred to a 1.5-mL tube.

6) The 100 μL are lyophilized to about 3–5 μL, and water is added to a final volume of 6 μL. Two microliters of dimethyl sulfoxide (DMSO; References 5 and 6), 2 μL of Sequenase buffer and 1 μL of sequencing primer (10 pmol) are added. For annealing the primer to the template DNA, the tube is placed in a 65°C temperature block for 5 min, and the block is then placed on the bench for gradual cool-

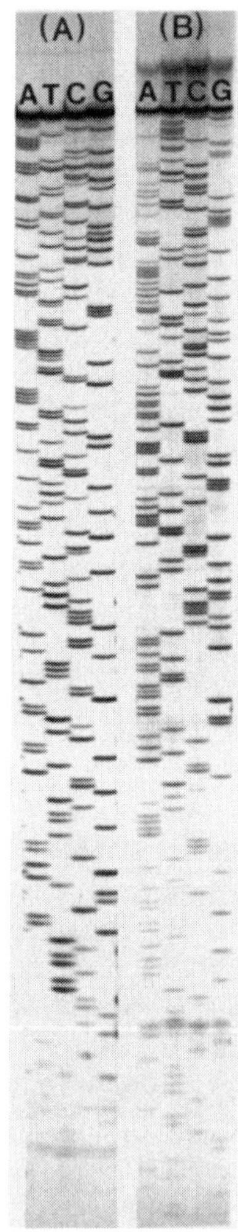

Figure 1. Autoradiograms of sequencing gels of the biotinylated, immobilized ssDNA on Dynabeads (A) and also the non-biotinylated, free, complementary DNA strand (B). This procedure is easier than the method using streptavidin-agarose (3) and circumvents the need for accurate neutralization of the NaOH following separation of the non-biotinylated strand (1).

ing to room temperature (40 min).

7) The 11 µL of annealed mixture are collected in the bottom of the tube by a 10-s spin. One microliter of dithiothreitol (DTT, 100 mM), 2 µL of 1:5 diluted labeling mixture, 0.5 µL of α-^{35}S-dATP (DNA sequencing grade) and 2 µL of diluted enzyme mixture (1:8 dilution of Sequenase Version 2.0 T7 DNA polymerase and 1:16 dilution of pyrophosphatase) are added on ice. Without incubation, 3.7-µL portions of the reaction mixture are transferred to 1.5-mL tubes containing 2.5 µL of termination mixture of each dideoxynucleotide (on ice). The tube (6.2 µL total volume) is incubated at 48°C for 15 min. Four microliters of stop solution are then added.

8) The samples are heated to 70°C for 5 min, and 3 µL portions are loaded on lanes of a 7 M urea, 6% polyacrylamide wedge (1:5) gel (50 × 30 cm) and electrophoresed for 3–4 h at 50 W. The gel is fixed for at least 2 h in 10% acetic acid, 10% methanol, dried under vacuum at 80°C and exposed overnight to X-Omat RP film (Eastman Kodak, Rochester, NY, USA).

ACKNOWLEDGMENTS

This work was supported by a Korean Government Overseas Scholarship to Y.-H. Lee and NIH Grant 12986 to V.D. Vacquier.

REFERENCES

1. **Hultman, T., S. Bergh, T. Moks and M. Uhlén.** 1991. Bidirectional solid-phase sequencing of *in vitro*-amplified plasmid DNA. BioTechniques *10*:84-93.
2. **Hultman, T., S. Stahl, E. Hornes and M. Uhlen.** 1989. Direct solid phase sequencing of genomic and plasmid DNA using magnetic beads as solid support. Nucleic Acids Res. *17*:4937-4946.
3. **Mitchell, L.G. and C.R. Merril.** 1989. Affinity generation of single-stranded DNA for dideoxy sequencing following the polymerase chain reaction. Anal. Biochem. *178*:239-242.
4. **Sanger, F., S. Nicklen and A.R. Coulson.** 1977. DNA sequencing with chain-terminating inhibitors. Proc. Natl. Acad. Sci. USA *74*:5463-5467.
5. **Schuurman, R. and W. Keulen.** 1991. Modified protocol for DNA sequence analysis using Sequenase® 2.0. BioTechniques *10*:185.
6. **Winship, P.R.** 1989. An improved method for directly sequencing PCR amplified material using dimethyl sulfoxide. Nucleic Acids Res. *17*:1266.

Update to:

A Method for Obtaining High-Quality Sequences from the Non-Biotinylated, Free ssDNA Remaining after Solid-Phase Sequencing

Youn-Ho Lee

California Institute of Technology, Pasadena, CA, USA

We have found that the protocol presented in the original paper works consistently and so no efforts have been made to change or improve it. Members of Dr. Vacquier's lab are still using the protocol as it is.

A survey of the literature shows that no significant improvements to our work have been made. In fact, many researchers use only the biotinylated strand for solid-phase sequencing and simply discard the other non-biotinylated strand; this procedure proves to be a waste of time and ultimately makes the sequencing cost twice as much. I believe that the method presented in our original article is still useful, especially for anyone who has limited money and time.

Sequencing Homopolymer Tracts and Repetitive Elements

Christiane M. Robbins, Elvin Hsu and Patrick M. Gillevet

National Center for Human Genome Research and the National Institutes of Health, Bethesda, MD, USA

ABSTRACT

We investigated the use of Taq *dye primer and* Taq *terminator sequencing chemistry to optimize the quality of sequence data obtained from templates containing homopolymer tracts and repetitive elements. In direct side-by-side comparisons using the Applied Biosystems Model 373A Fluorescent Sequencer, the* Taq *terminator sequencing chemistry gave much cleaner and more consistent results on long homopolymer tracts and dinucleotide repeats. We also investigated various thermal cycling conditions and determined that higher annealing temperatures and longer denaturation times improved the ability to sequence through these problem templates.*

INTRODUCTION

The Nucleic Acid Core Facility (NACF) was established at the National Center for Human Genome Research (NCHGR) as a resource for routine sequencing of templates for the new intramural program. The NACF is using Applied Biosystems (ABI) Model 373A Automated DNA Sequencers, which are based on real-time detection of electrophoresed DNA bands by means of laser-induced fluorescence (1,4) to sequence a variety of double-stranded (ds) templates. Each DNA sample is labeled with four different fluorescent dyes, which is then run in one lane on the Model 373A Sequencer (2). The vast majority of these templates are derived from projects investigating microsatellite detection or exon trapping and thus contain various repetitive elements or homopolymer tracts. The difficulties associated with sequencing these templates on the Model 373A Fluorescent Sequencer are well documented (3) and can be very problematic. Furthermore, it is difficult to standardize the choice of a vector, methods for library construction, techniques for insert selection and template preparation in a core situation. Core sequencing facilities are now being established routinely (5), and many of these sites have encountered similar methodological problems as the NACF and

(Reprinted from BioTechniques 20:862-868, 1996)

have had the same difficulties in educating the local community on the use of the core and the technical limitations of the fluorescent sequencing technology. In this work, we describe the methodology used by the NACF to overcome many of the difficulties encountered in sequencing homopolymer tracts and repetitive elements, and we have presented examples of data from these problem templates on the World Wide Web (WWW) with the hope of helping to educate other users of similar services.

We initially used the ABI *Taq* dye primer sequencing chemistry with the standard ABI thermal cycling conditions for all reactions on ds templates, but observed problems when ds templates contained homopolymer tracts such as the poly T tracts found in the exon trapping vector system (pAMP-1). We subsequently investigated the use of the *Taq* terminator chemistry to alleviate noisy background and premature termination of the signal from these problem templates. Finally, we modified the thermal cycling conditions to accommodate very long repetitive elements, and now we routinely use a higher annealing temperature and longer denaturation times for all fluorescent sequencing of ds templates done at the NACF.

MATERIALS AND METHODS

All of the fluorescent sequencing reagents were purchased from Perkin-Elmer/Applied Biosystems Division (PE/ABI, Foster City, CA, USA). Unlabeled primers were purchased from Genosys (The Woodlands, TX, USA). The thermal cycling protocols were conducted on a Model PTC-100-96™ Thermal Cycler (MJ Research, Watertown, MA, USA). Excess ddNTPs were removed from the *Taq* terminator reactions using Centri-Sep™ columns purchased from Princeton Separations (Adelphia, NJ, USA).

DNA Preparation

Double-stranded plasmid templates having the host strains DH5α or XL1-Blue were prepared by picking a single colony from a streaked LB plate and incubated overnight in a shaker-incubator at 37°C in 5 mL of LB broth from Advanced Biotechnologies (Columbia, MD, USA) with appropriate antibiotic. The following day, DNA was prepared from these cultures by using either the Wizard™ Miniprep kit (Promega, Madison, WI, USA) or the QIAwell™-8 plasmid kit (Qiagen, Chatsworth, CA, USA). The DNA was accurately quantitated by checking the optical density (OD) value on a Model 650 DU®-Spectrophotometer (Beckman Instruments, Fullerton, CA, USA) using a capillary tube system to read the DNA solution directly without dilution. The control template (vector without insert) was incubated and prepared in parallel with the sample DNA as a quality control for the template DNA preparation.

Taq Dye Primer Sequencing Reactions

In the fluorescently tagged primer A and C reactions, we used 200 ng plasmid DNA, 3.2 pmol labeled primer, 1 mL dNTP mixture, 1 mL 5× cycle sequencing buffer (400 mM Tris-HCl, pH 9.0, 10 mM $MgCl_2$) and 1 µL diluted *Taq* DNA polymerase (0.57 U/µL) in a final volume of 5 µL. These amounts were doubled for the G and T reactions to accommodate the lower quantum yield from the TAMARA and ROX fluorescent tags. The sequencing reactions were run with the recommended ABI thermal cycling conditions in a Model PTC-100-96 thermal cycler (initial denaturation at 95°C for 3 min, 95°C for 30 s, 55°C for 30 s, 70°C for 1 min and then back to the denaturation step for 14 more cycles). The reactions were then finished by 14 cycles of 95°C for 30 s and 70°C for 1 min. The reactions were then stored at 4°C until further processing. The sequencing reactions were then concentrated by adding 80 µL of 95% ethanol with 1.5 µL of 3 M sodium acetate (pH 5.3) in a clean tube, combining all four sequencing reactions into this mixture and precipitating the DNA fragments at -70°C for 1 h. The samples were then spun for 30 min at 15000 rpm (Eppendorf® Model 5145C Centrifuge; Brinkmann Instruments, Westbury, NY, USA), the supernatant was decanted, and the pellets were washed with 250 µL of 70% ethanol. The samples were re-spun for 10 min, the supernatant was discarded and the pellets were dried in a SpeedVac® (Savant Instruments, Holbrook, NY, USA) for 1–3 min. The samples were then resuspended in 4 µL of DNA loading buffer (deionized formamide with 50 mM EDTA), heated at 90°C for 1 min and loaded onto the Model 373A Fluorescent Sequencer (PE/ABI).

Taq Terminator Reactions

The *Taq* terminator reactions used 1 µg of plasmid DNA, 5 pmol unlabeled primer and 9 µL of *Taq* terminator mixture. We initially used the ABI standard thermal cycling conditions (96°C for 3 min, 96°C for 10 s, 50°C for 5 s, 60°C for 4 min and back to the denaturing step for 24 more cycles on the Model PTC-100-96 thermal cycler). We subsequently used a modified cycle that consisted of 93°C for 3 min, 94°C for 30 s, 55°C for 5 s, 60°C for 4 min and back to the 30-s denaturing step for 24 more cycles on the Model PTC-100-96. The sequencing reactions were placed over the Centri-Sep columns to remove excess dye terminators (following the Centri-Sep column protocol), dried down on the SpeedVac and then denatured in 4 µL of loading buffer as above. The reactions were loaded completely on the Model 373A Sequencer (PE/ABI) and run under standard electrophoretic conditions.

Quality Assessment

The quality of the data was determined by the number of ambiguous base calls (number of n's) between bases 50 and 350. Less than 5% ambiguous calls was considered a successful run. Furthermore, the quality of the chromatographic data coming after a homopolymer stretch or repeat was considered when optimizing the thermal sequencing conditions. The data were aligned using the "Sequencing Management Project" program (DNA Star Inc., Madison, WI, USA) to compare the analyzed data from the various conditions run on these samples. We did not manually edit any of the data. Large amounts of sequence data are generated by the Core Facility. These data were moved to a Sun SPARCstation® 10 (SUN Microsystems, Mountain View, CA, USA) for both analysis and archival purposes. Through the Genetic Data Environment (GDE) and custom programs, it was possible to take the ABI sequence data and calculate a success rate with a 5% error criteria. This information was then taken and published on the WWW (http: //uranus. nchgr.nih.gov/NACF_html).

RESULTS

We sequenced a series of plasmid clones using both the ABI *Taq* dye primer and ABI *Taq* terminator chemistry and directly compared the chromatographic profiles as well as the analyzed sequence data. We specifically picked clones that contained problem regions such as homopolymer tracts and dinucleotide repeats. We enumerated the number of ambiguous base calls between bases 50 and 350 of the analyzed data and compared the clarity of chromatographic profile following the problem region in the clone. In all cases, we observed that the *Taq* terminator chemistry gave superior results (see below).

In Figures 1A, we show an example of a chromatographic profile from a region of a double-stranded clone sequenced using *Taq* dye primer chemistry that contains a (GT) repeat from position 269 to position 299 in the analyzed data. The chromatographic profile for the *Taq* dye primer reactions had low

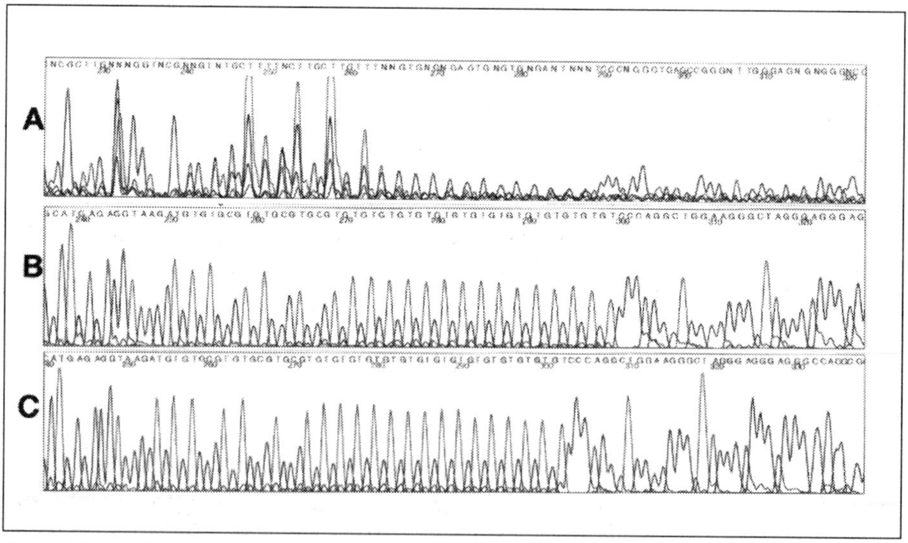

Figure 1. Plot of chromatographs. Plot of chromatograph of the dye primer data (A) and *Taq* terminator data (B) using the ABI thermal cycling conditions. These chromatographs show the sequence of a GT repeat 269 bases into the sequence. (C) Plot of chromatograph of *Taq* terminator data using the NACF's thermal cycling conditions.

peak heights with much higher background noise, and the signal dropped off after about 270 bp in the middle of the (GT) repeat. There were 30 ambiguous calls between 250 and 370 bp, and the data were noisy before the repeat as well. It may be that the template itself is difficult to denature due to the high GC content of the repeats, resulting in the mispriming of the primer and premature termination of the chain extension reaction.

In contrast, the terminator reactions produced a clear chromatographic profile with no ambiguous calls between positions 250 and 370 (Figure 1B). The peak heights were high throughout the first 350 bases with baseline peak resolution. In addition, the data analysis was clean through the (GT) repeat, and readable sequence was obtained out to 390 bases.

We also show the sequence from the same plasmid clone using the *Taq* terminator kit with the modified thermal cycling conditions. The analyzed results from the chromatographic profile (Figure 1C) had a 1.5-fold stronger average signal (G 248:161, A 161:93, T 97:72, C 83:75) than the corresponding *Taq* terminator chemistry. The peak height was much higher and distinct for the modified cycling conditions both through the repeat region and around the repeat regions. We saw baseline resolution of the peaks throughout the region and obtained a readable sequence out to 500 bases. Thus, the modified thermal cycling conditions greatly improves the overall sequence data through difficult homopolymer tracts and dinucleotide repeats.

Figure 2 shows the analyzed data from a clone that contains a (TCTG) repeat and a (TC) repeat in tandem sequenced with the *Taq* terminator chemistry and the modified thermal cycling protocol. We were able to sequence through two distinct repeats maintaining high signal strength, baseline peak resolution and no ambiguous base calls in the problem region from the first 50–250 bp. This sequencing run had 3% ambiguous calls over the readable sequence out to 360 bases.

DISCUSSION

It should be noted that about 20% of the templates submitted to the core contain homopolymer tracts and repetitive elements. The use of *Taq* terminator reactions is highly recommended over the use of the *Taq* dye primers for all sequencing in order to get through difficult templates, such as the homo-

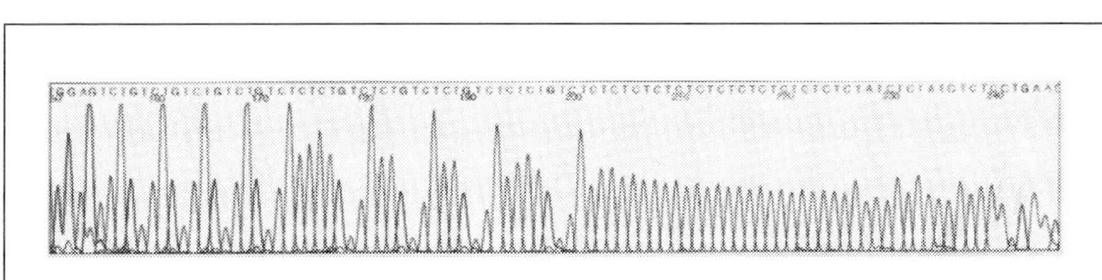

Figure 2. Plot of chromatograph *Taq* terminator data using the NACF's thermal cycling conditions showing a 100-bp repeat sequence starting at position 150 into the sequence.

polymer tracts. Furthermore, the use of the *Taq* terminator chemistry obviously facilitates the routine use of custom primers, which is crucial in the studies conducted at NCHGR as, in many instances, clones must be "walked on" to obtain the sequence flanking dinucleotide repeats for micro-satellite analysis.

Our thermal cycling conditions gave better sequence data when compared to the thermal cycling conditions of ABI, though they were subtle on some clones. In our experience, our thermal cycling conditions have allowed us to get through some very long repeats and poly-T tails, as well as given us excellent data on routine clones without repetitive elements. We hypothesize that the higher annealing temperature used in our thermal cycling conditions allows for more primer specificity and less slippage in the repeat or homopolymer tract during the thermal cycling, which would theoretically assist in high-fidelity copying through the full repetitive element. In addition to using a higher annealing temperature, we also used a longer denaturation step at a lower temperature, which is less harsh on *Taq*, while allowing the higher annealing temperature to keep these problem templates denatured longer.

The methodology presented in this report should help other core facilities to overcome many of the systematic problems associated with sequencing ds templates obtained from many investigators, especially with respect to templates with inserts of human origin. Furthermore, it is hoped that the information presented will help educate the various local communities on the use of the core and the technical limitations of fluorescent sequencing technology. In light of the latter point, we have established a home page on the WWW (http://uranus.nchgr.nih.gov/NACF_html/ NACF.Home.html) where we present many examples of problems routinely encountered in a core situation along with their resolution and a troubleshooting flow chart for core facilities.

REFERENCES

1. **Ansorge, W., B. Sproat, J. Stegemann, C. Schwager and M. Zenke.** 1987. Automated DNA sequencing: ultrasensitive detection of fluorescent bands during electrophoresis. Nucleic Acids Res. *15*:4593-4602.
2. **Connell, C., S. Fung, C. Heiner, J. Bridgham, V. Chakarian, E. Heron, R. Jones, S. Menchen, W. Mordan, M. Raff, M. Recknor, L. Smith, J. Springer, S. Woo and M. Hunkapiller.** 1987. Automated DNA sequence analysis. BioTechniques *5*:342-348.
3. **Perkin-Elmer/Applied Biosystems Division.** 1991. Applied Biosystems User Manual 18, October. Foster City, CA.
4. **Pon, R.T., G.A. Buck, R.L. Niece, M. Robertson, A.J. Smith and E. Spicer.** 1994. A survey of nucleic acid services in core laboratories. BioTechniques *17*:526-534.
5. **Prober, J.M., G. Trainor, R. Dam, F.W. Hobbs, C.W. Robertson, R.J. Zagursky, A.J. Cocuzza, M.A. Jensen and K. Baumeister.** 1987. A system for rapid DNA sequencing with fluorescent chain-terminating dideoxynucleotides. Science *238*:336-341.

PCR with Random Primers to Obtain Sequence from Yeast Artificial Chromosome Insert Ends

Jeff Swensen

University of Utah, Salt Lake City, UT, USA

ABSTRACT

A technique that can be used to isolate vector/insert junctions from clones in vectors, such as yeast artificial chromosomes and P1s, and to sequence plasmid inserts more rapidly has been developed. A vector primer is combined with single, randomly chosen oligonucleotides in PCRs, to create pools of products. With 12–24 random primers used in separate reactions, a given insert junction can frequently be isolated. For plasmid inserts, multiple products are created that can be sequenced from their random-primed ends to provide internal coverage for a clone. It is often possible to sequence a significant portion of an insert with one set of reactions. The speed and simplicity of the method in each case and its use of existing techniques and reagents make it appealing.

INTRODUCTION

A method for isolating yeast artificial chromosome (YAC) and P1 vector/insert junctions and for sequencing plasmid inserts was developed during the process of searching for the recently discovered *BRCA1* gene (4).

An important step in the search was to create a physical map of the localized region (6) using overlapping YAC clones assembled into a contig (7). To do this, it was desirable to obtain sequence from the YAC insert ends so sequence-tagged sites (STSs) could be designed as probes for the most efficient possible walking steps (1). As previously reviewed (9), there were several techniques available for isolating stretches of DNA around the vector/insert junctions. The simplest was likely Alu-vector (A-V) polymerase chain reaction (PCR) (5). This was successful in approximately 50% of the cases, but the remaining ends still had to be obtained. Other techniques that were tried, including subcloning YAC DNA followed by probing for insert ends by hybridization, seemed time-consuming and cumbersome. The method

(Reprinted from BioTechniques 20:486-491, 1996)

described in this report was developed to provide a simple supplement to A-V PCR for isolating YAC insert junctions.

This new technique consists of pairing nested YAC vector primers with single, randomly chosen primers in PCRs. Pools of products that occasionally contain a given YAC insert end can be created this way. If enough single random primers (RPs) are used in separate reactions (generally 12–24), it is often possible to obtain sequence for a particular YAC STS with minimal effort.

In subsequent stages of *BRCA1* physical mapping, contigs were assembled using vector types that were more convenient. P1s (8) were utilized most frequently for this project. STSs were again needed on the ends of the inserts, but with these vectors it was possible to sequence the clones directly. Occasionally, however, an end that required an STS contained repetitive sequence around which PCR primers could not be placed. As an alternative to sequencing beyond such areas, it was shown that the random-priming technique (RPT) could be used to create PCR products in the 0.5- to 1.5-kb range from the vector/insert junctions. The random-primed ends of these products could be sequenced to provide STSs that often bypassed the repeats in one step.

A recent report (3) describes a PCR-based technique known as thermal asymmetric interlaced PCR that can be used for the isolation of insert ends from YACs and P1s. This method also utilizes nested sequence-specific primers, but it uses a more complex set of PCR programs and requires short degenerate random primers rather than single primers of normal PCR length.

In the final stages of the *BRCA1* search, a large number of cDNA clones with lengths in the 1- to 3-kb range were isolated and needed to be sequenced quickly. Options for doing this included synthesizing oligonucleotides between sequence-walking steps and making nested deletion mutants (2) for each clone. Both methods were time-consuming, and while deletion mutants provided a way of sequencing an entire insert with one set of reactions, they were cumbersome to prepare when there were a large number of clones.

It was found that the RPT could provide a useful supplement to sequence walking. When vector-specific primers were used in reactions with the RPs, multiple products were produced that could be sequenced from the random-primed ends to provide internal coverage for the clones. If products were generated from both ends, it was often possible to sequence a significant portion of an insert at one time. In this paper, the RPT and some instances where it has proven useful are described.

MATERIALS AND METHODS

Primer Sets

Two RP sets of 12 primers each were chosen from stocks of oligonucleotides that remained from previous projects (Table 1). All had annealing

Table 1. Random Primers

	Oligonucleotide	Sequence
Set "A"	R1A	CTAGCCAATGGAAGCTGTGAC
	R2A	CCCAGCTCTGGAAATATGGATG
	R3A	GGGAGGGCCTGCACTCAG
	R4A	ACAGGTGCCGAACATGCAC
	R5A	CATAGATGATGCAGCATTCAC
	R6A	CTGATGTGTCTCCTAGCGTG
	R7A	CTGTGGAAGTCGGATGCAGG
	R8A	GAAGGTGGGAGCTGATGAC
	R9A	CCTACAAGTGTTGTAGTTTTAGG
	R10A	GAGAGAAAATTATAAAGTAAATGTGG
	R11A	TTGTCACTTTCACATGTATTGA
	R12A	AACAGAAATCAGTAATGAGCTG
Set "B"	R1B	CAGTGTATTGGTCAGGATTCTC
	R2B	TCCCTCCGCAGTTCTGGAAG
	R3B	GGTTTGAGGATTAAGGCTGTC
	R4B	ATTGAATTTAAGTCTGATATGACC
	R5B	ACGGCAAGTTAGGGGCAGA
	R6B	CTCCTCCTCCTTTCTTCCTTCA
	R7B	CTAGGAGGATGCATATATTCTGG
	R8B	CCACCTCAACTCACTGACTC
	R9B	CAGAGGCGCCATTTACACTG
	R10B	GTCCTTGAGTGTGCTAGTGG
	R11B	TTGTGTTTCTGCCATAGCTGG
	R12B	TACCTTGGCTTCCCACAGTC

temperatures of 55°C. Selection was at random with a limited visual inspection of the primer sequences to increase the variety of sequence patterns chosen. The oligonucleotides were suspended at 5 µM and arrayed in two rows on a 96-well plate to allow rapid aliquoting with a 12-channel pipet. Sets of nested primers suitable for PCR were designed around the cloning sites of YAC and P1 vectors (Table 2). Single primers (Table 2) were placed on the T3 and T7 sites of the pBluescript® II (Stratagene, La Jolla, CA, USA) plasmid that was most commonly used.

PCR

PCRs were performed in a 20-µL volume. Reaction mixtures contained 200 µM deoxyribonucleoside triphosphates (dNTPs), 0.5 µM each primer, buffer (10 mM Tris-HCl, pH 8.3, 50 mM KCl, 1.5 mM $MgCl_2$, 0.001% gelatin) and 0.5 units of *Taq* DNA polymerase (Perkin-Elmer, Norwalk, CT, USA).

Table 2. Vector Primers

Vector	Location	Oligonucleotide	Sequence
YAC	Right arm, external	YRJ-E	TCTGGGAAGTGAATGGAGAC
	Right arm, nested	YRJ-N	G AGTCGAACGCCCGATCTC
	Left arm, external	YLJ-E	CATGGCGACCACACCCGTC
	Left arm, nested	YLJ-N	CAAGTTGGTTTAAGGCGCAAG
P1	T7-External	T7-PE	CATTGTAGGACTATATTGCTCTAA
	T7-Nested	T7-PN	CCGCTAATACGACTCACTATAG
	SP6-External	SP6-PE	TCCTCCCGAATTGACTAGTGG
	SP6-Nested	SP6-PN	GCCGTCGACATTTAGGTGAC
pBluescript	T7	T7-BLS	TACGACTCACTATAGGGCGAA
	T3	T3-BLS	GCGCAATTAACCCTC ACTAAAG

With YACs and P1s, a two-stage amplification was necessary. For sequencing plasmids, only a single stage was required. In the initial amplification, 10 ng of DNA were added per reaction. Amplification was performed with a GeneAmp™ PCR System 9600 Thermal Cycler (Perkin-Elmer) using 96-well trays. Each reaction mixture was aliquoted out into a row of 12 tubes on a tray; then a multichannel pipet was used to add the RPs. The reaction was carried out with 60 cycles of denaturation at 94°C for 10 s, annealing at 55°C for 15 s, and elongation at 72°C for 2 min with YACs and P1s, or for 20 s to 1 min with plasmids depending on the insert length and the length of the products desired. Denaturation time was kept to a minimum to preserve the viability of the *Taq* DNA polymerase.

If plasmids were being sequenced, the entire volume of each reaction was electrophoresed on a 1.5% agarose gel. Bands of interest were recovered and cleaned using a Geneclean® kit (Bio 101, La Jolla, CA, USA). For YACs and P1s, it was necessary to reamplify the first set of reactions to differentiate specific bands from the large number of nonspecific bands. In the second round of PCR, nested primers were used in place of the vector oligonucleotides to eliminate products that arose from nonspecific annealing. Since it was not possible to put nested primers within the RPs, products primed nonspecifically on both ends by a random primer reamplified as well. To identify them, a second reaction mixture was prepared for each template that contained no vector oligonucleotides. Since only the RPs were present, all products seen in these controls were primed nonspecifically.

For the next set of PCRs, a multichannel pipet was used to dilute aliquots of products from the first set of reactions 1:100 into a second 96-well plate for use as templates. The mixtures were set up as previously stated. The mixture with the nested primer was dispensed in one row, and the mixture that was to contain only RPs was aliquoted in a separate row below. RPs and 2 µL of 1:100 template were added with the multichannel pipet. These reactions were amplified for 40 cycles of 94°C for 10 s, 55°C for 15 s, and 72°C for 2 min.

The entire volume of each reaction was electrophoresed on a 1.5% agarose gel that was loaded so that nested primer reactions were run next to their RP-only mates. Bands common to both lanes were disregarded as nonspecific. Products present only in the nested-primer lanes were taken as candidates for insert ends. These were recovered from the gel and cleaned as previously described.

Sequencing

The cleaned products were resuspended in 5–20 µL of TE buffer (5.0 mM Tris-HCl, pH 8.0, 0.1 mM EDTA). Cycle sequencing was performed with a Cyclist™ Kit (Stratagene) using 2–5 µL template. YAC insert ends were verified by identifying appropriate vector sequence on the products. P1 insert ends were confirmed by comparison of sequence from the vector ends of products with that obtained by direct sequencing of the clones.

RESULTS AND DISCUSSION

This technique was developed to simplify the isolation of YAC insert ends that could not be obtained by A-V PCR. It was observed that in the absence of the appropriate template, PCR often produced nonspecific products that electrophoresed as distinct bands. Provided these artifacts did not contain the same priming site on both ends, they could be sequenced with the primers

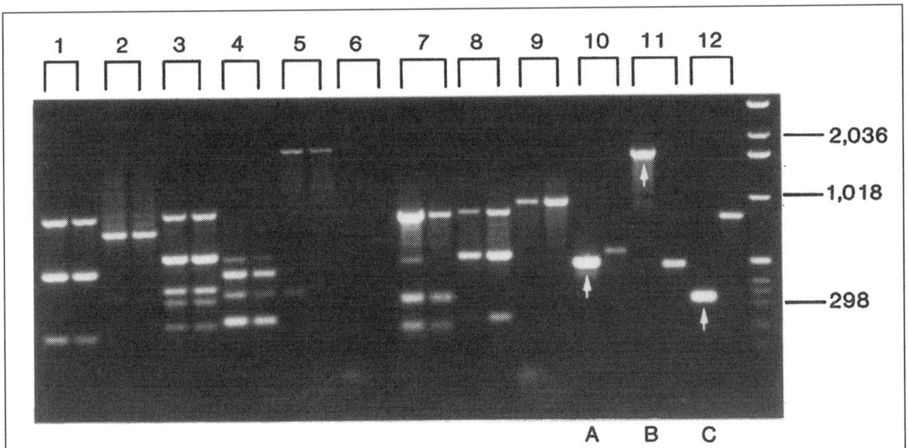

Figure 1. Reamplified random priming reactions from the right-arm end of a YAC insert. The numbers at the top are the random primer numbers (in this case the "A" set). Each primer is represented by two adjacent lanes on the gel. The left reaction contains both vector and random oligonucleotides, while the right reaction contains only the random oligonucleotide. With this gel, bands (A) and (B) were recovered and shown upon sequencing to be insert ends. (C) is a false band that appears with this primer on other YACs. The size marker is a 1-kb DNA ladder (Life Technologies, Gaithersburg, MD, USA).

Table 3. Results of Random Priming Trial on YAC Vector/Insert Junctions

ARM	YAC	Random Oligonucleotide Set	Products Analyzed	True End Products	Lengths of End Products (kb)	Oligonucleotide Priming End Product
Right	1	A	3	1	0.2	1A
	2	A	2	2	0.5/1.6	10A/11A
	3	A	1	0		
		B	2	1	0.7	12B
	4	A	0	0		
		B	3	1	1.5	12B
	9	A	1	0		
		B	1	1	1.3	7B
	10	A	2	0		
		B	2	1	0.9	8B
	11	A	0	0		
		B	0	0		
	12	A	1	1	1.3	10A
Left	5	A	2	1	0.6	5A
	6	A	4	3	0.4/0.7/1.0	2A/5A/10A
	7	A	1	1	0.2	7A
	8	A	1	1	1.3	8A
	9	A	2	2	0.4/0.45	1A/3A
	10	A	0	0		
		B	2	0		
	11	A	0	0		
		B	1	1	0.7	8B
	12	A	2	2	0.25/0.8	7A/10A

from which they were generated. Since this type of product required two different primers to anneal near to each other nonspecifically, it was postulated that if a true priming site were present for one oligonucleotide, then a randomly chosen second oligonucleotide might anneal close enough to the first to create a usable product. This would presumably be facilitated by the specific primer enriching the reaction with single copies of the desired template early in the reaction, making it more likely that an RP would anneal where a product could be formed.

After some success, a limited number of experiments were performed with PCR conditions to increase the production of specific products in the first set of reactions. The most significant factor found was the number of cycles performed. With 40 cycles, the initial set of reactions would produce a few faint bands from a given template. However, with 60 cycles the bands would become more numerous and much more apparent. This was especially important in sequencing plasmids, where only a single round of amplifications was performed.

Lowering the annealing temperature of the reactions was also attempted, but it was found on an experimental template that specific products that could be obtained at a higher annealing temperature did not appear. Presumably this was because the lower temperature decreased the specificity of the vector primers. Limited tests involving the use of multiple RPs in a reaction were less successful as well. The extra oligonucleotides interacted not only with the vector primers and themselves but also with each other; this caused a significant increase in the number of nonspecific products present. This effect was accentuated if shorter RPs were used. Keeping the stringency of the reactions high was important if small numbers of distinct products were to be obtained from a template. Overall, in the tests completed, normal PCR conditions with extra cycles consistently worked the best.

YACs

Since the technique evolved as it was developed, a new set of YACs was evaluated to determine its efficiency. A-V PCR had always proven to be the easiest way to obtain YAC insert ends, and it seemed that the RPT would be most valuable when used as a backup. Consequently, four vector/insert junctions that had failed to yield A-V products in a recent mapping project were chosen for testing on each YAC arm. Eight different YACs were represented (Table 3, YACs 1–8). Four additional YACs were selected at random (YACs 9–12) for which isolation of both ends would be attempted.

Each YAC was initially amplified using only the "A" primer set (Table 1). The "B" set was subsequently used on those that failed to yield ends. Products from 14 of 16 insert junctions were obtained and could be sequenced clearly (Figure 1, Table 3). Three of the right-arm ends were acquired with the "A" set and four with the "B" set. Six left-arm junctions were obtained with the "A" set and one was acquired with the "B" set. It was noted that two of the products had lengths that were near 200 bp. When vector and RP sequence were disregarded, approximately 100 bp where STSs could be placed remained. It was also noted that oligonucleotides "12A" and "1B" used with the right-arm primers often produced strong bands in their nested lanes that were of different sizes from those in the control lanes of their gels (Figure 1). These products appeared to be good candidates for insert ends, but because they were present on multiple YACs, they were recognized as artifactual and were not included for further analysis. In addition, since experience showed that prominent bands were the most likely to be useful, they were preferentially sampled, and some faint products were not evaluated.

Because the number of YACs in this sample is quite small, it is difficult to accurately predict how often the technique will work. To improve it, RPs can be chosen more carefully so artifacts such as those present with oligonucleotides "12A"and "1B" are eliminated. Primers that generate products more frequently can also be utilized preferentially. For example, oligonucleotide

"10A" primed 4 of the 19 confirmed YAC end products (Table 3). Additional sets of RPs can be used on YACs that do not yield ends with the first two sets.

If a large number of specimens are to be dealt with, it might prove useful to add a screening step so false products can be identified prior to sequencing. Aliquots of the cleaned PCR products can be dotted onto hybridization membranes. Oligonucleotides that anneal to vector sequence internal to the nested PCR primers can then be hybridized to these membranes to determine which products contain true insert ends.

P1s

When applying this technique to P1 ends, a reamplification step was again necessary. Although P1s were considerably less complex than YACs, there were still unacceptable numbers of nonspecific bands present after the first set of reactions. Using the SP6 and T7 nested primer sets on a trial of four P1s, an average of three products in the 0.5- to 1.5- kb range was produced on each end with the "A" RP set. Several smaller bands were also present. With P1s, the number of false products after a second amplification was minimal. Of 24 bands examined, all were found to be specific.

Plasmids

When the RPT was applied to DNA inserts in plasmids, most products present after the initial amplification were specific (Figure 2). The occasional exception was when the vector oligonucleotide annealed within the insert and created a product with the same primer on both ends. These products were easily recognized because they created a constant band across the gel. To avoid them, a different vector primer was used in the reactions.

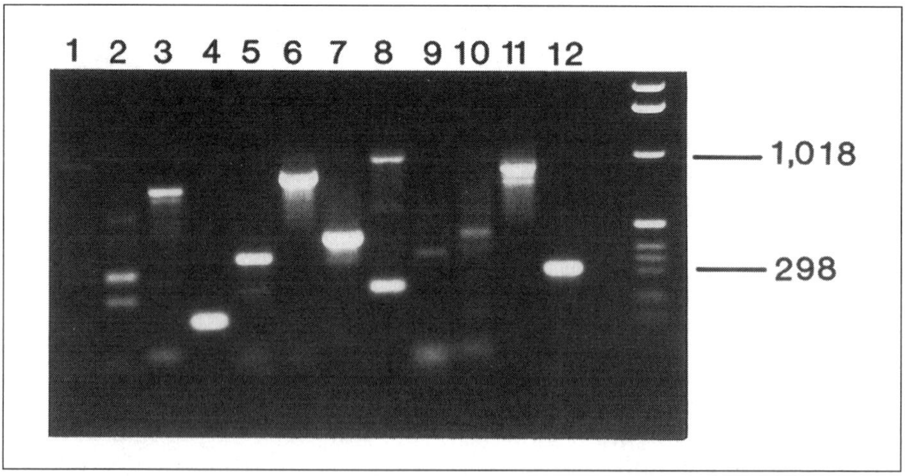

Figure 2. Random-primed products from a 1.5-kb cDNA insert cloned in pBluescript. The "A" random primer set was used along with the T3-BLS vector primer. The elongation step of the PCR was 45 s. A number of products could be recovered from the gel for sequencing

Using a set of 12 RPs with one vector oligonucleotide, it was generally possible to obtain two to six bands from a given plasmid that were distinct enough to be sequenced. It was important to consider the size of the inserts, since there were often random products long enough to reach the vector on the other end of the clone. While the elongation time of the PCR could be altered accordingly, for short inserts often only a percentage of the products were useful. Also, unlike deletion mutants, the products were generated randomly so they did not always give evenly spaced coverage across a clone. In practice they sometimes clustered. This could be an advantage in areas of complex sequence, because multiple products were often obtained that had different starting points and orientations. One useful method found to cover sequence gaps was to use the RPT on a random-primed product from a previous reaction that contained the desired region. This localized the sequence gap and often gave coverage in the desired area without having to synthesize oligonucleotides.

The RPT can be used for a number of purposes. Particularly appealing is its simplicity and its use of existing techniques and reagents. Advantage is taken of a PCR by-product that is normally considered a nuisance. While at least 36 reactions are generally required to obtain a particular YAC or P1 vector/insert junction, the use of a multichannel pipet allows these reactions to be set up with considerable speed. The repetitive nature of the technique should also easily lend itself to automation. It seems reasonable to assume that the method can be applied to vectors other than those described in this report.

More can be done to optimize the RPT. PCR conditions such as template concentration and buffer composition require further experimentation. In addition, specifically designed primers might be valuable, although an advantage of the method presented here is that leftover oligonucleotides can be utilized.

ACKNOWLEDGMENTS

I would like to thank Mark Skolnick, Yoshio Miki, Alexander Kamb and Arnold Oliphant for valuable discussions and helpful criticism during the preparation of this manuscript. This work was supported by NIH Grants CA-48711 and CA-55914.

REFERENCES

1. **Green, E.D. and M.V. Olson.** 1990. Chromosomal region of the cystic fibrosis gene in yeast artificial chromosomes: a model for human genome mapping. Science *250*:94-98.
2. **Henikoff, S.** 1984. Unidirectional digestion with exonuclease III creates targeted breakpoints for DNA sequencing. Gene *28*:351-359.
3. **Liu, Y.-G. and R.F. Whittier.** 1995. Thermal asymmetric interlaced PCR: automatable amplification and sequencing of insert end fragments from P1 and YAC clones for chromosome walking. Genomics

25:674-681.

4. **Miki, Y., J. Swensen, D. Shattuck-Eidens, P.A. Futreal, K. Harshman, S. Tavtigian, Q. Liu, C. Cochran, et al.** 1994. Isolation of a strong candidate for the 17q-linked breast and ovarian cancer susceptibility gene, *BRCA1*. Science *266*:66-71.
5. **Nelson, D.L., A. Ballabio, M.F. Victoria, M. Pieretti, R.D. Bies, R.A. Gibbs et al.** 1991. Alu-primed polymerase chain reaction for regional assignment of 110 yeast artificial chromosome clones from the human X chromosome: identification of clones associated with a disease locus. Proc. Natl. Acad. Sci. USA *88*:6157-6161.
6. **Neuhausen, S.L., J. Swensen, Y. Miki, Q. Liu, S. Tavtigian, D. Shattuck-Eidens et al.** 1994. A P1-based physical map of the region from D17S776 to D17S78 containing the breast cancer susceptibility gene *BRCA1*. Hum. Mol. Genet. *3*:1919-1926.
7. **Olson, M.V., J.E. Dutchik, M.Y. Graham, G.M. Brodeur, C. Helms, M. Frank et al.** 1986. Random-clone strategy for genomic restriction mapping in yeast. Proc. Natl. Acad. Sci. USA *83*:7826-7830.
8. **Pierce, J.C., B. Sauer and N. Sternberg.** 1992. A positive selection vector for cloning high molecular weight DNA by the bacteriophage P1 system: improved cloning efficacy. Proc. Natl. Acad. Sci. USA *89*:2056-2060.
9. **Silverman, G.A.** 1993. Isolating vector-insert junctions from yeast artificial chromosomes. PCR Methods Appl. *3*:141-150.

Update to:

PCR with Random Primers to Obtain Sequence from Yeast Artificial Chromosome Insert Ends

Jeff Swensen

University of Utah, Salt Lake City, UT, USA

Since the original paper was published, I have received informal comments from researchers who have had success with the technique. One criticism was that the nested right arm YAC primer (YRJ-N) was so close to the *Eco*RI cloning site that it was difficult to see the vector sequence required to confirm the specificity of a product. It might prove useful to move this primer farther from the cloning site.

I have recently used the RPT to obtain internal sequence from large inserts. The insert DNA was digested with a restriction endonuclease, then specific priming sites for the RPT were ligated onto the mixed fragments. The digests were cleaned, and then added to ligation mixes along with a fragment of DNA (a plasmid worked well) that had a compatible end and a known site for a PCR primer. Treatment of this fragment with calf intestinal phosphatase (CIP) was necessary. The RPT was performed on a 1:100 dilution of the ligation mixture by specifically priming off the ligated fragment. A number of products were obtained that could be sequenced. Generally, attempts to obtain PCR products across mixed inserts in plasmids result in products from the smaller inserts overrunning the reactions. The RPT gave products independent of the size of the inserts.

DNA Sequencing from Single Phage Plaques Using Solid-Phase Magnetic Capture

Shouwen Wang, Marie Krinks and Malcolm Moos, Jr.

Food and Drug Adminstration, Rockville, MD, USA

ABSTRACT

Many operations encountered in molecular cloning are labor-intensive and time-consuming. One case that is often troublesome is the subcloning of cDNA clones from λgt11 phage into plasmid vectors. In situations where several clones have been isolated, time could be saved by a means of assessing insert size and sequence unambiguously without subcloning, particularly where degenerate PCR or low-stringency hybridization approaches are taken to identify multiple members of a gene family. We describe a simple and reliable strategy for efficient sequencing of small amounts of λ phage DNA, lysates, or individual phage plaques. The strategy combines the advantages of universal λ phage primers, rapid air thermal cycling, streptavidin magnetic bead capture of highly purified single-stranded templates and the unparalleled clarity of T7 DNA polymerase sequence. We routinely obtain 350–500 bases of unambiguous sequence from each reaction. It takes only hours from lifting phage plaques to finishing the sequencing reactions. The method provides an alternative to thermal cycle sequencing that has comparable sensitivity and affords sequence data of much higher clarity.

INTRODUCTION

Subcloning of cDNA inserts from phage vectors into plasmids usually precedes DNA sequencing. When many clones must be characterized, this approach represents a formidable commitment of time and resources. Thermal cycle sequencing (5,8) has allowed analysis of small amounts of relatively impure template; bacterial colonies or phage plaques have sufficed in many cases. Nevertheless, the quality of the data and the length of the sequencing ladders generated in this way have seldom compared with those obtained with modified T7 polymerase (Sequenase®) and single-stranded templates.

A recently introduced commercial kit for sequencing plasmid DNA or

(Reprinted from BioTechniques 18:130-135)

bacterial colonies (3,4) is based on streptavidin magnetic beads. In this procedure, a biotinylated and non-biotinylated vector primer are used to amplify part of an individual colony. The product is immobilized on magnetic beads and washed, and the non-biotinylated strand is eluted with NaOH. Either the eluted or adsorbed strand may be used for single-stranded sequencing with conventional protocols. This method, while slightly more labor-intensive than colony cycle sequencing, routinely yields much cleaner signals and sequencing ladders of nearly double the length obtained with most cycle sequencing protocols. We reasoned that an analogous procedure for phage plaques, lysates or DNA might provide similar advantages.

In preliminary experiments, sequencing from purified λ phage DNA proved straightforward. Phage lysates required some purification in order to yield useful sequencing data. An attractive method for sampling phage plaques with fragments of nylon hybridization membrane was introduced by Wang and Cutler (9) and gave extremely high-quality data with a minimum of hands-on time. These elements were combined into an efficient strategy for characterizing multiple clones isolated while searching for novel growth factors by degenerate PCR. The strategy includes *(i)* membrane plaque-lifting, *(ii)* amplification of cloned inserts using biotinylated and non-biotinylated universal λ phage primers, *(iii)* preparation of solid-phase sequencing templates using streptavidin magnetic beads and *(iv)* sequencing using universal λ phage primers or gene-specific primers and modified T7 DNA polymerase (Sequenase).

MATERIALS AND METHODS

Sequencing Templates

Several λgt11 clones of a bone morphogenetic protein gene isolated from *Xenopus* were used as templates. Phage DNA was prepared from plate lysates by the QIAEX™ ion-exchange procedure (Qiagen, Chatsworth, CA, USA). Phage lysates were plated using top agarose to obtain individual plaques where necessary, including well-isolated plaques from tertiary screens. DNA was purified from phage lysates using a modified Qiagen procedure as follows: Three hundred microliters of phage lysate were treated with 4 µL (10 U/µL) of RNace-It™ ribonuclease cocktail (Stratagene, La Jolla, CA, USA) at 37°C for 30 min. Then 2 µL of proteinase K (20 mg/mL; Sigma Chemical, St. Louis, MO, USA) and 1/10 volume of 10× buffer (0.1 M Tris-HCl, pH 8.0, 50 mM EDTA, 5% sodium dodecyl sulfate [SDS]) were added, and the reaction was incubated at 50°C for 30 min. The protease was inactivated by incubating at 80°C for 15 min, and the phage DNA was purified using the QIAEX Gel Extraction Kit following Qiagen's instructions. With phage titers of 10^9 plaque-forming units per mL (pfu/mL), very pure DNA can be obtained that is sufficient for at least 20 sequencing reactions.

We recommend this method only with high phage titers (>10^9 pfu/mL). Plasmid DNA was prepared using conventional methods (7).

Phage Plaque Lifting

Single or confluent phage plaques were lifted using 1- × 1-mm to 2- × 2-mm pieces of nylon membrane (Boehringer Mannheim, Mannheim, Germany) by letting the membrane sit on a plaque for 15 s and transferring the membrane into microcentrifuge tubes containing the PCR mixture. Both round- and square-shaped nylon membranes (which can be cut using a clean razor blade) may be used. Membrane pieces are easy to handle using a sharp-tipped forceps (#5; A. Dumont & Fils, Montignez, Switzerland). To avoid contamination between samples, we recommend dipping the forceps in ethanol and flaming three times between liftings. The membrane pieces were then incubated in the PCR mixture at 94°C for 5 min before thermal cycling; the PCR solutions were transferred into 10- or 50-µL glass capillary tubes (Idaho Technology, Idaho Falls, ID, USA). A similar technique was used for lifting bacterial colonies.

PCRs and Preparation of Sequencing Template

Cloned DNAs in λgt11 phage were amplified using universal primer set A: biotinylated forward (5′-TGGCGACGACTCCTGGAGCCCG) and non-biotinylated reverse (5′-TGACACCAGACCAACTGGTAATGG); primer set B: non-biotinylated forward (5′-TCCAGTTCAACATCAGCCG) and biotinylated reverse (5′-TGACACCAGACCAACTGGTAATGG). Ten-microliter PCRs contained 1 µL of 10× buffer (500 mM Tris-HCl, pH 8.3, 2.5 mg/mL bovine serum albumin, 5% Ficoll® 400 [Pharmacia Biotech, Piscataway, NJ, USA], 10 mM tartrazine, 2 mM $MgCl_2$), 0.2 mM dNTPs, 10 pmol each of forward and reverse primers, 0.5 U of *Taq* DNA polymerase diluted in 1 µL of *Taq* diluent (2.5 mg/mL bovine serum albumin, molecular biology grade; Boehringer Mannheim) and phage templates. After preheating with the membrane plaque lifts, the PCRs were transferred to 10-µL glass capillary tubes, which were flame-sealed at both ends. Thermal cycling was carried out by preheating at 94°C for 1 min, cycling at 94°C for 0 s, 55°C for 0 s, 72°C for 2 min for 30 cycles, with a 5-min 72°C extension after the last cycle. The reactions can also be performed in 50-µL capillary tubes, but the denaturation and annealing times are increased from 0 to 10 s. A setting of 0 s is used to indicate that the samples are brought to the specified temperature momentarily before immediately proceeding to the next step in the cycle. Plasmid DNA inserts were amplified using the LacZ primer sets supplied in the Dynabeads® Template Preparation Kit (Dynal AS., Oslo, Norway). Aliquots of 5-µL PCR product were separated and visualized on a 1% agarose minigel containing 1 µg/mL of ethidium bromide in 1× TAE buffer (7). The gels were photographed using a Polaroid MP-4 camera and type 55 film (Polaroid,

Cambridge, MA, USA). Sequencing templates were prepared with 40 μL of PCR product using Dynabeads M-280 Streptavidin (Dynal AS.) as instructed by the manufacturer.

Non-recombinant phage contamination may occur when amplifying from phage plaques or lysates; there will be a small band of about 100 or 300 bp as well as the desired insert band. Some templates may give other nonspecific products that are often much smaller than the desired product. We therefore used spin columns (CLONTECH Laboratories, Palo Alto, CA, USA) to purify the PCR products to remove excess primers or nonspecific small PCR products before the magnetic bead separation step when small nonspecific product was present. The appropriate spin column to use depends on the size of the desired product and the presence or absence of undesired small products as assessed with an analytical minigel prior to the magnetic separation step.

Sequencing Reactions and Sequencing Gels

The λgt11 forward (5′-ATTGGTGGCGACGACTCC) and reverse (5′-CCAACTGGTAATGGTAGC) sequencing primers and gene-specific sequencing primers (unpublished data) were used for sequencing. For plasmid templates, which were inserted into either Bluescript® SK+ or pCR-Script™ (Stratagene), M13 forward and reverse sequencing primers were used (United States Biochemical, Cleveland, OH, USA). The 10-μL annealing reactions contained the magnetic bead immobilized template, 2 μL 5× Sequenase reaction buffer (United States Biochemical), and 2 pmol of sequencing primer and were performed in the PTC-100 thermal cycler by heating the reactions at 65°C for 2 min and then slowly cooling to 30°C over 40 min. The sequencing reactions were carried out by adding 4 μL of ^{35}Sequetide™ (Du Pont NEN, Boston, MA, USA) and 2 μL of Sequenase Version 2.0 (1:8 diluted in dilution buffer; United States Biochemical), incubating at room temperature for 3–4 min, aliquoting 3.5 μL of the above reactions into tubes containing 2.5 μL termination mixtures (United States Biochemical) and incubating at 37°C for 5 min. The reactions were stopped by adding 4 μL stop solution (United States Biochemical). Sequencing was done with isotopes no older than three weeks. If the labeled nucleotide was over 3–4 weeks, we noted poor sequence ladders. The samples were separated on 7%–8% Long Ranger™ gels (AT Biochem, Malvern, PA, USA) containing 1× TBE buffer, 6 M urea (enzyme grade; Life Technologies, Gaithersburg, MD, USA) and 25% deionized formamide (Oncor, Gaithersburg, MD, USA). The gels were run at about 50 watts, so that the running temperature was maintained between 41° and 45°C. Two loadings separated by 4–6 h may be performed in order to get more sequencing data from each set of sequencing reactions. The gels were fixed in a solution of 33% ethanol and 10% acetic acid for 30 min, vacuum-dried for 1–2 h and exposed to x-ray films (Kodak

XAR, Eastman Kodak Company, Rochester, NY, USA). Readable signal was usually obtained by exposing 12–16 h.

RESULTS

Optimization of Primer Concentrations

Since unincorporated biotinylated primers may compete with amplified template for binding sites on the streptavidin Dynabeads, it is important to minimize primer excess while maintaining efficient amplification. We tested several primer concentrations from 0.1 to 5 μM (Figure 1). The 0.5 μM concentration gave near maximal amplification in most instances; 1 μM was for subsequent experiments and has performed well in the experience accumulated to date.

Successful Amplification of Cloned DNA from Various λ Phage Preps

Initial attempts to amplify inserts from λ DNA gave variable results, often characterized by efficient amplification of nonspecific low molecular weight products and little or no amplification of the desired insert

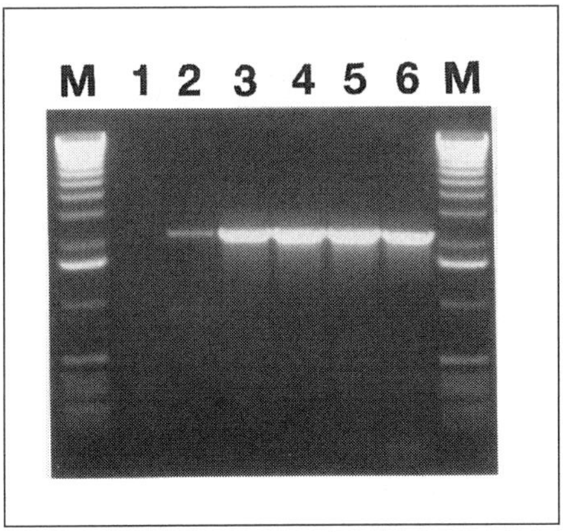

Figure 1. Optimization of λgt11 amplification primer concentration. The concentrations of each of the forward and reverse primers are 0 μm, 0.1 μm, 0.5 μM, 1 μM, 2.5 μM and 5 μM in lanes 1, 2, 3, 4, 5 and 6, respectively. Lane M is a 1-kb DNA ladder (Life Technologies); the amplified product is approximately 2.4 kb.

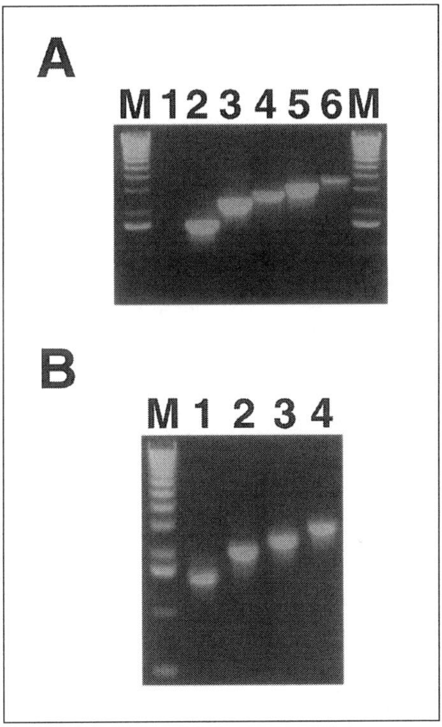

Figure 2. Universal amplification from various λ phage preps. (A) Products from a set of purified phage DNA templates. Lane 1 is a no-template control. (B) Products from several phage plaques. The templates used in lanes 2–5 in (A) and lanes 1–4 in (B) are from the same clones, allowing direct comparison of purified phage DNA and phage plaques as sequencing templates. Lane M in each panel is a 1-kb DNA ladder.

despite the efforts to vary Mg^{++}, annealing temperature, etc. We evaluated an air thermal cycler because of a report of improved specificity (10) and found it to produce good amplification from phage DNA consistently (Figure 2A). Inserts of up to 3.7 kb were amplified successfully.

We explored approaches to sequencing several types of templates encountered during a cDNA cloning project. Amplification of inserts by simply boiling the phage lysate was not reproducibly successful. The method described here allowed at least 20 sequencing reactions to be prepared from 0.3 mL of phage lysate with a titer of 1×10^9 pfu/mL (data not shown). Though detailed comparisons were not done, our experience suggests that lower phage titers do not yield enough product to sequence. It is also feasible to amplify the inserts using λgt11 universal primers and then use the amplified product as template for sequencing when very small amounts of phage DNA are available (data not shown). The amplifications using nylon membrane-lifted plaque as template were as successful as using purified λ phage DNA (Figure 2B).

Sequence Data

The data from various types of phage prep (Figure 3, Band C) were as good as those from single-stranded DNA prepared from purified plasmids with the Dynal Template Preparation Kit for sequencing (Figure 3A). The clarity, reproducibility and readability of data from phage plaques were consistently superior to those obtained with conventional double-stranded templates. With two loadings, we could read at least 350 bases and often determined up to 500 bases of sequence from each reaction. Even in cases where analytical electrophoresis revealed contaminating bands in the PCR prior to spin column treatment, very clean sequence was obtained (data not shown).

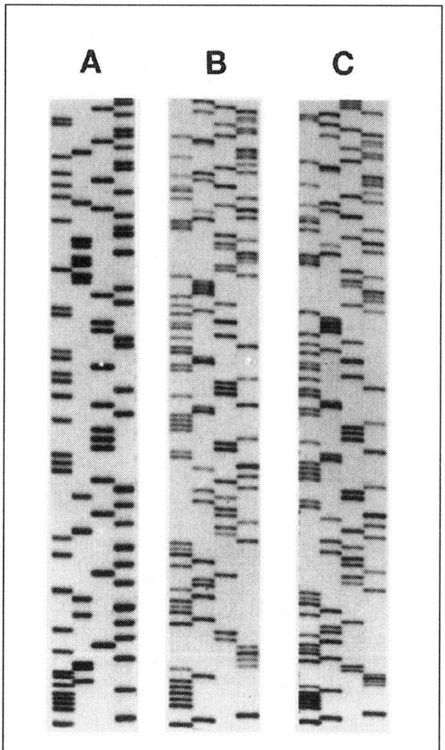

Figure 3. Sequence data from plasmid DNA and λ phage preps. An identical gene-specific sequencing primer was used for both plasmid and phage sequencing. Sequences shown are over 200 base pairs from the primer and the four lanes are in G-A-T-C order. (A) Sequence from plasmid DNA. (B) Sequence from purified phage DNA. (C) Sequence from phage plaques. The differences between sequences obtained with an identical primer reflect the presence of closely similar alternate alleles present in this pseudotetraploid organism.

DISCUSSION

We have extended the concept of magnetic capture, solid-phase DNA sequencing to the analysis of phage DNA, lysates and plaques. The key elements of our procedure are the use of the air thermal cycler, which enhanced amplification specificity and yield; spin columns to eliminate excess primers and unwanted nonspecific products; and rapid but effective methods to prepare PCR templates from phage lysates or plaques.

Since the operations described here can be carried out in parallel with many samples simultaneously, tremendous savings in time and labor can be realized, even if only a few clones need to be analyzed. In practice, the PCR amplifications are performed and an aliquot of each is run on an analytical gel. Attention can be focused on clones containing inserts of the desired size. If sufficient sequence information is available (e.g., from degenerate PCR experiments), gene-specific sequencing primers can establish the identity or lack of identity of the various clones and their membership in a particular gene family immediately. Universal λ phage sequencing primers can resolve more ambiguous situations. If gene-specific primers are available, they can be used in conjunction with λ phage primers to begin sequencing the entire clone from four points of extension, enabling acquisition of over 1000 bases from the initial set of reactions. Though the utility of universal amplification primers based on vector sequences is limited by decreasing amplification efficiency with increasing insert size, we were nevertheless able to amplify cloned inserts of up to 3.7 kb. With the long-distance PCR method reported recently (1), it is possible that this limitation will be overcome.

Rapid air thermal cycling not only shortens the time of amplification by decreasing the denaturation and annealing times to near zero, but may also increase the specificity and efficiency of some amplifications (10).

Magnetic capture of the biotinylated sequencing template (3,4) efficiently removes materials that may interfere with the sequencing reactions. The single-stranded templates obtained facilitate sequencing with T7 DNA polymerase and consistently yield sequence data of the highest quality.

Recently developed methods for thermal cycle sequencing based on native or modified thermostable polymerases have allowed rapid sequencing of small amounts of relatively crude templates (5,7). The data published thus far indicate that although these methods are rapid and sensitive, read length and clarity of data rarely, if ever, equal those obtained with modified T7 polymerase and single-stranded templates (4), and method-related artifacts may occur (6). Using the strategy described here, it takes as little as 4–5 h from lifting the plaques to finishing the sequencing reactions. Although the time requirement slightly exceeds the time for cycle sequencing methods, the advantages far outweigh the additional time spent. Since we can obtain approximately twice the data per reaction set, with clarity which matches that of the best classical methods, even for difficult templates, we sequence as described here whenever possible. Preliminary results indicate that related strategies will allow routine sequencing of PCR products such as those generated by rapid amplification of cDNA ends (RACE) (2) procedures.

ACKNOWLEDGMENTS

The authors wish to thank John Bishop and Mary Whiteley of CBER for their critical review of the manuscript.

REFERENCES

1. **Barnes, W.M.** 1994. PCR amplification of up to 35-kb DNA with high fidelity and high yield from lambda bacteriophage templates. Proc. Natl. Acad. Sci. USA *91*:2216-2220.
2. **Frohman, M.A., M.K. Dush and G.R. Martin.** 1988. Rapid production of full length cDNAs from rare transcripts: amplification using a single gene-specific oligonucleotide primer. Proc. Natl. Acad. Sci. USA *85*:8998-9002.
3. **Hultman, T., S. Stahl, E. Hornes and M. Uhlen.** 1989. Direct solid phase sequencing of genomic and plasmid DNA using magnetic beads as solid supports. Nucleic Acids Res. *17*:4937-4946.
4. **Hultman, T., S. Bergh, T. Moks and M. Uhlén.** 1991. Bidirectional solid-phase sequencing of *in vitro*-amplified plasmid DNA. BioTechniques *10*:84-93.
5. **Krishnan, B.R., R.W. Blakesley and D.E. Berg.** 1991. Linear amplification DNA sequencing directly from single phage plaques and bacterial colonies. Nucleic Acids Res. *19*:1153.
6. **Lee-Jackson, D., J. L. Abbruzzese and M. Bar-Eli.** 1993. Artifactual frame-shift *p*53 mutation at codon 249 detected with the Cyclist™ DNA sequencing method. BioTechniques *15*:363-364.
7. **Sambrook, J., E.F. Fritsch and T. Maniatis.** 1989. Molecular Cloning: A Laboratory Manual, 2nd Edition. Cold Spring Harbor Laboratory Press, Cold Spring Harbor, NY.
8. **Sears, L.E., L.S. Moran, C. Kissinger, T. Creasey, H. Perry-O'Keefe, M. Roskey, E. Sutherland and B.E. Slatko.** 1992. CircumVent™ thermal cycle sequencing and alternative manual and automated DNA sequencing protocols using the highly thermostable $Vent_R$® (exo^-) DNA polymerase. BioTechniques *13*:626-633.
9. **Wang, H. and A.J. Cutler.** 1992. A simple, efficient PCR technique for characterizing bacteriophage plaques. PCR Methods Appl. *2*:93-95.
10. **Wittwer, C.T. and D.J. Garling.** 1991. Rapid cycle DNA amplification: Time and temperature optimization. BioTechniques *10*:76-83.

Update to:

DNA Sequencing from Single Phage Plaques Using Solid-Phase Magnetic Capture

Shouwen Wang, Marie Krinks and Malcolm Moos, Jr.
Food and Drug Administration, Rockville, MD, USA

The ability to compare numerous clones without subcloning has afforded major savings in time and effort. Extended experience with the procedure allows us to make some additional recommendations.

After lifting phage plaques, the membrane pieces may be put into 1× PCR buffer or water instead of complete PCR mixture. The samples can then be heated at 94°–100°C for 5–10 min and stored at -20°C if necessary. This may save time if multiple sequencing reactions are to be done from the same template.

For most amplifications, spin column purification is not required. Nonspecific amplification products can usually be avoided with a primer concentration of 0.5 μM and appropriate amplification parameters.

PCR efficiency varies with template and other factors. If 30 cycles of amplification do not give enough product for sequencing, the number of cycles can be increased.

The technique has worked equally well for a wide variety of phage/primer combinations (e.g., λgt11, λZAP, λSHlox) with appropriate vector primers. With certain templates, we have observed stop artifacts (i.e., bands in all four lanes). A recent publication (1) has suggested that this difficulty may be ameliorated with carrier protein to prevent adsorption of the polymerase used in the sequencing reaction. While we have not yet evaluated this recommendation, the concept seems sound.

REFERENCE

1. **Lawson, V.A., D.A. McPhee and N.J. Deacon.** 1996. Elimination of sequence ambiguities by a single-step modification of a solid-phase, single-stranded sequencing protocol. BioTechniques *21*:356-358.

FINS (Forensically Informative Nucleotide Sequencing): A Procedure for Identifying the Animal Origin of Biological Specimens

Sylvia E. Bartlett and William S. Davidson

Bio-ID Corp. Ltd. and Department of Biochemistry, Memorial University of Newfoundland, St. John's, Newfoundland, Canada

ABSTRACT

The keys to identifying different species normally rely heavily on morphological characteristics. However, when an animal has been killed for food or sport, these markers are often destroyed or intentionally removed from the animal. This presents a problem for government agencies who are involved in determining the species origin of an animal or products derived from it in order to enforce conservation and/or health-related regulations. The problem is compounded if the meat of the animal has been processed in any way. We have developed a procedure called FINS (Forensically Informative Nucleotide Sequencing) that overcomes these problems. FINS has four components. First, methods have been developed that can isolate DNA from a wide range of biological samples including processed foods (e.g., canned, partially cooked, pickled, salted or smoked). Second, a specific segment of DNA is amplified using PCR. Third, the nucleotide sequence of the amplified segment of DNA is determined. Fourth, this nucleotide sequence is subjected to a phylogenetic analysis using a database, and the most closely related species is identified. FINS is a rapid, reliable and reproducible procedure that is based on established techniques. This procedure fills the need for an accurate method of determining the species identity of a specimen when this is not possible by conventional means.

INTRODUCTION

Many government agencies require a means of identifying the species origin of meat or other biological specimens. For example, law enforcement agencies need to know if a blood stain comes from a human or animal. Food inspectors check meat for accurate labeling for economic (e.g., selling cheap-

(Reprinted from BioTechniques 12:408-411, 1992)

er meats such as horse for beef) and religious (e.g., pork for beef) reasons. Wildlife officers must be able to identify the species type of meat to detect animals that have been obtained illegally (i.e., hunted without a permit, out of season or the quota exceeded). In addition, customs officers require methods for identifying restricted, endangered species or products derived from them. There is, therefore, a need for a rapid, accurate and reliable method for determining the species origin of a biological sample.

Currently there are three standard procedures for identifying the animal origin of meat, but each has major drawbacks. The most commonly recommended method (1) involves protein electrophoresis. The soluble proteins are detected by general protein or specific enzyme activity stains after electrophoresis using starch gels, polyacrylamide gels or cellulose acetate strips (12). This method only works with fresh or frozen tissue. Immunological procedures rely on antibodies raised against soluble proteins. Simple Ouchterlony analyses usually suffer from cross-reaction of proteins from related organisms. This procedure may indicate the family of organisms to which the specimen belongs, but rarely can it identify the species. Enzyme-linked immunosorbent assays have been developed that can detect pork or poultry in cooked and canned meat products (5). However, the polyclonal antibodies raised against the chicken heat-resistant antigens could not discriminate between turkey and chicken products. The last method depends on the ability to separate water-soluble muscle proteins by high-performance liquid chromatography (2,3). This procedure has the advantage of being able to estimate the relative proportions of raw meats in mixtures (2,3), but it is only applicable to fresh or freshly frozen meats.

The disadvantages of each of the methods outlined above are a consequence of their dependence on the ability to characterize proteins. Many proteins are heat-labile, lose their biological activity soon after death, are subject to modification in different cell types and their presence is a function of the cell type being examined. It would be preferable to analyze DNA rather than proteins for identifying the species origin of an unknown sample. DNA carries an organism's genetic information. It is the same in all cell types of an animal. Thus, it does not matter if one isolates DNA from a blood stain, a muscle extract or a liver sample. The information content of DNA is also greater than that of proteins. Information is lost as one goes from DNA to protein because of the degeneracy of the genetic code. Moreover, electrophoretic, immunological and chromatographic methods do not detect all the amino acid differences that may occur between different proteins (e.g., electrophoresis will only detect changes in the net charge of two proteins, and immunological techniques require different antigenic determinants to be present on the proteins under investigation). DNA is a remarkably stable molecule (17). It has been isolated from animal skins from the nineteenth century (10), mummified tissue (15) and a 7000-year-old human brain (16) (see Reference 17 for review). There are sensitive techniques available that permit the isolation of small amounts of crude preparations of nucleic acids from many diverse sources. In conjunction with PCR (19), specific segments of DNA can be amplified and their nucleotide sequences determined (13). Here we describe a procedure called FINS (Forensically Informative Nucleotide Sequencing). FINS offers a novel approach that can rapidly, reliably and reproducibly determine the species origin of a biological specimen, especially when traditional, protein-based methods are inadequate.

MATERIALS AND METHODS

Materials

Ground beef, ground pork, lamb chops, chicken breasts, a frozen turkey, partially-cooked battered cod, chicken nuggets and canned salmon were purchased from a local supermarket. Fresh cod fillets, smoked salmon, smoked mackerel, pickled herring and salt cod were obtained from Fishery Products International (St. John's, Newfoundland). Other biological samples (e.g., blood, tissues not mentioned above, formalin-fixed specimens, skins, etc.) were found in the Faculty of Science, Memorial University, St. John's, Newfoundland.

DNA Extraction

The sample was cut into small pieces and homogenized directly at room temperature with a glass-Teflon® homogenizer in a solution containing 4 M guanidinium thiocyanate, 25 mM sodium citrate, pH 7.0, 0.5% Sarkosyl®, and 0.1 M 2-mercaptoethanol. For every 10 mL of homogenate, 1.5 mL 2 M sodium acetate, pH 4.1, were added, and the solution was mixed with 10 mL of Tris-saturated phenol and 4 mL of chloroform/isoamyl alcohol (24:1, vol/vol). After shaking, the mixture was left on ice for 15 min and then centrifuged at 10 000× *g* for 20 min at 4°C. The aqueous phase was transferred to a fresh tube containing 10 mL of isopropanol. Solutions were mixed and left at -20°C for at least 1 h. Precipitated nucleic acids were obtained by centrifuging at 10 000× *g* for 20 min at 4°C. The pellet was washed with cold 70% ethanol, dried briefly under vacuum, dissolved in 10 mM Tris-HCl, 1 mM EDTA, pH 8.0, and stored at 4°C. This crude nucleic acid preparation was suitable for DNA amplification.

Synthesis of Oligonucleotide Primers

The oligonucleotide primers were synthesized using cyanoethyl phosphoramidite chemistry in the Biochemistry Department, Memorial University on a MilliGen/Biosearch cyclone DNA synthesizer (Burlington, MA, USA) in the dimethoxytrityl (DMT)-off mode. It was assumed that all steps in the synthesis and recovery of an oligonucleotide had efficiencies of 100%. It was not necessary to purify the oligonucleotides prior to use. The primers cyt bH and cyt bL are 5′CCCCTCAGAATGATATTTGTCCTCA3′ and 5′CCATCCAACATCTCAGCATGATGAAA3′, respectively (13).

Amplification of the Mitochondrial Cytochrome b Gene

The procedure for amplifying part of the mitochondrial cytochrome b gene from a crude nucleic acid extract obtained as described above was adapted from the method of Kocher et al. (13). Double-stranded amplifications were carried out in a final volume of 25 µL in 0.5-mL microcentrifuge tubes in a reaction cocktail containing 67 mM Tris-HCl, pH 8.8; 2 mM magnesium chloride; 10 mM 2-mercaptoethanol; 0.2 mM each of dATP, dCTP, dGTP and dTTP; 10 pmol of cyt bH and cyt bL; and 1 unit of AmpliTaq DNA Polymerase (Perkin-Elmer Cetus, Norwalk, CT, USA). Promega *Taq* polymerase from Fisher Scientific (Halifax, Nova Scotia, Canada) works equally well using buffer conditions suggested by the supplier. A drop of mineral oil was added to cover the reaction mixture, and 30 cycles of amplification (92°C for 45 s, 50°C for 45 s and 72°C for 90 s) were carried out using a Perkin-Elmer Cetus DNA Thermal Cycler. Fifteen microliters of the reaction mixture were electrophoresed at 4°C through a 2% NuSieve® (FMC BioProducts, Rockland, ME, USA) agarose gel made up and run in a 40 mM Tris, 30 mM sodium acetate buffer titrated to pH 7.4 with HCl. The DNA was visualized by staining with ethidium bromide, and the amplified fragment of DNA was excised from the gel and melted in 100 µL of sterile, deionized water. The sample was heated to 65°C for 5 min and mixed vigorously. Five microliters of this solution were used to prepare a single-stranded template by asymmetric amplification (9). The reaction was carried out in 100 µL of the same reaction cocktail and under the same cycling conditions as those used for the double-stranded amplification (see above) except that 40 pmol of the cyt bH primer were used and only 0.4 pmol of the cyt bL was present. Five microliters of the reaction were sufficient to visualize double- and single-stranded DNA bands following electrophoresis through a 2% agarose gel and staining with ethidium bromide. Residual nucleotides, primers and buffer salts were removed from the single-stranded template by centrifugal dialysis using Centricon-30 filters (Amicon Ltd., Oakville, Ontario, Canada). Two milliliters of water were added to the reaction mixture, and the solution was applied to the filtration apparatus and centrifuged at 3500× *g* for 20 min at 20°C. An additional 2 mL of water were added, and the centrifugation repeated. Another 2 mL of water were added, and the centrifugation was increased to 1 h until the volume in the chamber was approximately 60 µL. Single-stranded DNA template was obtained in this final volume, and 7 µL were taken for DNA sequencing.

DNA Sequencing Reactions

Nucleotide sequencing of the single-stranded template was carried out using Sequenase® and a commercial DNA sequencing kit purchased from United States Biochemical, Cleveland, OH, USA. One picomole of cyt bL was used as the sequencing primer and ^{35}S dATP was used as the label. The sequencing reactions were performed according to the manufacturer's instructions except that the elongation mixture was diluted 50-fold. The sequencing products were separated on 6% polyacrylamide denaturing gels (18). The primers amplify a 307-bp (excluding size of primers) segment of the cytochrome b gene (13), and these procedures routinely allowed the sequence of 290 nucleotides of the non-coding strand to be determined.

RESULTS AND DISCUSSION

FINS is composed of four steps. If FINS is to be used to identify the species origin of a specimen, conditions must be found that allow each step to be carried out routinely and reproducibly. The steps are as follows: first, DNA is extracted from the biological sample; second, a specific segment of the isolated DNA is amplified; third, the nucleotide sequence of the amplified DNA is determined; and fourth, a phylogenetic analysis is carried out on the nucleotide sequence to determine its closest relative(s) in a database.

Extraction of DNA

The most common procedure for isolating DNA uses protease digestion in the presence of the anionic detergent sodium dodecyl sulfate (SDS), followed by organic extraction (phenol and/or chloroform) and then ethanol precipitation of the DNA. This method has been very successful, and high molecular weight DNA is usually obtained from fresh tissue. However, many of the specimens that one encounters in forensic work are far from fresh, and meat inspectors and game wardens often have to deal with meat that is "well hung" or has been processed. We have recently tested several procedures for isolating "food quality" DNA from fish tissues (4). The one that routinely yielded DNA capable of being amplified by PCR was a modification of a method designed to isolate RNA (8). A modification of this method has been used to prepare DNA for fingerprint analysis (23). In the samples that we examined, the nucleic acids in the final preparation were usually less than 5 to 7 kbp in length (estimated by ethidium bromide staining after agarose gel electrophoresis). The protocol outlined above is easily scaled down for small amounts of sample. It is rapid, and amplifiable DNA was obtained from a wide variety of samples including processed meats (Table 1).

Which DNA Segment to Amplify?

Several factors had to be considered before a specific segment of DNA was selected for amplification. This region of the genome has to accept

Table 1. The FINS Procedure Was Successful in Determining the Species Identity of Biological Samples Processed in the Following Ways

Process	Samples Tested
Freshly frozen	Many mammals, birds, fish
Ground meat	Beef, pork
Smoked	Salmon, mackerel
Pickled	Herring
Canned	Salmon
Salted	Cod
Partially cooked	Battered cod, chicken nuggets
Formalin-fixed	Salmonids
Ethanol preservation	Salmonids
Blood smears on slides	Birds
Preserved skins	Mammals, birds

mutations quickly enough so that closely related organisms have different sequences, but not so rapidly that the amount of variation within a species is substantial. The length of segment had to be long enough to permit the detection of sequence differences between congeneric species, but short enough that its nucleotide sequence could be determined from a single loading of a standard DNA sequencing gel. A protein coding region of a gene was preferred because amplification and/or sequencing errors could be detected by translating the nucleotide sequence and comparing it with known amino acid sequences of the gene. A knowledge of the pattern of mutations in the gene would be an asset as would information on the structure and function of the corresponding protein, especially invariant amino acids. This would also facilitate comparing the nucleotide sequence from a sample with those in a data bank. Moreover, the creation of a data bank is a lengthy process, and so it was desirable to take advantage of a gene whose sequence had been determined in many diverse organisms. The mitochondrially encoded gene for cytochrome b satisfied these criteria admirably.

Part of the nucleotide sequence of the gene for cytochrome b has been determined for many vertebrates (13) including mammals (11), birds (6) and fish (4,7,14). Thus, a substantial database already exists. Moreover, the cytochrome b gene has become very popular with population geneticists and molecular systematists. For this reason the amount of sequence information on this gene is expanding rapidly, and this is likely to continue. The pattern of nucleotide substitutions has been examined in mitochondrial genes coding for proteins (22), and it has been shown that the cytochrome b gene follows the typical pattern with silent changes predominating and transitions being more abundant than transversions (11,13). The amino acid sequence of cytochrome b is highly conserved, but because of the degeneracy of the genetic code, the genes for cytochrome b differ by at least a few nucleotides even in very closely related species. Inter- and intra-specific studies have also shown that the amount of variation within a species is less than between species (4,6,14,20). A mitochondrially encoded gene has the additional advantage of behaving as if it were a haploid locus, and therefore the problem of heterozygosity is avoided (heteroplasmy being rare in mitochondria). Another major factor in favor of the cytochrome b gene was the observation by Kocher et al. (13) that one pair of primers could be used to amplify a 309-bp segment of the cytochrome b gene from any vertebrate that they tested. Thus, primer design was not a problem. The cytochrome b gene corresponding to amino acids 33 to 134 is the segment of choice for FINS.

Nucleotide Sequence Analysis

No single nucleotide sequence of the cytochrome b gene is diagnostic of a particular species. This is because of the nature of mutations in mitochondrial genes coding for proteins and intra-specific variations (e.g., see References 4, 6 and 7 for intra-specific variation in tuna, thick-billed murre and cod, respectively). This must be taken into account in FINS because it will be impossible to sample enough individuals from all the species that one may wish to have in a data bank to cover all the genetic variation within a species. Therefore, the procedure that is used to overcome this potential problem relies on a phylogenetic analysis. This is a form of pattern recognition and makes use of all the information in a sequence. Fortunately, this field of research is well established in molecular evolutionary studies, and computer programs are available (e.g., PAUP) (21). This procedure also allows reversions and new mutations to be taken into account. Moreover, if the species to which the specimen belongs is not represented in the database, the phylogenetic analysis will reveal the most closely related species. At this point it would be possible to obtain additional standard species and to expand this section of the data bank. It has been our experience that the amount of intraspecific variation does not preclude the identification of even closely related species (Bartlett, Birt, Birt-Friesen, McVeigh and Davidson, unpublished observations). However, there may be some very closely related organisms that are regarded as separate species based on morphological characteristics, which are not genetically distinguishable at the molecular level. In these instances, it will not be possible to use FINS or any other biochemical genetic method to distinguish members of these species from one another.

Advantages of FINS

FINS is a rapid, accurate and reproducible procedure that is based on established techniques. It is not subject to operator bias and can be performed independently in any lab equipped to carry out standard molecular biology. Thus, an analysis could be confirmed in any other diagnostic lab. FINS produces genetic evidence, and this should be acceptable to the legal system.

Court cases that have previously been hampered by the lack of genetic evidence can now proceed. For example, FINS allows the discrimination between bluefin tuna (a regulated species) and yellowfin tuna (an unregulated species), whereas this was not achieved by protein electrophoretic procedures (4). Therefore, FINS fills the need for a reliable method for determining the biological identity of a specimen when this is not possible by conventional means.

ACKNOWLEDGMENTS

We thank the following for helpful discussions and for sharing their results with us: T. Birt, V. Birt-Friesen, S. Carr, M. Cutler, J. Goodier, G. Hughes, D. Marshall, C. McGowan and H. McVeigh. Financial support for this work came from the Department of Fisheries and Oceans, the Centre for Fisheries Innovation and the Natural Sciences and Engineering Research Council of Canada.

REFERENCES

1. **AOAC.** 1984. Official Methods of Analysis, 14th Edition, 18.108-18.113, p. 349. Association of Official Analytical Chemists, Inc., Arlington, VA.
2. **Ashoor, S.H., W.C. Monte and P.G. Stiles.** 1988. Liquid chromatographic identification of meats. J. Assoc. Off. Anal. Chem. *71*:397-403.
3. **Ashoor, S.H. and M.A. Osman.** 1989. Liquid chromatographic quantitation of chicken and turkey in unheated chicken-turkey mixtures. J. Assoc. Off. Anal. Chem. *71*:403-405.
4. **Bartlett, S.E. and W.S. Davidson.** 1991. Identification of *Thunnus* tuna species by the polymerase chain reaction and direct sequence analysis of their mitochondrial cyctochrome b genes. Can. J. Fish. Aquat. Sci. *48*:309-317.
5. **Berger, R.G., R.P. Mageau, B. Schwab and R.W. Johnston.** 1988. Detection of poultry and pork in cooked and canned meat foods by enzyme-linked immunosorbent assays. J. Assoc. Off. Anal. Chem. *71*:406-409.
6. **Birt-Friesen, V.L., W.A. Montevecchi, A.J. Gaston and W.S. Davidson.** 1992. Genetic structure of thick-billed murres (*Uria lomvia*) populations examined using direct sequences analysis of amplified DNA. Evolution *46*:267-272.
7. **Carr, S.M. and D. Marshall.** 1991. Detection of intra-specific DNA sequence variation in the mitochondrial cyctochrome b gene of Atlantic cod (*Gadus morhua*) by the polymerase chain reaction. Can. J. Fish. Aquat. Sci. *48*:48-52.
8. **Chomczynski, P. and N. Sacchi.** 1987. Single step method of RNA isolation by acid guanidinium thiocyanate-phenol-chloroform extraction. Anal. Biochem. *162*:156-159.
9. **Gyllensten, U. and H. Erlich.** 1988. Generation of single-stranded DNA by the polymerase chain reaction and its application to direct sequencing of the HLA-DQA locus. Proc. Natl. Acad. Sci. USA *85*:7652-7656.
10. **Higuchi, R.G., B. Bowman, M. Freiberg, O.A. Ryder and A.C. Wilson.** 1984. DNA sequences from the quagga, an extinct member of the horse family. Nature *312*:282-284.
11. **Irwin, D.M., T.D. Kocher and A.C. Wilson.** 1991. Evolution of the cytochrome b gene of mammals. J. Mol. Evol. *32*:128-144.
12. **Kim, H. and L.A. Shelaf.** 1986. Characterization and identification of raw beef, pork, chicken and turkey meats by electrophoretic patterns of their sarcoplasmic proteins. J. Food Sci. *51*:731-735.
13. **Kocher, T.D., W.K. Thomas, A. Meyer, S.V. Edwards, S. Pääbo, F.X. Villablanca and A.C. Wilson.** 1989. Dynamics of mitochondrial DNA evolution in animals: amplification and sequencing with conserved primers. Proc. Natl. Acad. Sci. USA *86*:6196-6200.
14. **McVeigh, H.P., S.E. Bartlett and W.S. Davidson.** 1991. Polymerase chain reaction/ Direct sequence analysis of the cytochrome b gene in *Salmo salar*. Aquaculture *95*:225-233.

15.**Pääbo, S.** 1985. Molecular cloning of ancient Egyptian mummy DNA. Nature *314*:644-645.
16.**Pääbo, S., J.A. Gifford and A.C. Wilson.** 1988. Mitochondrial DNA sequences from a 7000-year old brain. Nucleic Acids Res. *20*:9775-9787.
17.**Pääbo, S., R.G. Higuchi and A.C. Wilson.** 1989. Ancient DNA and polymerase chain reaction. The emerging field of molecular archaeology. J. Biol. Chem. *264*:9709-9712.
18.**Saiki, R.K., D.H. Gelfand, S. Stoffel, S.J. Scharf, R. Higuchi, G.T. Horn, K.B. Mullis and H.A. Erlich.** 1988. Primer-directed enzymatic amplification of DNA with a thermostable DNA polymerase. Science *239*:487-491.
19.**Sanger, F., S. Nicklen and R. Coulson.** 1977. DNA sequencing with chain-terminating inhibitors. Proc. Natl. Acad. Sci. USA *74*:5463-5467.
20.**Smith, M.F. and J.L. Patton.** 1991. Variation in mitochondrial cytochrome b sequence in natural populations of South American akodontine rodents (Muridae: Sigmodontinae). Mol. Biol. Evol. *8*:85-103.
21.**Swofford, D.S.** 1989. PAUP: phylogenetic analysis using parsimony, Version 3.0b. Illinois Natural History Survey, Champaign, IL.
22.**Thomas, W.K. and A.T. Beckenbach.** 1989. Variation in salmonid mitochondrial DNA: evolutionary constraints and mechanism of substitution. J. Mol. Evol. *29*:233-245.
23.**Turner, B.J., J.E. Elder and T.F. Laughlin.** 1989. DNA fingerprinting of fishes: a general method using oligonucleotide probes. Fingerprint News *4*:15-16.

Update to:

FINS (Forensically Informative Nucleotide Sequencing): A Procedure for Identifying the Animal Origin of Biological Specimens

William S. Davidson and Sylvia E. Bartlett

Bio-ID Corp. Ltd. and Department of Biochemistry, Memorial University of Newfoundland, St. John's, Newfoundland, Canada

The initial driving force behind the development of the FINS (Forensically Informative Nucleotide Sequencing) technique (4,5) was the difficulty of enforcing the quotas and regulations governing the bluefin tuna fishery off the east coast of Canada. Four commercially important tuna species in the genus *Thunnus* are caught in the Atlantic near Nova Scotia. The harvest of bluefin (*T. thynnus*) is regulated whereas those of bigeye (*T. obesus*), yellowfin (*T. albacares*) and albacore (*T. alalunga*) are not. With the value of an individual bluefin tuna ranging from $1,000 to $20,000 in 1990 on the Japanese sushi market, it was not surprising that a black market in illegally caught bluefin tuna arose, especially as it is very difficult to tell a small bluefin from the other three species. The usual characters used to differentiate among these species are skeletal and visceral in nature and these identifying features are removed once the fish has been landed and gutted. Protein studies using isoelectric focussing (IEF) of water soluble muscle proteins revealed identical patterns for bluefin and yellowfin, which differed from bigeye and albacore, which were also identical (3). We therefore turned to PCR/direct sequencing analysis of a segment of the mitochondrially encoded cytochrome b gene as a means of distinguishing these tuna species in the genus *Thunnus*. Although variation was observed within each of the four tuna species, distinct differences that characterize each of the species were found. These species-specific markers in the nucleotide sequence of the cytochrome b gene make it possible to use this procedure and a phylogenetic analysis to determine unambiguously the species identity of muscle, or other tissue, from an individual tuna. This method was adopted by Canada's Department of Fisheries and Oceans and, after a few test cases, fishermen were made aware of the procedure and the illegal fishing of bluefin tuna essentially stopped.

FINS offered a novel approach that could rapidly, reliably, and reproducibly determine the species origin of a biological specimen when traditional protein-based methods are inadequate. FINS is composed of four steps.

First, DNA is extracted from the sample. Second, a specific segment of the isolated DNA is amplified using PCR and universal cytochrome b primers (12) or another pair designed specifically for the species of interest. The third step is the determination of the nucleotide sequence of the amplified DNA, and the fourth is a comparison of this sequence with standard sequences in a database. FINS is a fast, accurate procedure that is not subject to operator bias and can be performed independently in any lab equipped to carry out simple standard molecular biology. This means that there is no need for standards to be run alongside the sample. The method has been accepted by the court system in Canada and is the method of choice for Canada's Department of Fisheries and Oceans and the Canadian Wildlife Service if there is any question concerning the species identification of a biological sample.

The FINS procedure was used to examine a supposed emu shish kebab purchased from a local restaurant (10). The cytochrome b sequence obtained from the DNA isolated from the "emu" shish kebab bore little resemblance to that of authenticated emu (<70% identity) and a comparison of this sequence with cytochrome b sequences from the 14 species most commonly encountered in the meat trade in Australia indicated that the "emu" shish kebab was most likely made from buffalo meat (>97% identity). The FINS process has also been used to determine the molecular genetic identification of whale meat and dolphin products from commercial markets in Korea and Japan (1,2). Instead of the cytochrome b gene, these researchers decided to use a 550-bp fragment of the mitochondrial DNA control region. This is the most variable segment of the mitochondrial genome and the rationale for choosing this region rather than the cytochrome b gene was that it could potentially reveal the populations from which the samples were derived in addition to the species. This proved to be the case in some instances. A total of 34 sequences were obtained from 31 products and at least eight species were represented. Some of the species identified in the products from the Japanese and Korean commercial markets were not consistent with recent catch records made available to the International Whaling Commission. The results of this survey raised questions concerning the legality of some of the products tested and point to the need for a standardized form of commercial labeling of whale products. It also suggested that the international and domestic management of whale products was not being practiced as it should be. The FINS process is regarded as an important tool for monitoring the regulations covering the international trade in whale products and the domestic sale from unregulated local whaling or fisheries by-catch. As with the bluefin tuna problem that existed in Canada, the realization that there is now a reliable method for detecting cheating in the whaling industry might prove useful in conserving these mighty mammals of the sea.

The identification of the species in canned tuna presents a special problem. Unlike canned salmon that is usually smoked and then canned without much in the way of heat treatment, tuna is first cooked thoroughly and then autoclaved in the can. This leads to extensive degradation of the DNA in the sample, and DNA isolated from commercial cans of tuna is no larger than 300 bp (13), with the average size being about 100 bp (16). This precludes the use of the cytochrome b primers that flank the 307-bp region of the cytochrome b gene for use in the FINS process with these samples. A comparison of cytochrome b sequences from a wide variety of tuna and tuna-like species (6) provided the information for the design of primers that amplified a 59-bp fragment (123 bp including the primers) (16). This allowed the FINS procedure to be carried out on 30 commercially distributed cans of tuna and it was possible to determine a reliable sequence in each case. However, although this region of the cytochrome b gene provides unambiguous molecular markers for the identification of nine of the eleven tuna species commonly used by the canning industry, it was not able to distinguish between *Thunnus thynnus* (bluefin) and *Thunnus albacares* (yellowfin) tuna. This turns out to be a major impediment for the use of these primers as five of the 30 cans yielded the sequence corresponding to the indistinguishable species. Le Roux et al. (13) independently solved the problem of working with highly degraded tuna DNA from cans. They designed two pairs of primers that amplify a 125-bp fragment and a 101-bp fragment, respectively. The information from both of these sequences is sufficient to identify unambiguously all the species that one might expect to encounter in cans of tuna. These primers were successful when tested with skipjack tuna and yellowfin tuna that had been processed to extreme conditions of sterilization. These minor modifications of the original FINS process illustrate the widespread utility of this procedure.

Having to sequence the amplified product as part of the FINS process to identify the species from which a sample is derived is both time-consuming and expensive. It is, however, necessary if one wishes to obtain an unambiguous identification. Often it is sufficient to show the absence of a species in a product (e.g., no pork present). In other cases it is desirable to have methods that can be used as a rapid screen for detecting possible violations of regulations. By a judicious choice of primers, it should be possible to design primers that are specific for one species and only that species. This twist on the use of universal primers has been used to detect pork in heated meat products (14). As the absence of a product is taken as a positive result for no contamination of a sample by pork, it is necessary to have a control that can assess the potential of the extracted DNA from the sample to amplify in a PCR assay. A highly conserved 137-bp fragment corresponding to part of the 18S ribosomal RNA gene provided this important control and it is worth noting that this fragment is larger than the diagnostic segment. If a positive result is observed, the fragment can then be sequenced (as in the FINS process) to confirm that it is indeed pork and not some other spurious contaminant.

A similar but more extensive approach has been taken to determine the species designations of commercially available lots of caviar (9). The increasing demand for caviar has had a serious impact on the population sizes of commercial sturgeon species. The high cost of the product (> $2 per g) has increased the likelihood of poaching or the misrepresentation of one type of caviar for another. The traditional method for identifying the species of a particular shipment of caviar has relied upon such factors as egg size, appearance, smell, texture and color. The sequences of portions of the cytochrome b genes and 12S and 16S ribosomal RNA genes were compared from all extant sturgeons and paddlefishes. This permitted the design of pairs of primers specific for diagnostic nucleotide substitutions for each of the three main commercial Russian sturgeon species: beluga sturgeon (*Huso huso*); Russian sturgeon (*Acipenser guelenstaedt*); and sevruga (*Acipenser stellatus*). A survey of 23 commercially available lots of caviar purchased in New York found that five of them were misrepresentations. Three of these samples came from species that are threatened with extinction. The development of this procedure provides wholesalers with an alternative to the crude identification methods presently in use and should assist international conservation efforts to preserve what is left of commercial sturgeon populations. In addition, it should help protect the other species that are likely to be substituted as the commercial species disappear.

A PCR single-strand conformational polymorphism (PCR-SSCP) assay can also be used as a rapid first screen for species identification. It has been shown that this method is able to discriminate many species of fish using small samples such as eggs or larvae (11). PCR-SSCP has also been used for the species identification of canned tuna (15). Two different regions of the cytochrome b gene were selected for amplification, the 123-bp region described above (16) and another region that gave a product of 148 bp. Separation of the denatured 123-bp fragment on 7.5 or 10% polyacrylamide gels revealed species-specific patterns for six different tunas and bonitos that were different from the pattern for bluefin and yellowfin tuna, which, as expected, were identical. The universal application of PCR-SSCP has been demonstrated by Pharmacia Biotech (Piscataway, NJ, USA; Application Note 18-1108-90). In this report it was shown that the 307-bp segment of the cytochrome b gene can be used to resolve deer species, poultry, pork, beef and lamb, and fish species. PCR-SSCP provides a rapid screening method for the analysis of the species origin of a biological sample. The main disadvantage of this technique is that, like IEF of proteins, it is necessary to run reference standards and samples side by side on the same gel. The information content of a PCR-SSCP result is obviously less than that of the FINS procedure. However, it is possible to cut out bands from the PCR-SSCP gel and determine the nucleotide sequence of them after reamplification. By this means, the two methods complement one another. PCR-SSCP can also indicate that a sample contains mixtures from different organisms. Here again species identification would be possible by examining the DNA sequences of the multiple bands.

Patenting of the FINS procedure resulted in considerable controversy (7,8). In 1990 Davidson and Bartlett, in a joint venture with their employer

Memorial University, set up a company called Bio-ID Corporation to provide FINS and other DNA typing procedures as a service primarily for government agencies. It was decided that a patent should be filed on the FINS process and this was done in 1990. Bio-ID Corporation then entered into an agreement with Atlangene of Nantes, and this French company obtained the rights to market the FINS process in Europe. It should be noted that no royalties are ever sought for the use of patented processes in a purely research setting. Despite this, researchers at the Ministry of Agriculture Fisheries and Food (MAFF) Lab in Aberdeen, Scotland, who had obtained a grant from the European Community in 1994 essentially to reinvent the FINS work of Bartlett and Davidson (3,4) on tuna species, placed an Internet appeal for researchers of the world to unite to fight the Bio-ID FINS patent. The European FINS patent was awarded to Bio-ID Corporation in 1996 but MAFF, through the British government, continued to challenge the patent; eventually their financial and legal resources were too much for a small biotech company like Bio-ID and the patent was allowed to lapse. As a footnote to this highly publicized incident (7,8), it is interesting to note that the MAFF lab in Aberdeen was closed in late 1996.

REFERENCES

1. **Baker, C.S. and S.R. Palumbi.** 1994. Which whales are hunted? A molecular genetic approach to monitoring whaling. Science *265*:1538-1539.
2. **Baker, C.S., F. Cipriano and S.R. Palumbi.** 1996. Molecular genetic identification of whale and dolphin products from commercial markets in Korea and Japan. Mol. Ecol. *5*:671-685.
3. **Bartlett, S.E. and W.S. Davidson.** 1991. Identification of *Thunnus* tuna species by the polymerase chain reaction and direct sequence analysis of their mitochondrial cytochrome b genes. Can. J. Fish. Aquat. Sci. *48*:309-317.
4. **Bartlett, S.E. and W.S. Davidson.** 1992. FINS (forensically informative nucleotide sequencing): a procedure for identifying the animal origin of biological specimens. BioTechniques *12*:408-411.
5. **Bartlett, S.E. and W.S. Davidson.** 1992. Erratum. BioTechniques *13*:14.
6. **Block, B.A., J.R. Finnerty, A.F.R. Steward and J. Kidd.** 1993. Evolution of endothermy in fish: mapping physiological traits on a molecular phylogeny. Science *269*:210-214.
7. **Butler, D.** 1995. Patent row splits fish identity researchers. Nature *377*:9.
8. **Coghlan, A.** 1995. Fishy patent causes stink in academe. New Scientist *2 September*:7.
9. **DeSalle, R. and V.J. Birstein.** 1996. PCR identification of black caviar. Nature *381*:197-198.
10. **Forrest, A.R.R. and P.R. Carnegie.** 1994. Identification of gourmet meat using FINS (forensically informative nucleotide sequencing). BioTechniques *17*:24-25.
11. **Hara, M., M. Noguchi, E. Naito, K. Dewa and H. Yamanouchi.** 1994. Ribosomal RNA gene typing of fish genome using PCR-SSCP method. Bull. Jap. Sea Natl. Fish. Res. Inst. *44*:131-138.
12. **Kocher, T.D., K.K. Thomas, A. Meyer, S.V. Edwards, S. Paabo, F.X. Villablanca and A.C. Wilson.** 1989. Dynamics of mitochondrial DNA evolution in animals: amplification and sequencing with conserved primers. Proc. Natl. Acad. Sci. USA *86*:6196-6200.
13. **Le Roux, M-G., O. Pascal, A. Lostanlen, M. Etienne, H. Loreal, S.E. Bartlett, W.S. Davidson, J. Fleurence and J-P. Moisan.** Identification of tuna species in preserved canned samples. J. Sci. Food Agric. (In press).
14. **Meyer, R., U. Candrian and J. Luthy.** 1994. Detection of pork in heated meat products by the polymerase chain reaction. J. AOAC Int. *77*:617-622.
15. **Rehbein, H., I.M. Mackie, S. Pryde, C. Gonzales-Sotelo, R. Perez-Martin, J. Quinteiro and M. Rey-Mendez.** 1995. Fish species identification in canned tuna by DNA analysis (PCR-SSCP). Inf. Fischwirtsch. *42*:209-212.
16. **Unseld, M., B. Beyermann, P. Brandt and R. Hiesel.** 1995. Identification of the species origin of highly processed meat products by mitochondrial DNA sequences. PCR Methods Appl. *4*:241-243.

Strategies for Automated Sequencing of Human Mitochondrial DNA Directly from PCR Products

Romelle Hopgood, Kevin M. Sullivan[1] and Peter Gill[1]

The Forensic Science Service, London, England; [1]The Forensic Science Service, Birmingham, England

ABSTRACT

A rapid, robust and sensitive method has been developed for the amplification and direct sequencing of human mitochondrial DNA. A 403-bp hypervariable segment was amplified by two successive rounds of nested PCR. This was then sequenced by the dideoxy chain termination method using dye-labeled universal sequencing primers in conjunction with an automated DNA sequencer. This paper describes the assessment of four different strategies for this amplification and sequencing process. Optimal results were obtained by immobilizing the biotinylated PCR product on Dynabeads™ followed by solid-phase sequencing with Sequenase®. Degraded samples and those containing trace amounts of DNA such as extracts from hair shafts can be analyzed by this method.

INTRODUCTION

Various approaches for the direct sequencing of polymerase chain reaction (PCR) products have been described that are based on the Sanger dideoxy chain termination method. One method involves asymmetric PCR to produce a single-stranded DNA template (9). When used in conjunction with fluorescently labeled primers, these templates can be sequenced directly using an automated DNA sequencer (17). Other methods involve the production of single-stranded DNA template for direct sequencing by exonuclease digestion (11), or solid-phase separation (13,14). Phosphorylation and biotinylation of one DNA strand during PCR are required for each of these methods, respectively. An alternative approach is to sequence double-stranded PCR products utilizing *Taq* DNA polymerase using repeated PCR-like temperature cycling for the sequencing reactions (6).

(Reprinted from BioTechniques 13:82-92, 1992)

Table 1. Oligonucleotide Primers

Primer	Position	Sequence
A	L15926	5′-TCAAAGCTTACACCAGTCTTGTAAACC
B	H00580	5′-TTGAGGAGGTAAGCTACATA
C	L15997	5′-CACCATTAGCACCCAAAGCT
D	H16401	5′-TGATTTCACGGAGGATGGTG

This paper describes the assessment of four different amplification and sequencing strategies in order to develop a robust protocol for the automated analysis of mitochondrial DNA (mtDNA).

A large amount of sequence variation can be observed in the mtDNA of humans (2,5). mtDNA evolves at 5–10 times the speed of nuclear DNA resulting in regions of high variability (5). This is possibly due to the poorer proofreading activity of mtDNA polymerase as compared with nuclear DNA polymerase. The variable regions fall into two main areas within the non-coding control region (8,24). mtDNA is maternally inherited, so sequence variations can be used to determine the maternal lineage of individuals (7,19) and for forensic purposes (12,19,23). This approach has also been used to study the genealogical relationships of human subpopulations (24), and by using the process in conjunction with PCR, information has been obtained by amplifying trace amounts of DNA from ancient remains (10,20,21). Amplifying mitochondrial rather than nuclear DNA has proven more successful with ancient samples, presumably because the former is present in high copy numbers: single mammalian cells contain between 1000 and 10 000 mitochondrial genomes (4).

MATERIALS AND METHODS

DNA Preparation: Semen and Hair Shaft

Total DNA was extracted from semen samples as follows: In a 1.5-mL microcentrifuge tube, 200 µL liquid semen were added to an equal volume of sodium dodecyl sulfate (SDS) containing dithiothreitol (DTT) and proteinase K. This mixture was then incubated overnight at 37°C, extracted twice with phenol and then precipitated using absolute ethanol at room temperature. The DNA was then pelleted by centrifugation, washed in 70% ethanol, vacuum-dried briefly and resuspended in 50 µL distilled water (dH_2O). The amount of DNA present was then quantitated fluorometrically with a Hoefer TKO 100 Mini fluorimeter (Hoefer Scientific, San Francisco, CA, USA).

Hair shaft DNA was extracted by the following rapid method: Three centimeters of hair shaft (washed in distilled water) were placed in a microcen-

Table 2. Chimeric Oligonucleotide Primers

Primer	Position	Sequence
E	L15997	5′-TGTAAAACGACGACGGCCAGTCACCAT-TAGCACCCAAAGCT
F	H16401	5′-TGTAAAACGACGACGGCCAGTTGATTT-CACGGAGGATGGTG
G	L15997	5′-KINASE-TGTAAAACGACGACGGCCAGT-CACCATTAGCACCCAAAGCT
H	L15997	5′-BIOTIN-CACCATTAGCACCCAAAGCT

trifuge tube containing 200 µL rapid hair-digestion buffer comprising 10 mM Tris-HCl, pH 8.0, 0.9% polyoxyethylene 10 lauryl ether (Laureth 10), 50 µg/mL proteinase K and 35 mM DTT and incubated at 56°C for at least 0.5 h. The mixture was then subjected to vortex mixing and any undigested material was pelleted. The supernatant was used directly in PCR amplifications.

DNA Amplification

All oligonucleotide primers were prepared on an Applied Biosystems 381A DNA synthesizer (Applied Biosystems, Foster City, CA, USA) (see Table 1).

These primer sequences have been cited previously (19), with the position letter indicating the strand of mtDNA and the number relating to the reference sequence assignment (1) corresponding to the base at the 3′ end of the primer. Two chimeric primers combining the M13-21 primer sequence with

	STRATEGY 1: ASYMMETRIC PCR	STRATEGY 2: KINASED PRIMERS	STRATEGY 3: DOUBLE-STRANDED SEQUENCING	STRATEGY 4: SOLID PHASE SEQUENCING
PCR 1ST ROUND	SYMMETRIC AMPLIFICATION USING PRIMERS A & B CYCLING PARAMETERS - 94°C, 45s; 53°C, 1min; 72°C, 3mins for 32 cycles			
PCR 2ND ROUND	ASYMMETRIC AMPLIFICATION WITH PRIMERS C & F or D & E	SYMMETRIC AMPLIFICATION PRIMERS C & G (KINASED)	PRIMERS C & F	PRIMERS H (BIOTINYLATED) & F
	CYCLING PARAMETERS - 94°C, 45s; 50°C, 1min; 72°C, 5.5mins for 35 cycles			
TEMPLATE GENERATION	PCR PRODUCT ADDED DIRECTLY TO SEQUENCING REACTION WITH NO PURIFICATION STEP	KINASED STRAND REMOVED BY LAMBDA EXONUCLEASE DIGESTION. RESULTING PRODUCT PURIFIED PRIOR TO ADDITION IN SEQUENCING REACTION	PCR PRODUCT ADDED DIRECTLY TO SEQUENCING REACTION WITH NO PURIFICATION STEP	IMMOBILIZATION OF BIOTINYLATED PRODUCT ON STREPTAVIDIN-COATED SUPPORT, ELUTION OF NON-BIOTINYLATED STRAND BY NaOH TREATMENT
SEQUENCING METHOD	TAQ CYCLE SEQUENCING A: 10 ROUNDS, 2 TEMPERATURE		TAQ CYCLE SEQUENCING B: 10 ROUNDS, 3 TEMPERATURE FOLLOWED BY 10 ROUNDS 2 TEMPERATURE	TAQ CYCLE SEQUENCING B / SEQUENASE SEQUENCING

Figure 1. Different strategies for amplifying and sequencing a segment of the D-loop of mtDNA.

the above mt-specific sequences and a primer combining biotin with an mt-specific sequence were also synthesized. For one set of experiments, one of the chimeric primers was kinased (11) (see Table 2).

All PCRs were performed in a Perkin Elmer DNA Thermal Cycler (Norwalk, CT, USA) with temperatures and times quoted as programmed, using *Thermus aquaticus* (*Taq*) DNA polymerase (Cambio, Cambridge, England) and its associated amplification buffer (PARR). The amplification and sequencing strategies are summarized in Figure 1.

Strategy 1 - Asymmetric PCR:

First-stage amplification. Primers A and B above were used to amplify a 1333-bp segment of human mtDNA containing both the origin of replication and the D-loop region (Figure 1). PCR conditions were as follows: in a total volume of 25 µL were mixed 100 ng semen DNA (or 10 µL of nonquantitated hair shaft DNA) as template, primers at 1 µM, deoxynucleoside triphosphates (dNTPs) each at 200 µM, 1.25 U *Taq* DNA polymerase and 1× amplification (PARR) buffer. Cycling conditions were as follows for 35 cycles: 94°C for 45 s; 50°C for 1 min; and 72°C for 5.5 min. Amplification products were analyzed as follows: 4 µL of the reaction mixture were added to 1 µL of loading buffer (15) and loaded onto a 3% NuSieve® (FMC BioProducts, Rockland, ME, USA) agarose gel cast on a 75- × 50-mm glass slide. The samples were then electrophoresed in Tris-borate-EDTA buffer (15) at 50 V for 30 min in a Bio-Rad Mini Sub-Cell apparatus (Bio-Rad, Richmond, CA, USA). The PCR products were identified by staining with ethidium bromide and visualization with UV light (semen only). In preliminary experiments, 2 plugs were picked with a sterile Pasteur pipet from the 1333-bp major amplification product, taking care to minimize the exposure to UV light. The 2 plugs were added to 18 µL dH_2O in a microcentrifuge tube and melted by heating at 65°C for 15 min. Later experiments demonstrated that 0.5 µL of the first amplification product could be used directly in the second round of amplification with no purification step.

Second-stage amplification. The second round of amplification was asymmetric utilizing a nested set of primers covering approximately 400 bases of the non-coding region and 0.5 µL of the reaction product from the first round of amplification. The primers were used at unequal concentrations, thereby generating a single-stranded (ss) PCR product (9). Furthermore, two different pairs of primers were used in separate reactions to generate ssDNA of each strand of the target-amplified segment. Primer pairs were as follows: C (1 µM final concentration) with F (0.02–0.002 µM final concentration); and D (1 µM final concentration) with E (0.02–0.002 µM final concentration). This corresponds to primer ratios ranging from 50–500:1 and generates ss-DNA-containing sequence at the 3′ end complementary to the -21M13 universal primer. The other amplification reaction conditions were as follows: 2.5 U *Taq* DNA polymerase, 1× PARR buffer and 20 µM final concentration for each dNTP in a final volume of 50 µL. The temperature cycling parameters were as follows for 32 cycles: 94°C for 45 s; 53°C for 1 min; and 72°C for 3 min. Eight microliters of the reaction products were added to 3 µL of loading buffer and loaded in a 100 mL 3% NuSieve agarose gel cast in a 10- × 15-cm tray. The sample was then electrophoresed in Tris-borate-EDTA buffer at 50 V for 90 min in a Bio-Rad Mini Sub-Cell apparatus.

Sequencing reactions and analysis. For the dideoxyadenosine sequencing reaction, 1.5 µL of the PCR solution were dispensed in a 0.5 mL microcentrifuge tube to which the following were added: 0.8 pmol of -21M13 JOE primer (fluorescent dye; Applied Biosystems) (22); 2.5 U *Taq* DNA polymerase (AmpliTaq®, Applied Biosystems); 5× sequencing buffer to give a final concentration of 10 mM Tris-HCl, pH 8.0, at room temperature, 10 mM $MgCl_2$, 25 mM NaCl; 2 µL of 2′-deoxy/2′,3′-dideoxy-adenosine-5′–triphosphate mixture and dH_2O to a final volume of 15 µL. The 3 other sequencing reactions for dideoxycytosine, dideoxyguanosine and dideoxythymidine using FAM-, TAMRA- and ROX-labeled primers (Applied Biosystems), respectively, were set up in parallel and under effectively identical conditions except that all quantities were doubled for the dideoxyguanosine and dideoxythymidine reactions. The capped microcentrifuge tubes were placed in a thermal cycler and heated as follows for 10 cycles: 90°C for 1 min and 70°C for 2 min. The reactions were then pooled and precipitated with ethanol, resuspended in 5 µL deionized formamide and 1 µL 50 mM EDTA, pH 8.0, denatured for 2 min at 95°C and loaded directly onto a 6% acrylamide gel. Electrophoresis and sequencing analysis were performed with an Applied Biosystems Model 370A DNA sequencer, and some samples were later reanalyzed by 373A software. Sequence comparisons were performed using a DNASTAR package.

Strategy 2 - Kinased Primers:

Amplification. First-stage amplification was carried out according to strategy 1. The second round of PCR was symmetric using primer pairs in equal concentrations. Primers C with G (0.5 µM final concentration) were utilized to generate dsDNA, one strand of which was kinased and the other strand contained sequence at its 3′ end, which was complementary to the -21M13 universal primer. The other amplification reaction conditions and the cycling parameters were as outlined in strategy 1.

Sequencing reactions and analysis. The PCR product was treated with 4 U of lambda exonuclease in 50 µL 67 mM glycine/NaOH, pH 9.4, 2.5 mM $MgCl_2$ buffer for 30 min at 37°C. The lambda exonuclease digested the kinased strand leaving predominantly ssDNA that was used for the sequencing reactions. After exonuclease treatment, the reaction mixture was extracted with phenol, precipitated with ethanol and resuspended in 15 µL dH_2O. This solution was then sequenced by the method outlined in strategy 1 above.

Strategy 3 - Double-Stranded Sequencing:

Amplification. First-stage amplification was carried out according to strategy 1. The second round of PCR was symmetric using primer pairs in equal concentrations. Primers C with F (0.5 µM final concentration) were utilized to generate dsDNA, one strand of which contained sequence at the 3′ end, which was complementary to the -21M13 universal primer. The other amplification reaction conditions and the cycling parameters were as outlined in strategy 1.

Sequencing reactions and analysis. The sequencing reaction mixtures contained the same reagents as outlined in the method for strategy 1, except that 2.5 U *Taq* DNA polymerase were used in total, and the final volume of the mixtures was adjusted to 13 µL for reactions A and C and 26 µL for reactions G and T, as outlined in the PCR Sequencing Kit protocol from Applied Biosystems. The reaction mixtures were overlaid with 15 µL light mineral oil, the microcentrifuge tubes were capped, placed in a thermal cycler and heated as follows for 10 cycles: 95°C for 30 s, 55°C for 30 s and 70°C for 1 min. The microcentrifuge tubes were then heated as follows for a further 10 cycles: 95°C for 30 s and 70°C for 1 min. The reaction products were then pooled, precipitated, resuspended and loaded onto a sequencing gel as usual.

Strategy 4 - Solid-Phase Sequencing:

Amplification. First-stage amplification was carried out according to strategy 1. The second round of PCR was symmetric using primer pairs in equal concentrations. Primers H with F (0.5 µM final concentration) were utilized to generate dsDNA, one strand of which was biotinylated at the 5′ end and which contained sequence at the 3′ end that was complementary to the -21M13 universal primer. The other amplification reaction conditions and the cycling parameters were as outlined in strategy 1.

Sequencing reactions and analysis. The PCR product was added to 50 µL 6 M LiCl and 50 µL streptavidin-coated Dynabeads™ (Dynal A.S., Oslo, Norway) and incubated at 48°C for 15 min. This was to get optimum binding of the biotin-labeled PCR product to the streptavidin. The complex was then washed twice with 250 mM Tris-HCl/0.1% Tween 20 solution prior to denaturation in 0.15 M NaOH for 4 min at room temperature. The Dynabeads that were still attached to the biotinylated PCR product (now ss) were stabilized with the use of a magnet and the supernatant was removed. The Dynabead complex was then resuspended in 60 µL dH_2O and used directly in the following sequencing reactions: *(i) Taq* cycle sequencing as outlined in strategy 3 using 8 µL of the bead complex for the dideoxyadenosine and dideoxycytosine reactions and 22 µL of the bead complex for the dideoxyguanosine and dideoxythymidine reactions; or *(ii)* for the dideoxyadenosine sequencing reaction, 8 µL of the Dynabead complex were dispensed in a 0.5-mL microcentrifuge tube, stabilized and the excess liquid removed. To the microcentrifuge tube the following were then added: 0.4 pmol of -21M13 JOE primer; 5× Sequenase® buffer (United States Biochemical, Cleveland, OH, USA) and dH_2O to a final volume of 4 μL. The reactions were then annealed by heating for 2 min at 65°C, cooling to 25°C over 25 min and chilling on ice. One microliter of 2′-deoxy/2′,3′-dideoxy-adenosine-5′-triphosphate mixture was added, and the microcentrifuge tube was warmed to 37°C prior to the addition of 2 U Sequenase T7 DNA polymerase (United States Biochemical). The 3 other sequencing reactions for dideoxycytosine, dideoxyguanosine and dideoxythymidine, using FAM-, TAMRA- and ROX-labeled primers, respectively, were set up in parallel and under effectively identical conditions except that all quantities were doubled for the dideoxyguanosine and dideoxythymidine reactions. The solutions were then incubated at 37°C for 20 min prior to stopping the reactions with Stop/Salt solution and pooling. The sequenced products that were still attached to the Dynabeads were then separated from all unused reagents by the repeated use of the magnet. The Dynabead complex was then denatured by incubating in 4 μL deionized formamide for 4 min at 37°C, stabilized and the recovered liquid loaded onto a 6% acrylamide gel. Electrophoresis and sequence analysis were performed with an Applied Biosystems Model 373A (updated from Model 370A) DNA sequencer. Comparisons of DNA sequences were carried out using a SEQED® package (Applied Biosystems).

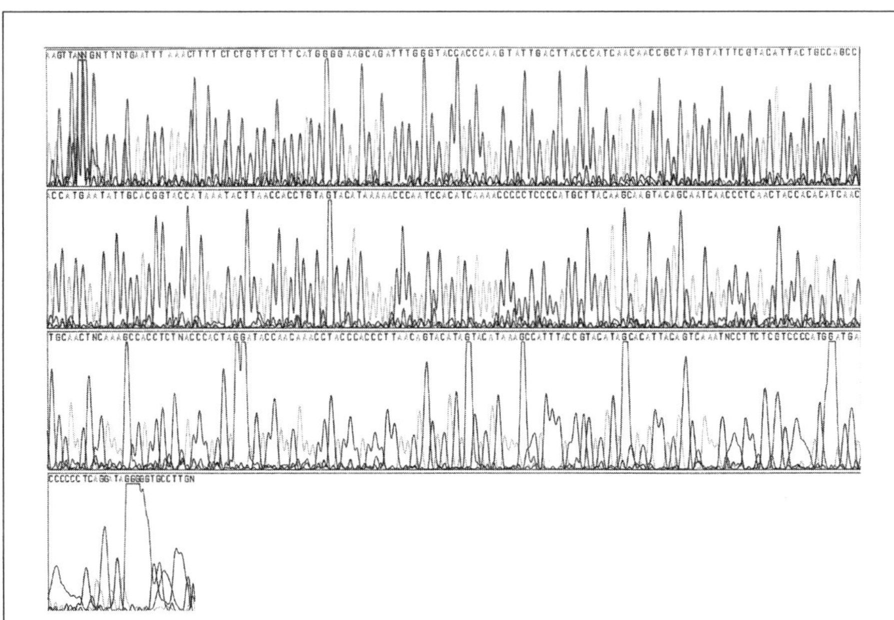

Figure 2. DNA sequence of the H strand of human mitochondrial D-loop, amplified from 100 ng of total semen DNA by strategy 1 - Asymmetric PCR.

RESULTS AND DISCUSSION

Strategy 1 - Asymmetric PCR

In the first round of amplification, 1333-bp amplification products were readily obtained from 100-ng samples of total DNA extracted from semen (data not shown). Because there was very little DNA present in hair shafts to begin with, no first amplification products were visible when electrophoresed into an agarose gel.

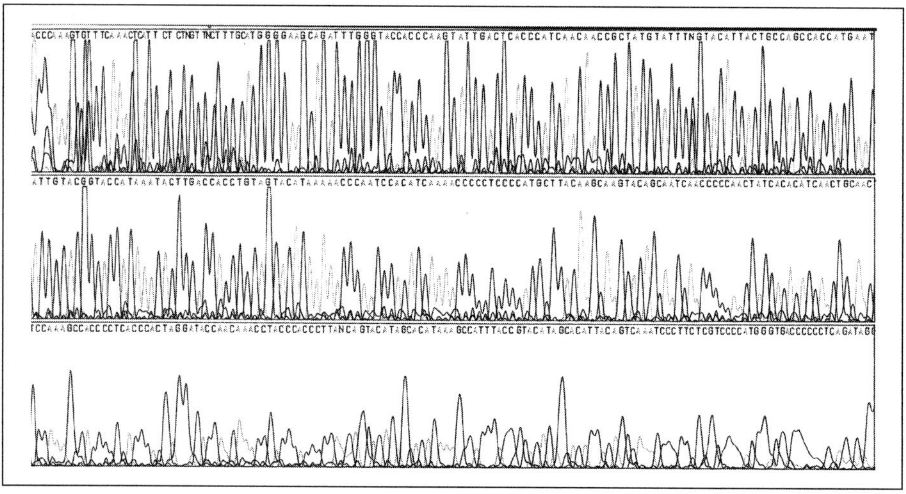

Figure 3. DNA sequence obtained by strategy 3. Double-stranded sequencing of the L strand of human mitochondrial D-loop, amplified from 100 ng of total semen DNA.

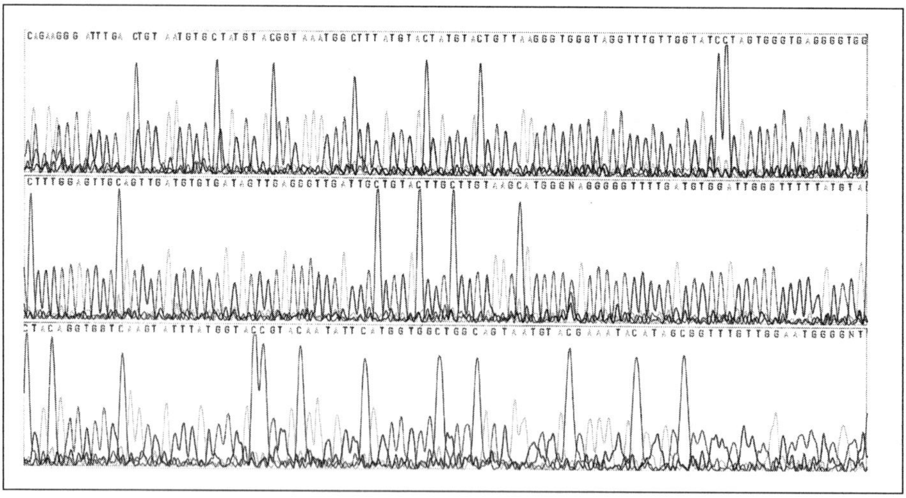

Figure 4. DNA sequence obtained by strategy 4. Solid-phase sequencing of the L strand of human mitochondrial D-loop, amplified from 3 cm of hair shaft.

In the second round of amplification, single-stranded DNA was generated with both sets of primers, with ratios ranging from 50:1 to 500:1. A final ratio of 200:1 was used in all subsequent asymmetric amplifications since this gave the optimal amounts of ssDNA as compared to dsDNA and other nonspecific products (results not shown). Introduction of the -21M13 universal primer sequence into the PCR products at this stage enabled commercially available universal fluorescent primers to be used in the sequencing reactions.

By performing 10 cycles of denaturation, annealing and extension for the sequencing reactions, we found a signal was obtained that could be analyzed from a small amount of template, thereby eliminating the need to concentrate the PCR product prior to sequencing. Raw data generated by the 370A sequencer were analyzed between the primer peak and the full-length product peak. An example of processed data is shown in Figure 2.

To resolve the identity of any ambiguities (bases not automatically resolved by the machine) present in the sequence, both strands of the amplified fragment were sequenced in single reactions from which any residual primer sequences were trimmed. Resultant sequences were initially compared using DNASTAR "COMPARE" computer program that enabled most sequence ambiguities to be resolved.

This method of generating ssDNA for sequencing was very unpredictable,

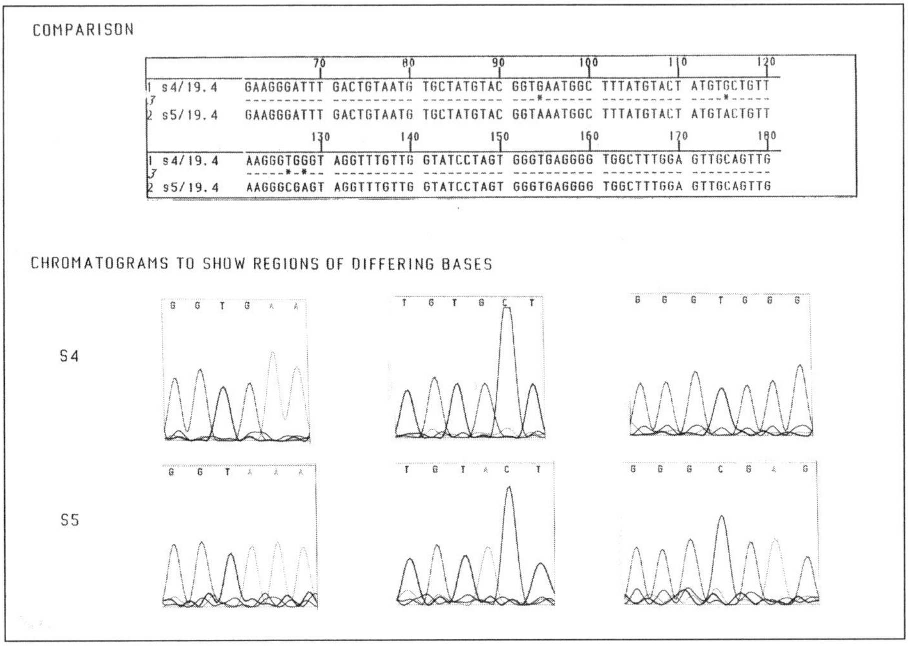

Figure 5. Comparison of 2 amplified mitochondrial sequences, S4 and S5, over a 120-bp section of the non-coding region. Amplification and sequencing are by strategy 4 using Sequenase. Dashes denote homologies, and asterisks denote differences between the sequences. The chromatograms show the regions of the differing bases.

especially for hair shaft DNA where the template concentration could not be quantitated. This frequently resulted in dsDNA being the only visible second amplification product even if the primer ratio was at the calculated optimum. However, if there was sufficient ssDNA present, the sequences obtained were of high quality with very few ambiguities. The sequences extended over a maximum of 410 bp for semen samples with 6 ambiguities or 380 bp for hair shafts with 47 ambiguities.

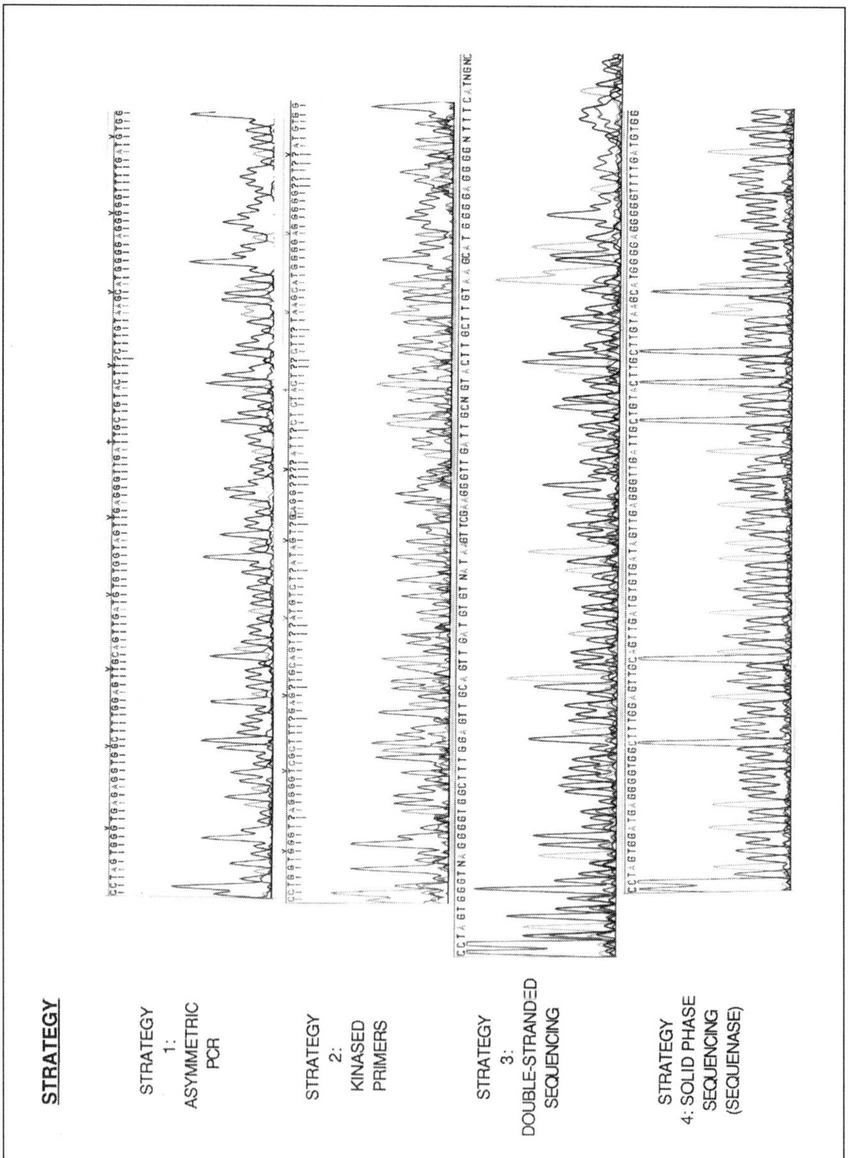

Figure 6. Comparison of the 4 main sequencing strategies using a "control" semen template. The chromatograms show the same 102-bp region of sequence [position 16173-16274 in the Anderson (1) reference sequence] in the same orientation.

Strategy 2 - Kinased Primers

First-amplification products were as described in strategy 1.

In the second round of amplification, 464 bp double-stranded products were readily obtained from both semen and hair shaft samples (data not shown).

In our hands, the method of generating ssDNA sequencing template by lambda exonuclease digestion of the kinased PCR product was more labor-intensive and produced more sequence ambiguities than strategy 1. The sequences obtained only extended over a maximum of 335 bp for semen samples with 46 ambiguities. Sequencing of hair shaft DNA by this method was not attempted.

Strategy 3 - Double-Stranded Sequencing

First and second amplification products were as described in strategy 2.

Producing ssDNA sequencing template from dsDNA was avoided by performing 10 sequencing cycles of a three-step PCR followed by 10 sequencing cycles of a two-step PCR. An example of the processed data is shown in Figure 3. Although the methodology of this strategy was relatively simple, sequences contained more ambiguities and did not extend as far as those obtained by strategy 1. A maximum of 366 bp for semen samples with 14 ambiguities and 370 bp for hair shafts with 21 ambiguities was produced. Difficulties in sequencing dsDNA, especially PCR products, are well documented (3,16,18) and may account for these problems.

STRATEGY	SAMPLE TYPE	AMPLIFICATION		SEQUENCING		Proportion of Automatically Assigned Bases (A-B/A x 100)
		1st 1333bp Amplification Product Present?	ss/ds 2nd Ampification Product	A Maximum Length of Sequence Obtained (bps)	B Number of Ambiguities in Obtained Sequence	
STRATEGY 1: ASYMMETRIC PCR	Semen	YES	ss and ds product present when primers at 200:1 ratio	410	6	98.5
	Hair Shaft	NO		380	47	87.6
STRATEGY 2: KINASED PRIMERS	Semen	YES	464bp ds products present	335	46	86.3
	Hair Shaft	NO		N/A	N/A	N/A
STRATEGY 3: DOUBLE-STRANDED SEQUENCING	Semen	YES	464bp ds products present	366	14	96.2
	Hair Shaft	NO		370	21	94.3
STRATEGY 4a: SOLID PHASE SEQUENCING (TAQ)	Semen	YES	464bp ds products present	330	44	86.7
	Hair Shaft	NO		178	8	95.5
STRATEGY 4b: SOLID PHASE SEQUENCING (SEQUENASE)	Semen	YES	464bp ds products present	421	0	100
	Hair Shaft	NO		347	7	98

Figure 7. Summary of the results of the 4 sequencing strategies.

Strategy 4 - Solid-Phase Sequencing

First and second amplification products were as described in strategy 2.

An example of a typical hair shaft sequence obtained by this method, sequencing with Sequenase, is shown in Figure 4. A very even signal was obtained across the whole length of the sequence that resulted in high-quality sequence with very few ambiguities present. A maximum of 421 bp for semen samples with 0 ambiguities and 347 bp for hair shafts with 7 ambiguities was produced by this strategy. Sequencing with *Taq* gave sequence comparable to sequences obtained previously with *Taq*. A maximum of 330 bp for semen samples with 44 ambiguities and 178 bp for hair shafts with 8 ambiguities was produced by this strategy.

In these experiments, comparisons between sequences were carried out on the computer program SEQED™ (Applied Biosystems) that enabled direct and rapid comparison of sequences. An example of the comparison between 2 semen samples, sequenced using Sequenase, is shown in Figure 5, and a summary of all results is given in Figures 6 and 7.

SUMMARY

This paper describes several approaches by which a hypervariable region of mitochondrial non-coding region could be rapidly and conveniently sequenced by employing PCR and automated DNA sequencing. A high level of automation would be required for the technique to be introduced into forensic analysis on a routine basis.

The best sequencing strategy utilized the Dynabead approach coupled with Sequenase sequencing because this appeared to be the most reliable, the most sensitive and the most consistent of the four methods attempted. This approach will be invaluable to the future analysis of evidential material such as hair shafts, which contain mtDNA but lack sufficient genomic DNA to be characterized by conventional DNA profiling.

Investigation of the feasibility of enhancing the discrimination power of the test by sequencing a second hypervariable region of the non-coding region is currently under way in our laboratory. Also, if the true statistical significance of the observed sequence variation is to be defined, a large population database will be required.

REFERENCES

1. **Anderson, S., A.T. Bankier, B.G. Barrell, M.H.L. de Bruijn, A.R. Coulson, J. Drouin, I.C. Eperon, D.P. Nierlich, B.A. Roe, F. Sanger, P.H. Schreier, A.J.H. Smith, R. Staden and I.G. Young.** 1981. Sequence and organisation of the human mitochondrial genome. Nature *290*:457-465.
2. **Aquadro, C.F. and B.D. Greenberg.** 1983. Human mitochondrial DNA variation and evolution: analysis of nucleotide sequences from seven individuals. Genetics *103*:287-312.
3. **Bachmann, B., W. Luke and G. Hunsmann.** 1990. Improvement of PCR amplified DNA sequencing with the aid of detergents. Nucleic Acids Res. *18*:1309.
4. **Bogenhagen, D. and D.A. Clayton.** 1974. The number of mitochondrial deoxyribonucleic acid genomes in mouse L and human HeLa cells. J. Biol. Chem. *249*:7991-7995.

5. **Cann, R.L., M. Stoneking and A.C. Wilson.** 1987. Mitochondrial DNA and human evolution. Nature *325*:457-465.
6. **Carothers, A.M., G. Urlaub, J. Mucha, D. Grunberger and L.A. Chasin.** 1989. Point mutation analysis in a mammalian gene: rapid preparation of total RNA, PCR amplification of cDNA and *Taq* sequencing by a novel method. BioTechniques *7*:494-499.
7. **Giles, R.E., H. Blanc, H.M. Cann and D.C. Wallace.** 1980. Maternal inheritance of human mitochondrial DNA. Proc. Natl. Acad. Sci. USA *77*:6715-6719.
8. **Greenberg, B.D., J.E. Newbold and A. Sugino.** 1983. Intraspecific nucleotide sequence variability surrounding the origin of replication in human mitochondrial DNA. Gene *21*:33-49.
9. **Gyllensten, U.B. and H.A. Erlich.** 1988. Generation of single-stranded DNA by the polymerase chain reaction and its application to direct sequencing of the HLA-DGA locus. Proc. Natl. Acad. Sci. USA *85*:7652-7656.
10. **Hagelberg, E. and B. Bykes.** 1990. Ancient bone DNA amplified. Nature *342*:485.
11. **Higuchi, R.G. and H. Ochman.** 1989. Production of single-stranded DNA template by exonuclease digestion following the polymerase chain reaction. Nucleic Acids Res. *17*:5865.
12. **Higuchi, R.G., C.H. von Beroldingen, G.F. Sensabaugh and H.A. Erlich.** 1988. DNA typing from single hairs. Nature *332*:543-546.
13. **Hultman, T., S. Bergh, T. Moks and M. Uhlen.** 1991. Bidirectional solid-phase sequencing of in vitro-amplified plasmid DNA. BioTechniques *10*:84-93.
14. **Kaneoka, H., D.R. Lee, K.-C. Hsu, G.C. Sharp and R.W. Hoffman.** 1991. Solid-phase direct DNA sequencing of allele-specific polymerase chain reaction-amplified HLA-DR genes. BioTechniques *10*:30-34.
15. **Maniatis, T., E.F. Fritsch and J. Sambrook.** 1982. Molecular Cloning: A Laboratory Manual. Cold Spring Harbor Laboratory, Cold Spring Harbor, NY.
16. **Mazars, G.R., C. Moyret, P. Jeanteur and C-G. Theillet.** 1991. Direct sequencing by thermal asymmetric PCR. Nucleic Acids Res. *19*:4783.
17. **Mcbride, L.J., S.M. Koepf, R.A. Gibbs, P. Nyugen, W. Salser, P.E. Mayrand, M.W. Hunkapiller and M.N. Kronick.** 1989. Automated DNA sequencing methods involving polymerase chain reaction. Clin. Chem. *35*:2196-2201.
18. **Mitchell, L.G. and C.R. Merril.** 1989. Affinity generation of single stranded DNA for dideoxy sequencing following the polymerase chain reaction. Anal. Biochem. *178*:239-242.
19. **Orrego, C and M.C. King.** 1990. Determination of familial relationships, p. 416-426. *In* M.A. Innis, D.H. Gelfand, J.J. Sninsky and T.J. White (Eds.), PCR Protocols: A Guide to Methods and Applications. Academic Press, London and San Diego.
20. **Paabo, S.** 1980. Ancient DNA: Extraction, characterization, molecular cloning and enzymatic amplification. Proc. Natl. Acad. Sci. USA *86*:1939-1943.
21. **Paabo, S., R.G. Higuchi and A.C. Wilson.** 1989. Ancient DNA and the polymerase chain reaction. J. Biol. Chem. *264*:9709-9712.
22. **Smith, L.M., J.Z. Sanders, R.J. Kaiser, P. Highes, C. Dodd, C.R. Connell, C. Heiner, S.B. Kent and L.E. Hood.** 1986. Fluorescence detection in automated DNA sequence analysis. Nature *321*:674-679.
23. **Stoneking, M., D. Hedgecock, R.G. Higuchi, L. Vigilant and H.A. Erlich.** 1991. Population variation of human mtDNA control region sequences detected by enzymatic amplification and sequence-specific oligonucleotide probes. Am. J. Hum. Genet. *48*:370-382.
24. **Vigilant, L., R. Pennington, H. Harpending and T.D. Kocher.** 1989. Mitochondrial DNA sequences in single hairs from a Southern African population. Proc. Natl. Acad. Sci. USA *86*:9350-9354.

Update to:

Strategies for Automated Sequencing of Human Mitochondrial DNA Directly from PCR Products

Gillian Tully

The Forensic Science Service, Birmingham, England

In the original paper, single-stranded sequencing of immobilized PCR products was shown to give the best quality of any of the automated sequencing strategies evaluated, and has been extensively used in research and forensic casework within our laboratory (2,3,7,8). The method is, however, labor-intensive, so additional work on evaluating alternative strategies for analyzing the human mtDNA major noncoding region has continued.

This update focuses on three major areas:

(1) alterations in methodology to deal with highly degraded samples;
(2) improvements in double-stranded sequencing technology; and
(3) alternatives to sequencing for distinguishing between individuals.

ALTERATIONS IN METHODOLOGY TO DEAL WITH HIGHLY DEGRADED SAMPLES

High product yield for sequence analysis can often be obtained from forensic samples by amplifying the entire non-coding region followed by nested PCRs in which each hypervariable region is separately amplified (2–4,7). PCR from extremely degraded samples is more efficient using closely spaced primers (e.g., 1,5,6), so single-round amplification of each hypervariable region in two overlapping fragments of less than 250 base pairs each has been applied to forensic mtDNA analysis (1,9,11,12).

IMPROVEMENTS IN DOUBLE-STRANDED SEQUENCING TECHNOLOGY

The introduction of new enzymes for cycle sequencing in conjunction with dye-labeled primers has produced a marked improvement in sequence quality. Figure 1 shows that sequence of quality approaching that obtained by single-stranded solid-phase sequencing using Sequenase® can now be obtained by double-stranded cycle sequencing using AmpliTaq® DNA Polymerase, FS (Perkin-Elmer/Applied Biosystems Division, Foster City, CA, USA). An advantage of cycle sequencing over Sequenase sequencing is that fewer false stops due to secondary structure formation in the template are encountered, and in particular, there is a higher success rate in sequencing through poly(G) tracts. Furthermore, automation of cycle sequencing reactions is more straightforward than automation of reactions involving magnetic separation; Figure 1b is an example of a sequence generated using the Catalyst™ 800 Molecular Biology Workstation (Perkin-Elmer/Applied Biosystems Division).

ALTERNATIVES TO SEQUENCING FOR DISTINGUISHING BETWEEN INDIVIDUALS

The distribution of sequence variability within the major non-coding region of human mtDNA is non-random, so a strategy whereby the most polymorphic nucleotides are determined first in a "screening" step enables rapid discrimination between many samples.

We have developed a novel technique, multiplex solid-phase fluorescent

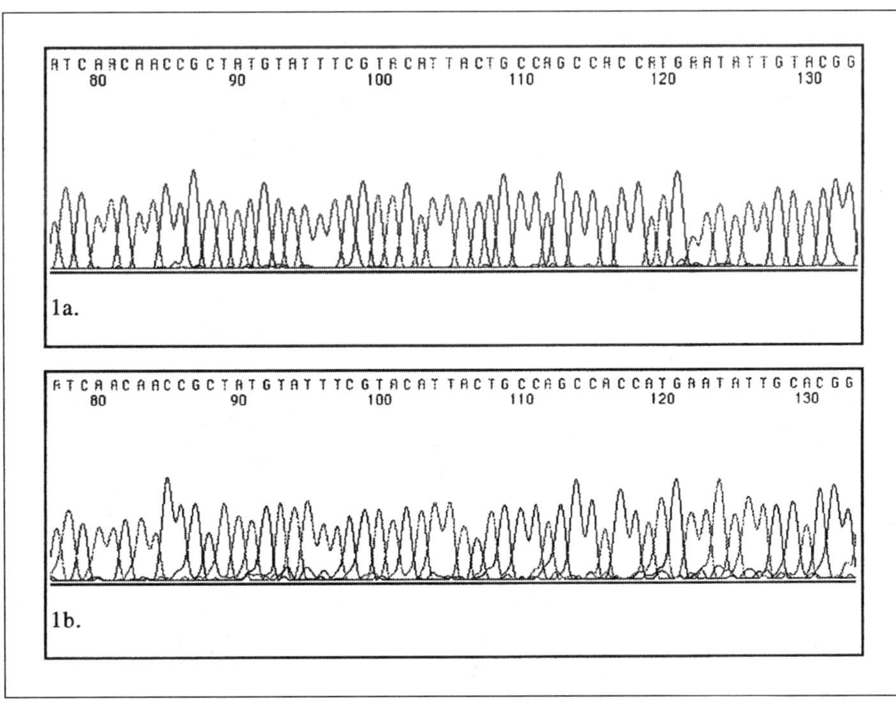

Figure 1. Comparison of sequencing strategies. (a) Solid-phase sequencing of single-stranded DNA using Sequenase as described (4). (b) Automated cycle sequencing of double-stranded DNA using AmpliTaq DNA Polymerase, FS. Products from nested PCR as described (4) were purified using DNA Select III columns (5 Prime → 3 Prime, Boulder, CO, USA) and sequenced using the Catalyst 800 Molecular Biology Workstation with a Dye Primer Cycle Sequencing Kit (Perkin-Elmer/Applied Biosystems Division) according to the manufacturer's instructions.

minisequencing, for the simultaneous detection of 10 substitution polymorphisms and 2 length polymorphisms within the mtDNA major non-coding region (10). Single-nucleotide extension of primers adjacent to polymorphic nucleotides is combined with four-color automated fluorescence detection (Perkin-Elmer/Applied Biosystems Division) to enable simultaneous characterization of multiple polymorphisms (Figure 2). The average probability of a random match between two unrelated individuals using the technique is 0.02–0.06 (10), enabling a significant reduction to be made in the number of samples that can only be distinguished by sequencing both hypervariable regions.

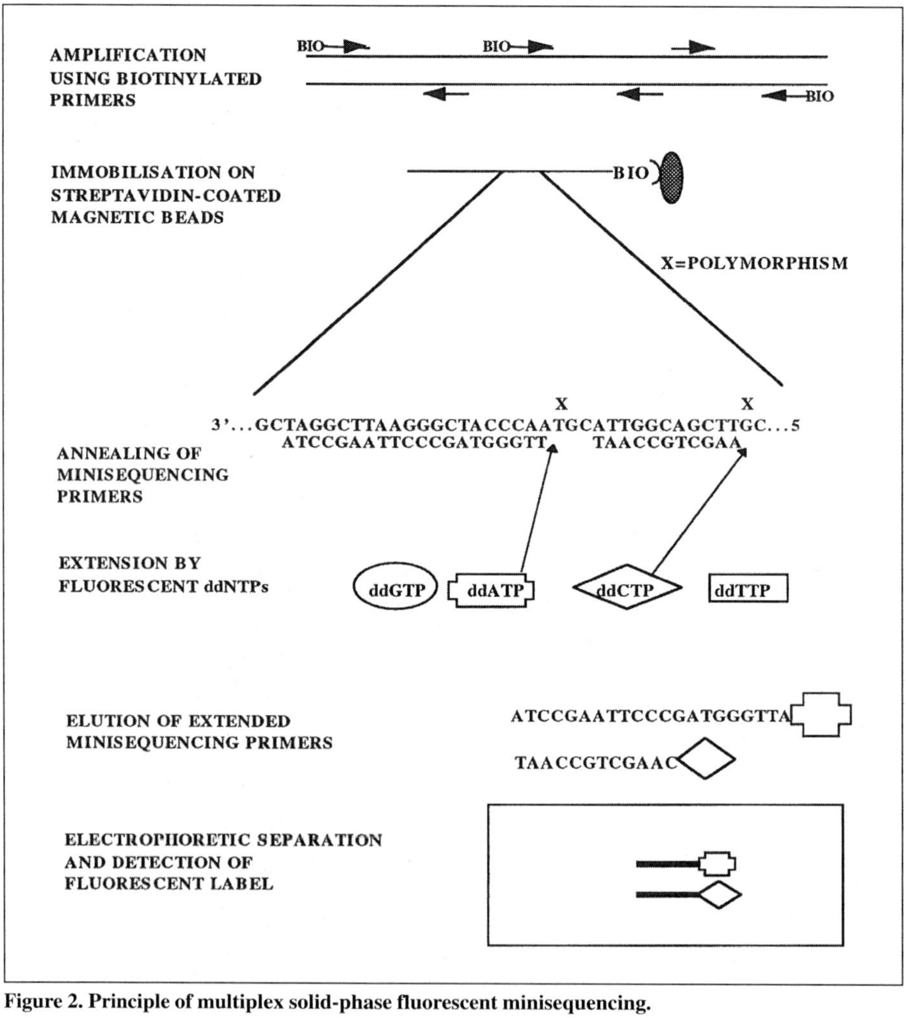

Figure 2. Principle of multiplex solid-phase fluorescent minisequencing.

ACKNOWLEDGMENTS

The input of K. Sullivan, R. Piercy, N. Woodyatt, A. Hopwood, J. Bark, C. Evans, A. Spriggs, R. Alliston-Greiner and L. Bark in developing the techniques described is gratefully acknowledged. This work was partially funded by the Police Research Group.

REFERENCES

1. **Fisher, D.L., M.M. Holland, L. Mitchell, P.S. Sledzik, A. Webb Wilcox, M. Wadhams and V.W. Weedn.** 1993. Extraction, evaluation, and amplification of DNA from decalcified and undecalcified United States Civil War bone. J. Forensic Sci. *38*:60-68.
2. **Gill, P., P.L. Ivanov, C. Kimpton, R. Piercy, N. Benson, G. Tully, I. Evett, E. Hagelberg and K. Sullivan.** 1994. Identification of the remains of the Romanov family by DNA Analysis. Nat. Genet. *6*:130-135.
3. **Gill, P., C. Kimpton, R. Alliston-Greiner, K. Sullivan, M. Stoneking, T. Melton, J. Nott, S. Barrit, R. Roby, M. Holland and V. Weedn.** 1994. Establishing the identity of Anna Anderson Manahan. Nat. Genet. *9*:9-10.
4. **Ginther, C., L. Issel-Tarver and M-C. King.** 1992. Identifying individuals by sequencing mitochondrial DNA from teeth. Nat. Genet. *2*:135-138.
5. **Hagelberg, E., L.S. Bell, T. Allen, A. Boyde, S.J. Jones and J.B. Clegg.** 1991. Ancient bone DNA: techniques and applications. Philos. Trans. R. Soc. Lond. B. Biol. Sci. *333*:399-407.
6. **Pääbo, S., J.A. Gifford and A.C. Wilson.** 1988. Mitochondrial DNA sequences from 7000-year old brain. Nucleic Acids Res. *16*:9775-9787.
7. **Piercy, R., K.M. Sullivan, N. Benson and P. Gill.** 1993. The application of mitochondrial DNA typing to the study of white Caucasian genetic identification. Int. J. Legal Med. *106*:85-90.
8. **Sullivan, K.M., R. Piercy, N. Benson, P. Ivanov, A. Mannucci and P. Gill.** 1994. Solid-phase sequencing of mitochondrial DNA: towards a fully automated forensic DNA test, p. 135-148. *In* M. Uhlén, E. Hornes and O. Ølsvik (Eds.), Advances in Biomagnetic Separation. Eaton Publishing, Natick.
9. **Sullivan, K.M., G. Tully, R. Alliston-Greiner, A. Hopwood, J.E. Bark and P. Gill.** 1996. A two-stage strategy for the automated analysis of mitochondrial DNA, p. 11-13. *In* A. Carracedo, B. Brinkmann and W. Bär (Eds.), Advances in Forensic Haemogenetics 6. Springer-Verlag, Berlin.
10. **Tully, G., K.M. Sullivan, P. Nixon, R.E. Stones and P. Gill.** 1996. Rapid detection of mitochondrial sequence polymorphisms using multiple solid-phase fluorescent minisequencing. Genomics *34*:107-113.
11. **Wilson, M.R., J.A. DiZinno, D. Polanskey, J. Replogle and B. Budowle.** 1995. Validation of mitochondrial DNA sequencing for forensic casework analysis. Int. J. Legal Med. *108*:68-74.
12. **Wilson, M.R., D. Polanskey, J. Butler, J.A. DiZinno and B. Budowle.** 1995. Extraction, PCR amplification and sequencing of mitochondrial DNA from human hair shafts. BioTechniques *18*:662-669.

Analysis of Heterogeneous Viral Populations by Direct DNA Sequencing

T. Leitner[1,2], E. Halapi[2], G. Scarlatti[2], P. Rossi[2], J. Albert[3], E.-M. Fenyo[2] and M. Uhlén[1]

[1]Royal Institute of Technology, [2]Karolinska Institute and [3]National Bacteriological Laboratory, Stockholm, Sweden

ABSTRACT

The ability of direct PCR sequencing to detect and quantify sequence polymorphisms was investigated using samples containing mixed populations of HIV-1. A part of the genome encoding the polymorphic variable region 3 of the envelope was directly sequenced to yield a consensus sequence of the virus population. The results were compared with sequences obtained by analysis of multiple clones derived from the same clinical samples. The results of five patients suggested that the direct sequencing method can be used as a rapid tool to analyze and quantify heterogeneous viral populations. Reconstitution experiments using cloned material demonstrated that it was possible to detect and quantify minor sequence variants present in as little as 10% of the total virus population. The use of the method for molecular diagnosis is discussed.

INTRODUCTION

DNA sequencing has become an important tool in biomedicine, including the analysis of genetic disorders and infectious diseases (10). The polymerase chain reaction (PCR) allows target regions of the genome to be amplified directly from clinical material (16). However, conventional DNA cloning of PCR products followed by sequencing is not suitable for such analysis due to the time and labor required. Thus, several alternative sequencing methods have been developed allowing direct sequencing of PCR products. These include asymmetric PCR (4), exonuclease protocols involving kinased primers (5), cycle sequencing protocols using *Taq* DNA polymerase (3) and solid-phase sequencing using magnetic beads (9).

For many clinical applications it is essential to have reproducible results with high quality. This is due to the fact that the analysis often involves heterozygotic loci, and thus the sequence analysis will yield mixed sequences at

(Reprinted from BioTechniques 15:120-127, 1993)

certain nucleotide positions. A good quality sequence with an even representation of bases will ensure correct interpretation of these alleles. Similarly, many viruses are genetically heterogeneous and should be described as populations of closely related genotypes, referred to as quasi-species (12). Obviously, the sequence data must be of good quality to allow quantification of this type of genetic polymorphism (20). Similarly, a great need for good quality sequencing exists in the field of cancer diagnosis involving direct genomic analysis of oncogenes and/or suppressor genes. To allow for direct analysis of biopsies, mutation "hot spots" must be evaluated. The signal from the specific mutant sequence is lowered beause it is often difficult to prepare cancer cells free from normal cells and/or the cancer genotype is often heterozygotic.

Recently, Hopgood et al. (6) evaluated different direct sequencing methods for the forensic analysis of human mitochondrial DNA using hair shafts and semen. The best results were obtained using magnetic beads as solid support and T7 polymerase (7,9). Solid-phase sequencing has also been shown recently to allow direct analysis of human heterozygotic alleles (18) and detection of mutations in the HIV-1 pol gene associated with development of AZT resistance (20). The question arises whether solid-phase sequencing can be used not only to detect but also to quantify the distribution in mixed sequences of a more complex nature.

To evaluate the efficiency and accuracy of solid-phase sequencing for the analysis of mixed populations, direct sequencing of the variable region 3 (V3) of the envelope gene of human immunodeficiency virus type 1 (HIV-1) has been carried out. The V3 region contains a cysteine loop recognized by human neutralizing antibodies (14,15) and seems to be under strong evolutionary pressure. The high variability (11,13) makes this region suitable for our study. Reconstitution experiments using cloned sequences as well as analysis of clinical samples directly obtained from patients showed that solid-phase sequencing can be used for rough quantification of highly polymorphic sequences.

MATERIALS AND METHODS

Clinical Samples and PCR

Five HIV-1-infected adult patients of European origin were studied. All patients were drug users or partners of drug users.

The DNA used was derived directly from the patient's uncultured peripheral blood mononuclear cells (PBMCs) (1) and amplified by PCR with nested primers, outer set: JA 9 and JA 12 and inner set: RIT 124 and RIT 128. The primers were specific for the V3 of the envelope glycoprotein (gp 120), as previously described (1,21), and the inner primer RIT 124 was biotinylated. In the first (outer) amplification, 24 cycles were done (94°C for 30 s, 50°C for 30 s and 72°C for 30 s, plus a final extension at 2°C for 5 min); then the product (diluted 1:100) was further amplified with the inner primer set for 30 cycles (with the same program). Negative (lysis buffer and samples from healthy blood donors) and positive controls (infected cells containing 10 HIV-1 viral DNA molecules) were included in each run. All PCR products were verified on ethidium bromide-stained 1.5% agarose gel electrophoresis.

Cloning

The PCR product amplified by the inner primers was purified using the Qiagen PCR Purification Spin Kit (Diagen, Hilden, Germany) and dissolved in 20 μL Tris-EDTA. The PCR fragments were treated with Klenow fragment of DNA polymerase I (Pharmacia Biotech, Uppsala, Sweden) as well as T4 polynucleotide kinase (Pharmacia Biotech). After precipitation with ammonium acetate, approximately 1/10 of the purified PCR product was ligated with 50 ng *Sma*I-digested pUC18 vector (Pharmacia Biotech) using the DNA ligating system (Amersham International, Little Chalfont, Buckinghamshire, England, UK). The ligated vector was used to transform competent cells *E. coli* JM105 (Pharmacia Biotech). Cells were transformed using the ligand vector for 90 s at 42°C. Positive colonies were picked, resuspended in 10 μL PCR buffer (10 mM Tris-HCl pH 8.3, 50 mM KCl and 0.01% Tween 20) and treated at 95°C for 5 min. To confirm the presence of the insert in the plasmid, amplification with pUC18-specific primers (RIT 28-29) were performed (8).

Sequencing

The PCR product used for direct sequencing was amplified with a biotinylated inner primer set (RIT 124 and RIT 128) (21) from the outer product of the nested PCR amplification.

All sequencing by either direct sequencing or indirect cloning was performed with a robotic and automated DNA sequencing system, as described in Reference 7. This system consists of a robotic workstation (Biomek™ 1000; Beckman Instruments, Fullerton, CA, USA) using a 96-well plate and an automated laser fluorescent DNA sequencer (A.L.F. DNA Sequencer; Pharmacia Biotech).

The workstation was configured to immobilize DNA fragments on streptavidin-coated magnetic beads (Dynabeads™ M-280; Dynal, Oslo, Norway) (9), binding to the biotinylated PCR primers (RIT 124 or RIT 29) and then melting the double-stranded DNA into two single-stranded DNA strands, one immobilized and one in the supernatant (22). In the next step, the workstation performed the sequencing reactions; these were done on both the immobilized and the supernatant strand using T7 DNA polymerase (Pharmacia Biotech) in a Sanger reaction (17). On the immobilized strand, the M13 universal sequencing primer was used; and on the supernatant strand, JA 10 F (when PCR primers RIT 124 and RIT 128 were used in the direct sequencing) or reverse primer (when PCR primers RIT 28-29 were used for sequencing clones) was used. The sequencing primers were fluorescein marked (FITC; Pharmacia Biotech) for laser detection of the fragments.

The generated sequence fragments were then loaded on the automated laser fluorescent sequencing apparatus. For separation, we used the Gel-Mix® 6 6% sequencing gel solution (Ultra Pure) (GIBCO/Life Technologies, Gaithersburg, MD, USA) of 0.5 thickness.

Heterogeneous Sample Reconstitution

Two clonal samples with diverse nucleotide sequences, clone 1 of patient 196 and clone 2 of patient 199, were used for mixing in different ratios. After PCR amplification (GeneAmp® PCR System 9600; Perkin-Elmer, Norwalk, CT, USA) with the biotinylated inner primer set (RIT 28-29) and immobilizing the fragments on streptavidin-coated magnetic beads, the double-stranded DNA was melted into single-stranded DNA. This left one strand immobilized and one strand in the supernatant. The supernatant strand was used to measure the amount of DNA created in the PCR amplification by optical density at 260 nm. The immobilized strands of the two clones were then slightly adjusted to keep their concentrations equal. They were mixed in different ratios (1:1, 1:3 and 1:9) and placed into a workstation performing the sequencing reactions (described under the subheading Sequencing). Duplicate samples of the clones and the mixtures were then sequenced on the same run.

Evaluation

The sequences were all processed with the A.L.F. Manager 2.2 program and manually edited for polymorphic sites, giving alternative bases, caused by the heterogeneous virus population. A subjective estimation of the distribution was made of the overlapping peaks at the polymorphic sites (Figures 1, 2 and 4) to decide major and minor forms. Peak areas and heights, compared with neighboring peaks of the same base, were considered in the evaluation. The IUPAC ambiguity codes (M = A or C; R = A or G; W = A or T; S = C or G; Y = C or T; K = G or T; V = A, C or G; H = A, C or T; D = A, G or T; B = C, G or T; and N is A, C, G or T) for uncertain bases were used to describe these positions; but here these should be regarded as alternative codes. The nucleotide sequence was then translated into an amino acid sequence and presented with major and minor forms (uppercase and lowercase letters, respectively) based on peak distributions (Figure 3).

RESULTS

Direct Sequencing of Clinical Samples

The V3 domain of the HIV-1 envelope gene was chosen as a model for this evaluation due to its high degree of variability and its important functional

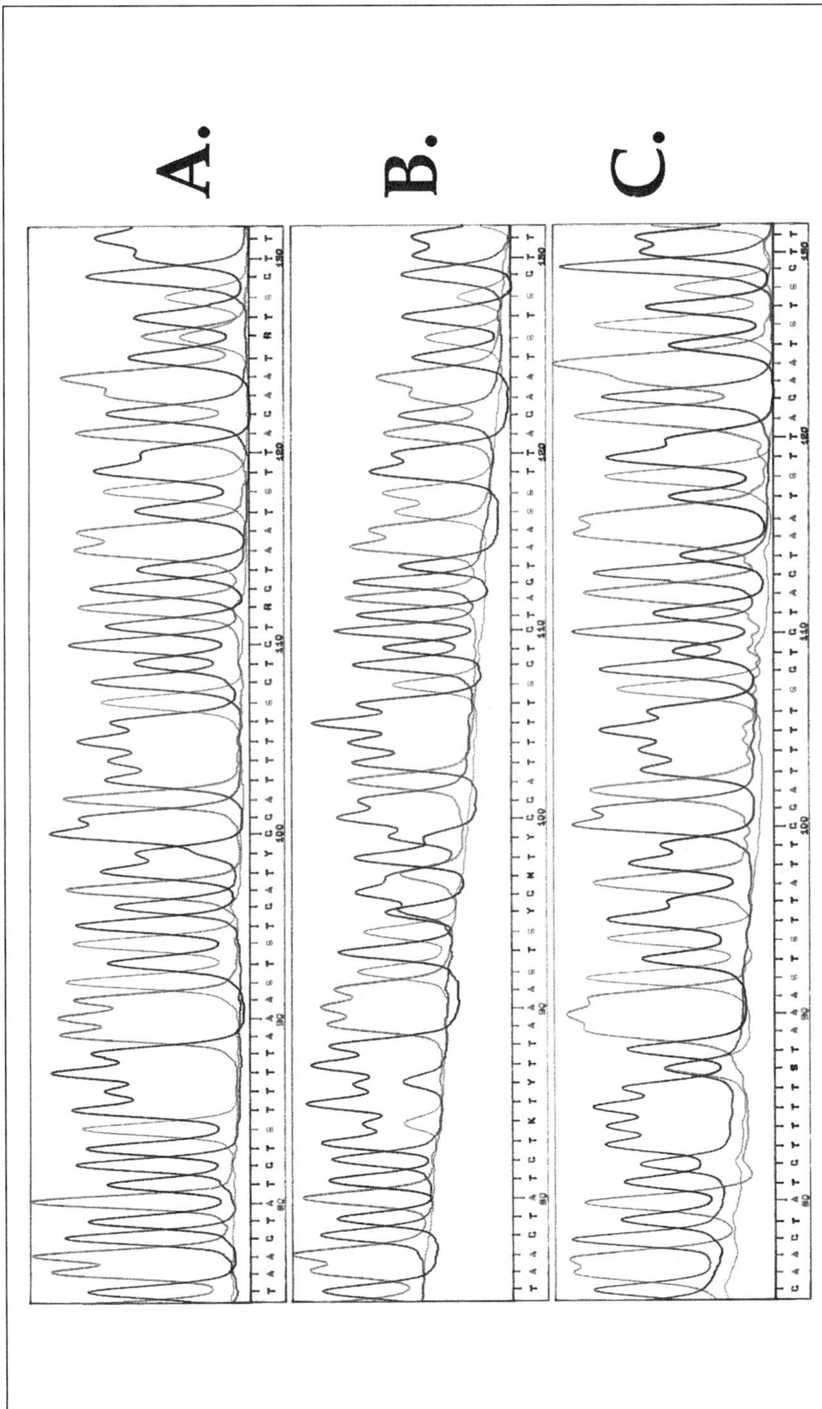

Figure 1. Examples of the raw data output of three independent clinical samples from the automated laser fluorescent DNA sequencer (A.L.F.) using direct solid-phase sequencing of a part of the HIV-1 genome encoding the V3 loop. Black letters indicate bases with possible mixed populations, as indicated by overlapping peaks at these positions. Note that the results of patient B suggest a highly heterogeneous virus population.

role. The robotic workstation used has been described in References 7 and 22 and the design and use of the V3-specific primers in Reference 21. A nested primer approach was used to decrease the background signal. No precipitation, filtration or centrifugation steps were needed throughout the whole procedure.

Three examples of data from the same region obtained by direct sequencing of uncultured PBMCs from HIV-1-infected patients are shown in Figure

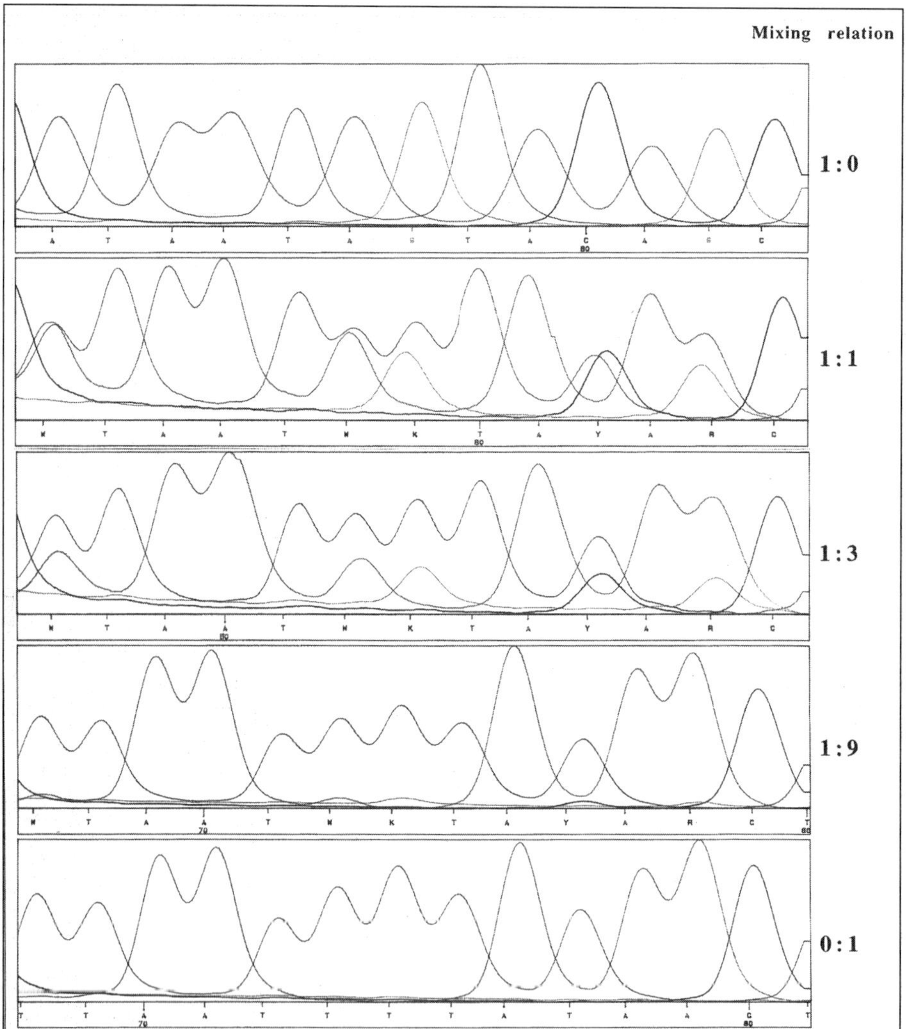

Figure 2. Quantification of polymorphism. A reconstitution of a heterogeneous sample was made in different ratios, as indicated in the figure. Two clones were mixed, patient 196 clone 1 and patient 199 clone 2; the mixtures and the pure clones were then sequenced and analyzed. The ambiguity codes were manually edited at the positions where the polymorphic sites occurred and should be regarded as alternative codes.

1. Clear sequences with very low background signal were obtained, thus facilitating the analysis of the heterogeneous sites. Note that the raw data output is used throughout the analysis to avoid artifacts due to software processing errors. Mixed peaks were obtained at positions in the sequences as indicated in IUPAC code with black letters. Note that some of these peaks, i.e., the approximately 50/50% peak at position 126 (A/G) in Figure 1A, position 95 (C/T) in Figure 1B and position 87 (C/G) in Figure 1C, are most likely significant, while it is more difficult to judge whether the minor peaks at other positions, i.e., positions 99 and 112 in Figure 1A, represent real polymorphism or simply background signal.

Direct Sequencing of Mixed Samples

A set of reconstitution experiments were performed to evaluate the direct sequencing method used to determine and quantify minor sequences. A 330-bp fragment flanking and including the V3 domain was amplified by PCR from two patients, and the resulting fragments were cloned into a plasmid vector and sequenced (data not shown). Two clones with several differences in the sequence were chosen for the analysis. The clones were subsequently mixed in different ratios, and solid-phase sequencing was performed using T7 polymerase. The non-processed results for samples containing 1:1, 1:3 and 1:9 proportions of the two templates are shown in Figure 2. By comparing the peak area and height to neighboring peaks of the same base, it was evident that the 1:1 mixture gave close to the expected 50% peak area/height. When the clones were mixed 1:3, a significantly lower signal for one of the bases in all of the polymorphic sites was observed (Figure 2). It is noteworthy that both clonal sequences can be deduced from this single sequence, since the major species has an approximately 75% peak area, while the minor species has a 25% area. In the case where 10% (1:9) of one template was used, only a signal slightly above background could be observed. Using the method described here, it is unlikely that sequence polymorphism present in 10% or less of the total viral genomes can be determined using the direct sequencing method.

Comparative Sequence Analysis of Genomic and Cloned Material

To further evaluate the direct sequencing method, analysis of the HIV population in five patients was carried out. The V3 loop was directly amplified from uncultured PBMCs in the patients, and the material was sequenced by direct solid-phase sequencing. The PCR material was also cloned into a plasmid vector, and individual clones were isolated and sequenced separately. The results are presented in Figure 3. The sequences reveal both highly heterogeneous and less variable viral populations.

Patient 190 had three positions with significant heterogeneity as indicated by direct sequence. A glutamic acid/lysine mixture was observed at position 1, while an aspartic acid/glutamic acid and a lysine/glycine heterogeneity

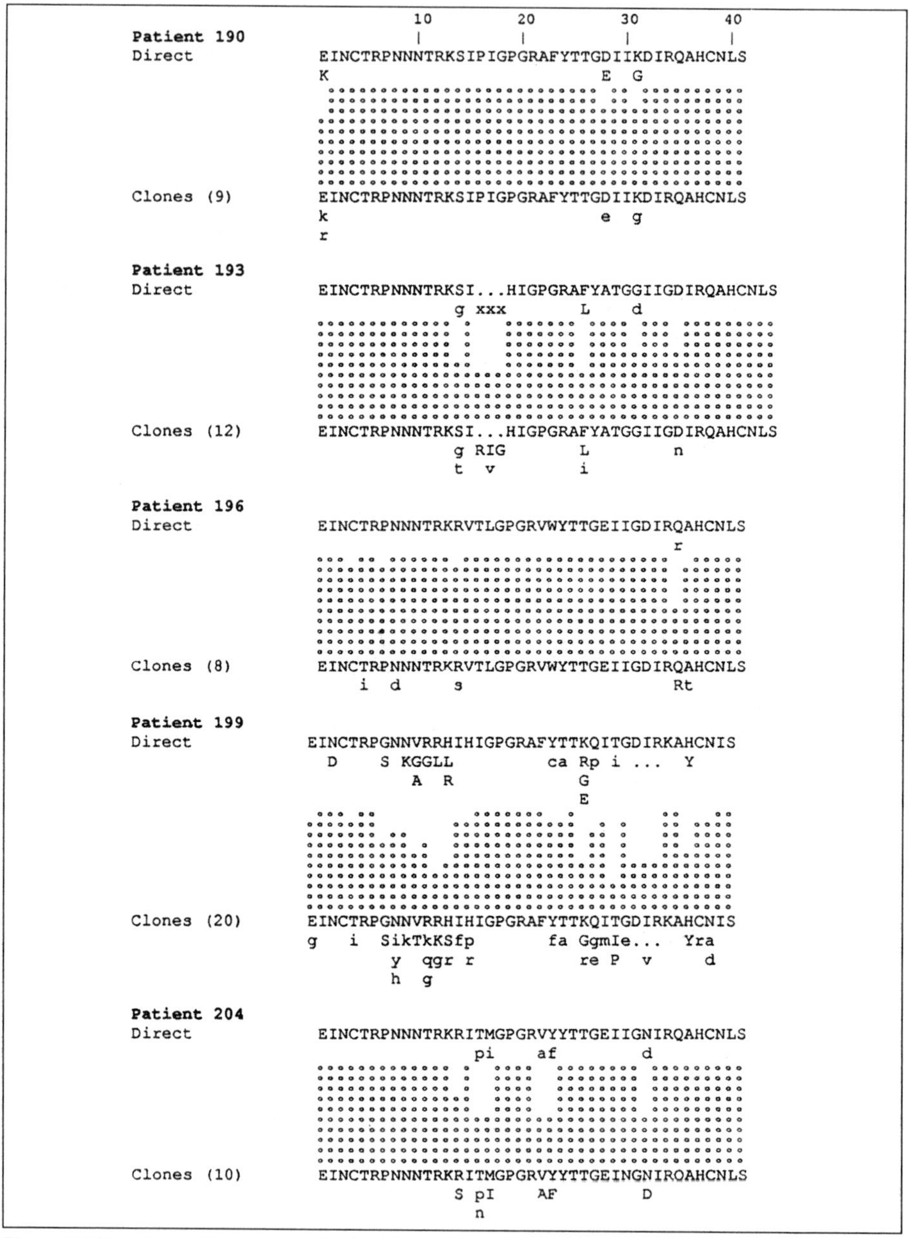

Figure 3. V3 amino acid sequences derived from direct sequencing and of clones from five patients. The direct sequence is presented at the top in each set as deduced amino acids with heterogeneous positions indicated by alternative residues below. Uppercase letters represent major amounts (>40%) and lowercase letters minor amounts (<40%) of the amino acid. Periods indicate insertions or deletions and x's are amino acids that were not possible to evaluate. The histograms are calculated from the sequences of the individual clones. The number of clones sequenced in each case is indicated. The consensus sequence is shown, presented in the same way as the direct sequence, and the circles in the histogram show the proportion of clones with the consensus sequence. Each circle represents 10% of the clones.

were observed at positions 28 and 31, respectively. Interestingly, the analyses of individual clones from the patient yield the same consensus sequence. In addition, the proportion of clones which carried the three mutations was in agreement with what was predicted by the direct genomic sequence.

Patient 204 was somewhat more heterogeneous as determined by the direct sequencing approach. The analysis revealed that five residues were variable, although in all cases a major form could be deduced. Note that none of these residues corresponds to the heterogeneous residues observed in patient 190. An example of the primary data from the electrophoresis unit is presented in Figure 4. From the results, it is possible to estimate that the R in position 88 is about 25% A and 75% G, and the R in position 92 had 50% of each base. The K at position 90 is predominantely G, but to a small propor-

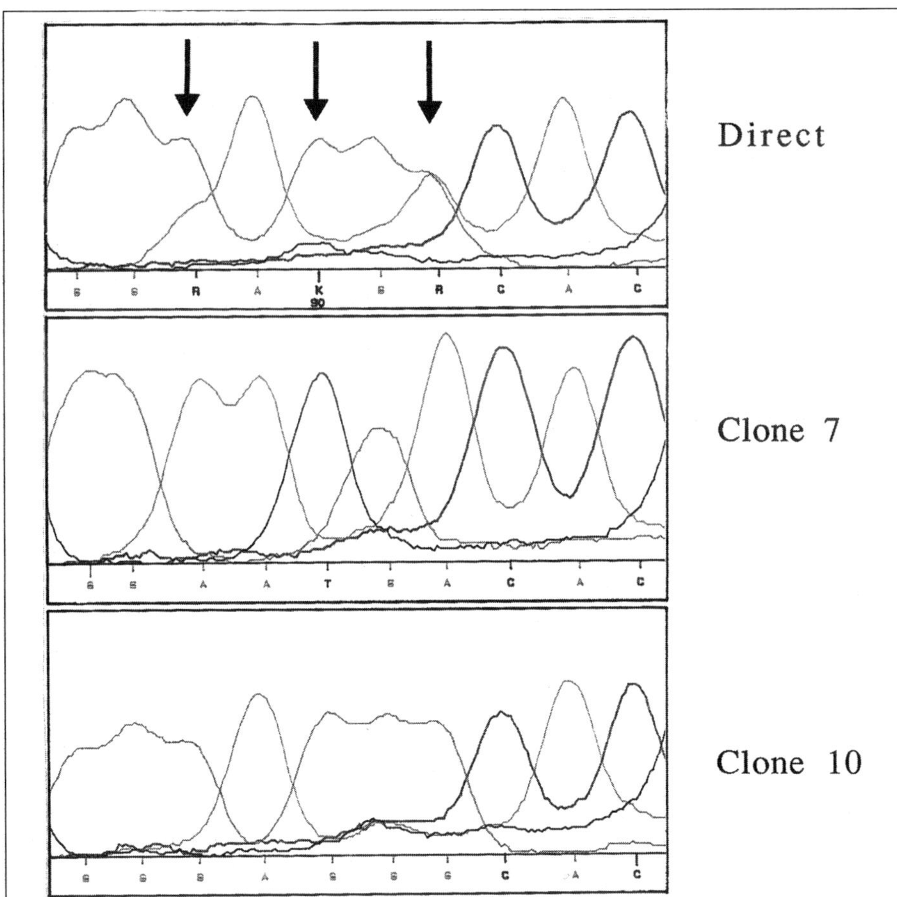

Figure 4. An example of a small part of the direct genomic sequence of the V3 loop of patient 204 compared with the sequence of two independent clones obtained from the same patient. Three positions with alternative nucleotides are presented as follows: R in position 88, K in position 90 and another R in position 92.

tion also a T. The cloning confirmed that all five residues observed by the direct method indeed correspond to mixed viral populations. However, the clones suggested an approximately equal mixture in the variable sites and an additional minor site (arginine/serine) in position 16 (Figure 3).

Patient 196 had a homogeneous population as detected by both the direct sequencing and the indirect cloning method. Only a glutamine/arginine mixture was observed for the direct method, while cloning showed minor variations (1 of 8) in 4 additional positions. These 4 residues probably constitute minor variants below the detection limit (ca. 10%) of the direct method. However, it is not unlikely that some or all of these rare mutations were due to *Taq* DNA polymerase errors introduced during the PCR amplification. Both methods suggest the same population with one certain variable amino acid in the studied region.

Patient 199 had a very heterogeneous viral population as detected by direct sequencing. The sequence had signs of point mutations as well as length variations. Populations of genomes with differences in length yield sequence data with complex patterns since the different genotypes are "out-of-frame" relative to each other. By sequencing the same region on both strands, a relatively clear sequence could be obtained throughout the whole region (data not shown). Sequences of clones, from the same patient, confirmed two major viral subpopulations with a length variation of 9 nucleotides (data not shown). The results of direct sequencing revealed large areas with various heterogeneity. In particular, residues 8 to 14 and 25 to 28 showed considerable variations. Interestingly, analysis of the cloned material showed variability in the same regions, with the same approximate relative proportions of the various residues (Figure 3). The deletion between residues 29 and 31 is observed in approximately half of the clones.

Patient 193 had a population with three variable amino acids but also a length variation. With direct sequencing, it was possible to read clearly until the length variation. After this point, as in patient 199, overlapping sequence was obtained (data not shown). The comparison with the cloned material showed good agreement with mixed populations in positions 14 (serine/glycine), 25 (phenylalanine/lysine) and 34 (glycine/aspartic acid). Note that the residues corresponding to the length variation (arginine-isoleucine-glycine) cannot be deduced easily by the direct approach, since this region will consist of mixed sequences when sequencing both strands. However, it is also noteworthy that the direct sequencing of this relatively complex pattern of quasi-species allowed us to determine and roughly quantitate the various major subpopulations.

DISCUSSION

Our results show that direct solid-phase sequencing can be used to analyze complex mixtures of heterogeneous viral populations. The more laborious cloning approach showed a close similarity in both the determination of the consensus sequence and the relative amounts of the different variants in the samples. The reconstitution experiments suggest that minor populations consisting of between 10% and 25% of the total population can be determined and analyzed. To obtain sequence data of high quality, several imortant points were considered. The PCR product was purified and concentrated by immobilization to a solid support, which gives a clean single-stranded template. The use of magnetic beads allowed us to use a semi-automated procedure involving robotic workstations that have a magnetic separator and a heating/cooling device (22). A key to the quality of the data is the use of a polymerase with high processivity that yields even peak heights, thus allowing quantification of mixed populations simply by measuring the peak heights/areas at positions with mixed nucleotides. Here, we used T7 polymerase using manganese as suggested by Tabor and Richardson (19).

The traditional way to analyze a heterogeneous population is by isolating many individual clones followed by large-scale sequencing. This procedure has many disadvantages as compared with the direct sequencing approach described here. Cloning is laborious and time-consuming, and the yield of recombinant clones is often variable. In addition, the cloning procedure may be selective, thus resulting in skewing of the data. The quantification is also performed in an indirect manner since the consensus sequence is obtained by statistical analysis of the various clones. When cloning PCR fragments, it can be difficult to determine whether random mutations that are introduced by the polymerase during the amplification step represent natural variations in the sample population or if they are PCR artifacts. This cannot be clarified without a detailed analysis of clones obtained by methods other than PCR. In the direct sequence, these PCR-induced mutations will only yield a "noise," which will not be detected. Note, however, that the linkage between mutations cannot be determined by direct sequencing of amplified fragments. If two or three juxtaposed nucleotide positions are variable, the translation might give the possibility of additional amino acid residues depending on the particular codon and the positions within it. Sequencing of individual clones can therefore be a valuable tool to determine such relationships.

An alternative technique to recombinant cloning procedures is to perform so-called limited dilution (2) in which the sample is diluted to allow amplification of single initial templates. Thus individual clones can be analyzed without the need for laborious cloning into recombinant vectors. However, such procedures are not suitable for large-scale routine analysis, since serial dilutions are needed and many PCRs have to be performed to obtain a sample with only one single molecule.

It is important to point out that the processing of the data using commercial instruments often involves adjustments of baselines and some lowering of "noise" signals. Since these background noises can be of interest to estimate minor populations, we prefer to use the raw data from the sequencing during the analysis. Processed data will have an adjusted baseline and peak treatment not giving full control of the separation and detecting events in the sequencing machine. As can be observed from Figures 1 and 4, the raw data output, using the automated solid-phase protocol (7,9) and the T7 polymerase, yields sequence data of high quality, thus simplifying the analysis of heteroegeous populations.

In conclusion, the direct sequencing protocol using magnetic beads used in this work is suitable for automation, thus opening up possibilities for large-scale routine analysis of complex viral and other heterogeneous genomic populations.

ACKNOWLEDGMENTS

This work was supported by U.S. Army Medical Research and Development Command under Grant DAMD17-89-Z-9038. Grants were also received from the Swedish Medical Research Council, the Swedish National Board for Technical Development and the Pediatric AIDS Foundation. We are grateful to T. Hultman and J. Wahlberg for their helpful discussion and critical comments.

REFERENCES

1. **Albert, J. and E.M. Fenyö.** 1990. Simple, sensitive and specific detection of HIV-1 in clinical specimens by polymerase chain reaction with nested primers. J. Clin. Microbiol. *28*:1560-1564.
2. **Brichman, J.E., J. Albert and F. Vartdal.** 1991. Few infected CD4+ T cells but a high proportion of replication competent provirus copies in asymptomatic human immunodeficiency virus type 1 infection. J. Virol. *65*:2019-2033.
3. **Carothers, A.M., G. Urlaub, J. Mucha, D. Grunberger and L.A. Chasin.** 1989. Point mutation analysis in a mammalian gene: rapid preparation of total RNA, PCR amplification of cDNA, and *Taq* sequencing by a novel method. BioTechniques *7*:494-499.
4. **Gyllensten, U.B. and H.A. Erlich.** 1988. Generation of single-stranded DNA by the polymerase chain reaction and its application to direct sequencing of the HLA-DQA locus. Proc. Natl. Acad. Sci. USA *85*:7652-7656.
5. **Higuchi, R.G. and H. Ochman.** 1989. Production of single stranded DNA templates by exonuclease digestion following the polymerase chain reaction. Nucleic acids Res. *17*:5865.
6. **Hopgood, R., K.M. Sullivan and P. Gill.** 1992. Strategies for automated sequencing of human mitochondrial DNA directly from PCR products. BioTechniques *13*:82-92.
7. **Hultman, T., S. Bergh, T. Moks and M. Uhlén.** 1991. Bidirectional solid-phase sequencing of *in vitro*-amplified plasmid DNA. BioTechniques *10*:84-93.
8. **Hultman, T., M. Murby, S. Ståhl, E. Hornes and M. Uhlén.** 1990. Solid phase *in vitro* mutagenesis using plasmid DNA template. Nucleic Acids Res. *18*:5107-5112.
9. **Hultman, T., S. Ståhl, E. Hornes and M. Uhlén.** 1989. Direct solid phase sequencing of genomic and plasmid DNA using magnetic beads as solid support. Nucleic Acids Res. *17*:4937-4946.
10. **Innis, M.A., D.H. Gelfand, J.J. Sninsky and T.J. White.** 1990. PCR Protocols. Academic Press, San Diego, CA.
11. **LaRosa, G.J., J.P. Davide, K. Weinhold, J.A. Waterbury, A.T. Profy, J.A. Lewis, A.J. Langlois, G.R. Dreesman, N. Boswell, P. Shadduck, L.H. Holley, M. Karplus, D.P. Bolognesi, T.J. Matthews, E.A. Emini and S.D. Putney.** 1990. Conserved sequence and structural elements in the

HIV-1 principal neutralizing determinant. Science *249*:932-935.

12. **Meyerhans, A., R. Cheynier, J. Albert, M. Seth, S. Kwok, J. Sninsky, L. Morfeldt-Månson, B. Asjö and S. Wain-Hobson.** 1989. Temporal fluctuations in HIV quasispecies in vivo are not reflected by sequential HIV isolations. Cell *58*:901-910.
13. **Myers, G., B. Korber, J.A. Berzofsky, R.F. Smith and G.M. Pavlakis.** 1992. Human retroviruses and AIDS. 1991. Los Alamos National Laboratory, Los Alamos, NM.
14. **Palker, T.J., M.E. Clark, A. Langlois, T.J. Matthews, K.J. Weinhold, R.R. Randall, D.P. Bolognesi and B.F. Haynes.** 1988. Type-specific neutralization of the human immunodeficiency virus with antibodies to env. encoded synthetic peptides. Proc. Natl. Acad. Sci. USA *85*:1932-1936.
15. **Rusche, J.R., K. Javaherian, C. McDanal, J. Petro, D.L. Lynn, R. Grimaila, A. Langlois, R.C. Gallo, L.O. Arthur, P.J. Fischinger, D.P. Bolognesi, S.D. Putney and T.J. Matthews.** 1988. Antibodies that inhibit fusion of human immunodeficiency virus infected cells bind to a 24 amino acid sequence of the viral envelope, gp 120. Proc. Natl. Acad. Sci. USA *85*:3198-3202.
16. **Saiki, R.K., D.H. Gelfand, S. Stoffel, S.J. Scharf, R. Higuchi, G.T. Horn, K.B. Mullis and H.A. Erlich.** 1988. Primer-directed enzymatic amplification of DNA with thermostable DNA polymerase. Science *239*:487-491.
17. **Sanger, F., S. Nicklen and A.R. Coulson.** 1977. DNA sequencing with chain terminating inhibitors. Proc. Natl. Acad. Sci. USA *74*:5463-5467.
18. **Syvänen, A.C., T. Hultman, K. Aalto-Setälä, H. Söderlund and M. Uhlén.** 1991. Genetic analysis of the polymorphism of the human apolipoprotein E using automated solid-phase sequencing. Genet. Anal. *8*:117-123.
19. **Tabor, S. and C.C. Richardson.** 1987. DNA sequence analysis with a modified bacteriophage T7 DNA polymerase. Proc. Natl. Acad. Sci. USA *84*:4767-4771.
20. **Wahlberg, J., J. Albert, J. Lundeberg, S. Cox, B. Wahren and M. Uhlén.** 1992. Dynamic changes in HIV-1 quasispecies from azidothymidine (AZT)-treated patients. FASEB *6*:2843-2847.
21. **Wahlberg, J., J. Albert, J. Lundeberg, A. von Gegerfelt, K. Broliden, G. Utter, E.M. Fenyö and M. Uhlén.** 1991. Analysis of the V3 loop in neutralization-resistant human immunodeficiency virus type 1 variants by direct solid-phase DNA sequencing. AIDS Res. Hum. Retroviruses *7*:983-990.
22. **Wahlberg, J., A. Holmberg, S. Bergh, T. Hultman and M. Uhlén.** 1992. Automated magnetic preparation of DNA templates for solid phase sequencing. Electrophoresis *13*:547-551.

From Field to Film: Rapid Sequencing Methods for Field-Collected Plant Species

Daniel L. Nickrent

Southern Illinois University, Carbondale, IL, USA

ABSTRACT

Methods that have proven effective in the rapid extraction of plant DNA, PCR amplification and sequencing of ribosomal and chloroplast genes are presented. Techniques that can be used under field conditions to preserve DNA for subsequent extraction are reviewed. Tissues that are fresh, heat-desiccated, silica gel-desiccated and cetyltrimethylammonium bromide (CTAB) buffer-preserved are compared for DNA quality and quantity. The "delayed CTAB" method yields high molecular weight DNA, thereby providing an alternative to preservation using silica gel. Optimized methodologies for PCR amplification and double-stranded sequencing of the product are detailed including gel purification of PCR products using DEAE membranes.

INTRODUCTION

With the growing interest in plant molecular phylogenetic studies and the advent of the polymerase chain reaction (PCR) (20), an increasing number of workers are using DNA sequencing as opposed to other approaches. This report describes a series of methods that have proven effective in the extraction of DNA, PCR amplification and sequencing of plant ribosomal and chloroplast genes. Given an interest in obtaining sequence information from a large number of plants from taxonomically diverse groups, methods have been selected that do not require long optimizations for any one species. Although not all of the methodologies compiled here are novel, each has been selected for convenience, speed and simplicity without sacrificing the final sequence quality. A review of the methods used to obtain DNA from plant samples that have been collected in remote locations and preserved in different ways, and a liquid preservation procedure—the "delayed hot CTAB" method—are discussed. The use of gel purification and direct sequencing of double-stranded

(Reprinted from BioTechniques 16:470-475)

PCR products is also presented.

PLANT DNA PRESERVATION AND EXTRACTION

The eventual use of the DNA ultimately determines one's requirements for amount and purity. Fresh or liquid nitrogen frozen tissues are obviously ideal; however, these options may not be available when collecting plants in remote areas (4,18,27). For this reason, a number of workers have experimented with alternate methods to preserve plant tissues prior to initiating DNA extractions. The limiting factors associated with obtaining good quantity and quality DNA from plants are tissue type, tissue amount (8,26), presence of secondary compounds such as polyphenolics (28), polysaccharides (6, 10), and extent of DNA degradation (when using other than fresh tissues). One of the first studies to examine such factors was by Rogers and Bendich (26) who tested the effect of the preservation method (fresh, frozen, dried, chemically fixed, etc.), tissue type (leaves, seeds, cotyledons, pollen, etc.) and tissue amount (ranging from milligram to gram amounts) on genomic DNA yield. Similarly, the study by Pyle and Adams (24) tested dried leaves as well as tissues stored in a number of chemical preservatives such as ethanol, Chlorox®, NaCl and formaldehyde. DNA was preserved for less than seven days in all solutions. Rogstad (27) had better success using a saturated NaCl–cetyltrimethylammonium bromide) (NaCl–CTAB) solution, at least for some of the plant species used.

One of the most effective and widely used methods to preserve plant tissue for subsequent DNA extraction is rapid drying at ambient temperatures using desiccants such as silica gel or $CaSO_4$ (W.A. Hammond Drierite, Xenia, OH, USA). Experiments using silica gel showed that this method often results in high-quality DNA that can be used for restriction endonuclease digestions, PCR and other applications (4,18). The most appealing aspect of this preservation method is that field-collected plant material can be quickly and easily preserved and shipped at ambient temperatures for later DNA extraction. It should be noted that

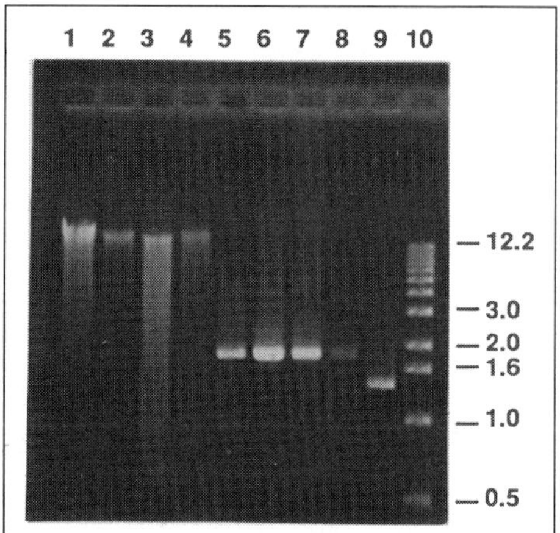

Figure 1. Genomic DNA (lanes 1–4), PCR-amplified 18S rRNA (lanes 5–8), PCR-amplified *rbc*L (lane 9) and 1-kb ladder (lane 10). Lane 1, fresh extracted *Viscum album* (Nickrent No. 2145). Lane 2, from silica gel-dried *Thesium impeditum* (No. 2845). Lane 3, delayed (4 weeks) CTAB-extracted *Notothixos subaureus* (No. 2790). Lane 4, from herbarium specimen of *Oncocalyx sulfurens* (No. 2850). Lanes 5–8, 1.8-kb PCR products amplified from genomic DNA shown in lanes 1–4. Lane 9, 1.4-kb PCR product amplified from delayed CTAB-extracted genomic DNA of *Opilia amentacea* (No. 2816).

extreme drying, e.g., over phosphorus pentoxide, removes structurally important water molecules from the DNA double helix, which makes bases more prone to damage (17).

Plant samples dried as herbarium specimens have also been used as sources of DNA. Such samples may be suitable for PCR amplification; however, the quantity and quality are often not sufficient for restriction fragment analyses. The factors that are important in obtaining DNA from traditionally dried samples are age of the specimen, drying temperature (gentle heat is optimal) and the taxonomic group. The use of ethanol or formaldehyde to combat fungal growth on specimens collected in moist, tropical areas adversely affects DNA quality and should be avoided if the samples are to be used for molecular work. Successful DNA extractions from herbarium material were obtained from several families of parasitic plants such as an 8-year-old specimen of *Ginalloa* (Viscaceae) and a number of genera of African Loranthaceae (e.g., *Oncocalyx*, Figure 1, lane 4). The latter are interesting since members of this family typically contain large amounts of secondary compounds that make DNA extractions difficult even with fresh tissue. Since the leaf samples were not discolored (darkened), it is assumed that they were dried slowly and with low heat.

Since silica gel drying may result in DNA degradation, an alternative method was developed called the "delayed hot CTAB" method. This procedure, a modification of that described by Rogstad (27), has also been used successfully to obtain DNA from 20 different genera of parasitic plants from Costa Rica and Australia. This method assumes no refrigeration is available and all of the following items can be packed into a small suitcase (carry-on size for most airlines): homogenizer (e.g., Polytron®; Brinkmann Instruments, Westbury, NY, USA), stirring hot plate and stir bar, miniature top-loading balance, thermometer, wash bottle, glass test tubes (25× 200 mm), cheesecloth, funnels, as many 50-mL sterile polypropylene centrifuge tubes with caps (e.g., Corning 25330; Corning, NY, USA) as the number of samples to be collected, scissors, beakers, 500-mL plastic bottle and the dry CTAB buffer components. The modified 2× CTAB buffer (8) is composed of 100 mM Tris-HCl, 1.4 M NaCl, 30 mM EDTA, 2% wt/vol CTAB, 5 mM ascorbic acid, 4 mM diethyldithiocarbamic acid (DIECA) and 2% wt/vol polyvinylpyrrolidone (PVP-40), the latter two ingredients added just before use.

The best DNA extracts come from young, recently collected material. For extraction, the buffer (ca. 25 mL for every 2–3 g of plant tissue) is heated to 95°C in a water bath. The sample is cut into small pieces, placed in the glass test tube and homogenized in the hot CTAB buffer. Homogenization should use only short bursts at low speeds to reduce shearing of DNA and minimize foaming. The extract is then strained through cheesecloth into the sterile 50-mL tube and incubated in a warm (70°–80°C) water bath for about 30 min with occasional swirling. The parasitic plant samples collected in Australia were processed to this point and stored at ambient temperature for four weeks prior to completing the extraction.

After returning to the laboratory, the samples are briefly centrifuged, and 20 mL of the supernatant are transferred to a clean 50-mL centrifuge tube. Chloroform:isoamyl alcohol (24:1) (0.7 volume) is added and the tubes are mixed for 5 min. They are then centrifuged at 8500 rpm for 15 min, and the aqueous phase is removed with a wide-bore pipet and placed in a clean centrifuge tube. The DNA is precipitated by the addition of a 0.7 volume of ice-cold isopropanol and incubated at -20°C for at least 1 h. The DNA pellet is collected by centrifugation for 20 min at 10 000 rpm. The pellet is allowed to dry at ambient temperature for 30 min and then resuspended in 3.0 mL of TE (1 mM Tris, 0.1 mM EDTA, pH 7.5) and 2.0 mL of 4.0 M ammonium acetate. The solution is extracted with an equal volume of Tris-saturated (pH 8.0) phenol and chloroform (1:1), and 2 volumes of ethanol are added to the aqueous phase. The tube is incubated at -20°C for at least 30 min, and the DNA is collected by centrifugation, dried and resuspended in TE. An RNase digestion can be performed at this time if required.

The above method differs from that of Rogstad (27) in that the buffer is not saturated in CTAB and NaCl, and it contains additional ingredients. Also, instead of simply cutting the leaf tissue and placing in the CTAB buffer, this method quickly homogenizes the tissue in buffer, which is similar to grinding on liquid nitrogen (the standard method when starting in the laboratory). Homogenization and heating following extraction may be important in allowing the buffer components to quickly begin binding polyphenolics and denaturing DNases. Figure 1 shows the results of genomic DNA extractions from fresh plant tissue or tissue preserved using silica gel, the delayed CTAB method and traditional drying (herbarium specimen). Although some DNA degradation is seen, the delayed hot CTAB method results in a sufficient quantity of high molecular weight DNA for PCR amplification and other procedures (Figure 1, *Notothixos*, lane 3).

PCR AMPLIFICATION OF PLANT GENES

Increasing attention has recently been focused on the direct sequencing of double-stranded PCR products. This method is preferred over asymmetric PCR (11) since it eliminates the second round of amplification (thereby saving time) and further reduces base incorporation errors inherent in the PCR process. It also overcomes the difficulty sometimes encountered when attempting amplification from an existing PCR product and allows determination of sequence from both strands of the same product.

In plant phylogenetic studies, the chloroplast gene *rbc*L has been the focus of a number of studies (1,22); however, nuclear 18S (and 26S) ribosomal RNA is now gaining increasing attention (12,21). PCR amplifications were performed in 0.5-mL microcentrifuge tubes in a Perkin-Elmer DNA Thermal

Cycler. The reaction mixture contained (final concentrations in a 100-mL reaction): 50 mM KCl, 10 mM Tris-HCl pH 8.8, 0.1% Triton® X-100, 2.5 mM $MgCl_2$, 1.25 mM of each deoxyribonucleoside triphosphate (dNTP), 1 µL of each primer at 125 µg/mL (ca. 20 pmol), 2.5 units of *Taq* DNA polymerase (M166; Promega, Madison, WI, USA), and about 10–30 ng of genomic DNA. The tubes are overlaid with mineral oil, centrifuged briefly and placed in the thermal cycler. An initial incubation (time delay) at 94°C for 3 min is linked to 35 cycles, each of which consists of 1 min at 94°C, 1 min at 50°C, and 2.5 min at 72°C. Two seconds are added to each extension step as the cycling progresses using the autoextension function. Figure 1 shows the results of PCR amplification of a 1.8-kb portion of 18S rDNA obtained from plant genomic DNAs (fresh and preserved in silica gel, CTAB buffer and an herbarium specimen) and a 1.4-kb product containing *rbc*L amplified from delayed CTAB-preserved *Opilia.*

In the PCR mixture, the optimal $MgCl_2$ concentration for the particular organism, primer combination, etc., must be determined experimentally and ranges from 1.0 to 3.0 mM (often on the high end of the range for plants). Oxidation of the dNTPs occurs following numerous rounds of freezing and thawing. This is circumvented by aliquoting small volumes of stock solution and storage at -70°C. High concentrations of RNA are known to suppress PCR amplifications (23). This is especially the case when large amounts of rRNA exist and one is attempting to amplify rDNA genes. This problem can be overcome by digesting the genomic DNA with RNase (e.g., RNase A, Type I-AS; Sigma Chemical, St. Louis, MO, USA) or by agarose gel purification (above). If amplifications are still not successful, other impurities (such as secondary compounds) should be suspected. Moderately impure DNA can often be amplified by further dilution; however, a rapid and reliable method involves purification of the genomic DNA on low melting temperature agarose (e.g., NuSieve®; FMC BioProducts, Rockland, ME, USA) containing ethidium bromide. A 1% gel is prepared in TBE or Tris-acetate. Genomic DNA (20 µL; ca. 5 µg) is mixed with 2 µL of Ficoll® loading dye (Pharmacia Biotech, Piscataway, NJ, USA) (25% Type 400 Ficoll, 40 mM Tris-HCl, pH 8.0, 20 mM EDTA, 0.05% bromophenol blue), loaded into the gel and electrophoresed. The genomic DNA band is visualized with long wavelength UV light, excised and excess agarose is removed. The band is then diluted 1:1 (by weight) with sterile water. For use in PCRs, the NuSieve band is heated to 65°C and about 1.0 µL used per 100 µL PCR. This method is more rapid and straightforward than purification using CsCl and has allowed amplification of numerous recalitrant plants.

GEL PURIFICATION OF PCR PRODUCTS USING DEAE MEMBRANES

Potential contaminants in a PCR mixture include remaining primers, *Taq*

DNA polymerase, unincorporated nucleotides, compounds introduced with the template DNA and mineral oil (if this was used as an overlay). These must be removed before proceeding to sequencing reactions. A number of methods to purify PCR products for sequencing are currently in use such as glass powder (GeneClean®; Bio 101, La Jolla, CA, USA) (25) or direct precipitation using ethanol or polyethylene glycol. Ultrafiltration devices, such as Centricon®-100 (Amicon, Beverly, MA, USA), are also quite effective in the removal of primers, mineral oil, and so on, from PCR products (15). Many of these methods, however, do not allow the isolation of a single band when multiple PCR products are obtained. Direct sequencing from PCR products excised from low melting point agarose gels has been reported (14), or the DNA in bands excised from gels may also be removed from the agarose using chaotropic agents combined with microconcentrator filter units (e.g., SpinBind™; FMC BioProducts). Although not tested, direct elution from an agarose gel, such as the method described by Hansen et al. (13), likely would produce similarly high-quality PCR product templates for sequencing.

The procedure described below involves agarose gel purification and recovery of the PCR product by binding onto a DEAE membrane (Grade NA45, No. 23400; Schleicher & Schuell, Keene, NH, USA). Similar recovery protocols from agarose and acrylamide gels have been reported that utilize DEAE-cellulose (9), but the former appears to be more durable, hence easier to use. This simple method of recovering nucleic acids up to 7 kb in length has been known for some time (19); however, its application to the purification of PCR products has not been widely reported. Template concentration, homogeneity and purity appear to be critical in obtaining good sequence from PCR products. We have found that higher yields are obtained using DEAE membranes than with spin columns or glass bead methods, possibly owing to retention of DNA on spin-column membranes or glass surfaces. The DEAE membrane method yields DNA of very high purity (29). Primers and nucleotides are eliminated owing to their faster migration rate on the agarose gel. Remnant mineral oil floats to the top of the gel buffer upon loading and is subsequently not carried over with the purified product.

The amount of PCR product to be made and purified depends on the amount required for subsequent sequencing reactions. As a starting point, we have found that 100 µL of product is usually sufficient for four primer/ template combinations. For purifying several 100-µL samples, a 20-× 6- × 1.2-cm gel with a 16-place comb is used. The comb teeth are 10 × 12 × 1.5 mm, which produce wells that each hold 100 µL of sample. A 1% Tris-acetate agarose gel (15 mM Tris-HCl, 33 mM acetate, pH to 8.3 with glacial acetic acid) with ethidium bromide (10 µg/100 mL gel) is prepared. The membranes are cut to fit the gel lanes (7.5 × 11 mm) and washed for 10 min in 10 mM EDTA, then 5 min in 0.5 M NaOH. The membranes are rinsed several times in distilled water and stored in water at 4°C until needed (up to several weeks). One hundred microliters of sample are mixed with 10 µL of Ficoll loading dye, loaded onto the gel and electrophoresed. When the band has migrated the desired distance (check progress with long wave UV light), the gel is removed from the electrode chamber, and an incision is made just ahead of the band of interest. The prepared DEAE membrane is placed in the incision, the gel is returned to the electrode chamber and electrophoresis is continued. After approximately 10 min, the DNA is bound to the membrane.

To elute the DNA from the membranes, they are removed from the gel and placed in a microcentrifuge tube. One milliliter of NET buffer (0.15 M NaCl, 0.1 mM EDTA, 20 mM Tris-HCl pH 8.0) is added, the tube is vortex mixed briefly to remove residual agarose and the buffer is eluted. For 2–3 membranes, 350 µL of high-salt NET buffer (1 M NaCl, 0.1 mM EDTA, 20 mM Tris-HCl pH 8.0) are added and the tubes incubated at 65°C for 30 min with occasional vortex mixing. The buffer is transferred to a clean tube. The membranes are washed with 50–75 µL of high-salt NET buffer and incubated at 65°C for 10 min, and then the buffers are pooled. The buffer is then extracted with 3 volumes of water-saturated *n*-butanol (BT-105; Sigma Chemical), vortex mixed, centrifuged for 5 min and the top phase removed. Two and one-half volumes of 100% ethanol are added, mixed thoroughly and precipitated at -70°C for at least 1 h. The pellet is collected by centrifugation for 20 min at 4°C and then rinsed in 70% ethanol. The ethanol is decanted, and the pellet is dried and resuspended in sterile, distilled water. The intensity of the bands in the gel should be used to determine the appropriate dilution, but usually 30 µL are used for every 100 µL of double-stranded DNA product loaded on the gel. Optionally, one can determine the DNA concentration spectrophotometrically, ideally using 0.5–1.0 µg per sequencing reaction set.

SEQUENCING REACTIONS

Many articles have appeared in recent years describing various methods for sequencing double-stranded DNA produced by PCR (3,7,16,25). The factors that appear most important in double-stranded sequencing are homogeneity and purity of the PCR product (to reduce false priming, seen as doublets on the sequencing ladder), efficiency of the primer/template annealing, and proper ratios of template DNA, nucleotides and primer. The template can be alkaline-denatured using sodium hydroxide (5,30) or denatured with heat. Since the two strands of the template DNA have a much higher binding affinity to each other than to the primer, it is essential to set the conditions so that primer annealing is favored. Rapid cooling at -70°C produces better results (judged by the clarity of sequencing ladders) than gradual annealing at warmer temperatures, even on molecules with a high degree of secondary structure (such as rRNA). The following protocol appears to work as well as alkaline denaturation and requires less time.

The double-stranded DNA (7 µL or ca. 0.5–1.0 pmol) is mixed with 1 µL of primer (125 μg/mL or 20 pmol) and the tube cap secured with lid locks. The primer/DNA mixture is then placed in a boiling water bath for 1 to 2 min, snap-cooled for 1–2 min in a -70°C ethanol bath and centrifuged briefly. Two microliters of 5 × Sequenase® reaction buffer (United States Biochemical, Cleveland, OH, USA) are then added. The reaction mixture is allowed to anneal at 37°C for 20 min. To the annealed mixure, the following are added (per set): 1 μL of Mn buffer, 1 μL of 0.1 M dithiothreitol (DTT), 2 μL Sequenase labeling mixture (diluted 1:5 in dH_2O), 0.5 μL [^{35}S]dATP and 2 μL of Sequenase (diluted 1:8 in enzyme dilution buffer). The reaction is mixed and incubated at room temperature for 5 min. Labeling mixture (3.5 μL) is added to each termination tube (A, C, G, T) containing 2.5 μL of the appropriate Sequenase termination mixture (see below). These are mixed and incubated at 37°C for 5 min. The reaction is terminated by the addition of 4 μL of stop solution and stored at -20°C.

For electrophoresis, the reaction tubes are heated to 85°C for 2 min, and 4 μL are loaded per lane in a standard (Tris-boric acid, EDTA) 6% or 8% acrylamide gel. Typical electrophoresis and autoradiography are then performed. Routinely, an 8% gel is used to read bases near the primer (2 h at 37 W), and a second loading on a 6% gel (7 h at 37 W) is used to resolve bases farther from the primer.

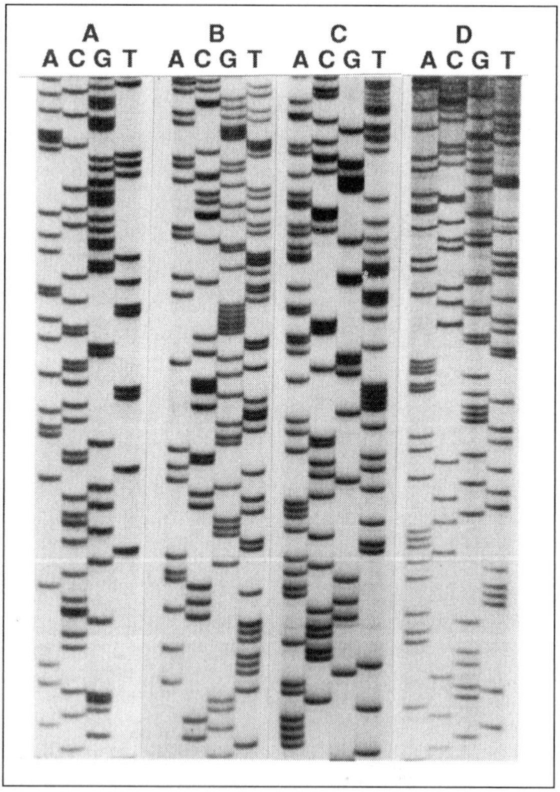

Figure 2. Autoradiograms derived from direct sequencing of double-stranded DNA purified using DEAE membranes. Set A, fresh extracted *Saruma henryi* 18S rDNA (1830 reverse primer). Set B, delayed CTAB-extracted *Opilia amentacea rbc*L (Z-967 forward primer). Set C, silica gel-preserved *Misodendrum* sp. *rbc*L (Z-674 reverse primer). Set D, *E. coli fur*, amplified directly from lysed cells (36 forward primer).

Experiments with primer concentrations in the range recommended in the Sequenase protocol manual (0.5–1.0 pmol) produced a very light sequence. Higher concentrations (e.g., 20 pmol) are required for double-stranded sequencing to optimize primer annealing. As previously noted (2,3), double-stranded sequencing is especially sensitive to the molar ratio of template DNA to nucleotides in the labeling reaction. Sufficient DNA template must be present to provide enough terminations close to the primer to visualize small products. When sequencing double-stranded PCR products, a high concentration of template is not a problem since this can be controlled by dilution, whereas insufficient DNA is more likely the cause of light bands on autoradiograms. To realize long reads using double-stranded sequencing, the extending mixtures described in the Sequenase protocol manual should be used. By using the 2.5× relative extension, long reads can be obtained without sacrificing fragments near the primer. Resolution of these small fragments is also increased by routinely using the Mn buffer, which does not appear to adversely affect long gels run from the same reaction. Also, by diluting the dGTP labeling mixture by 3.75 (11 µL dH_2O + 4 µL dGTP labeling mixture) instead of the recommended 5×, the readable sequence is maximized on the long gels. For sequencing rRNA genes where secondary structure may result in band compressions, inositol and deaza nucleotides (dITP or 7-deaza-dGTP) can be used in parallel with the dGTPs. With these modifications, 300–350 bp can routinely be read per sequencing reaction.

Figure 2 shows results of double-stranded sequencing of PCR products obtained by amplification of two different plant genes (18S rRNA and *rbc*L) and a bacterial gene (*fur*). These results demonstrate that the DEAE membrane purification method, combined with heat denaturation, can result in superior double-stranded sequencing reactions using a variety of organisms, genes and primer combinations.

ACKNOWLEDGMENTS

This work was supported by a grant from N.S.F. (BSR-8918385). Thanks go to G. Zurawski (DNAX Research Institute) and D. Soltis (Washington State University) who donated *rbc*L primers and to Laurie Achenbach for the *Escherichia coli fur* sequencing gel. Ellen Starr provided laboratory technical assistance.

REFERENCES

1. **Albert, V.A., S.E. Williams and M.W. Chase.** 1992. Carnivorous plants: phylogeny and structural evolution. Science *257*:1491-1495.
2. **Anderson, R.D., C.-Y. Bao, D.T. Minnick, M. Veigl and W. D. Sedwick.** 1992. Optimization of double-stranded sequencing for polymerase chain reaction products. US Biochem. Comments *19*:39-58.
3. **Casanova, J.-L., C. Pannetier, C. Javlin and P. Kourilisky.** 1990. Optimal conditions for directly sequencing double-stranded PCR products with Sequenase®. Nucleic Acids Res. *18*:4028.
4. **Chase, M.W. and H.H. Hills.** 1991. Silica gel: an ideal material for field preservation of leaf samples for DNA studies. Taxon *40*:215-220.
5. **Chi, H.-C., J.-C. Hsieh, C.-F. Hui and M.F. Tam.** 1988. Modified method for double-stranded DNA sequencing and synthetic oligonucleotide purification. Nucleic Acids Res. *16*:10382.
6. **Do, N. and R.P. Adams.** 1991. A simple technique for removing plant polysaccharide contaminants from plant DNA. BioTechniques *10*:162-166.

7.**Douglas, A.M., A.M. Georgalis and B.A. Atchison.** 1993. Direct sequencing of double-stranded PCR products incorporating a chemiluminescent detection procedure. BioTechniques *14*:824-828.

8.**Doyle, J.J. and J.L. Doyle.** 1987. A rapid DNA isolation procedure for small quantities of fresh leaf tissue. Phytochem. Bull. *19*:11-15.

9.**Dretzen, G., M. Bellard, P. Sassone-Corsi and P. Chambon.** 1981. A reliable method for the recovery of DNA fragments from agarose and acrylamide gels. Anal. Biochem. *112* :295-298.

10.**Fang, G., S. Hammar and R. Grumet.** 1992. A quick and inexpensive method for removing polysaccharides from plant genomic DNA. BioTechniques *13*:52-54.

11.**Gyllensten, U.B. and H.A. Erlich.** 1988. Generation of single stranded DNA by the polymerase chain reaction and its application to direct sequencing of the *HLA-DQA* locus. Proc. Natl. Acad. Sci. USA *85*:7652-7656.

12.**Hamby, K.R. and E.A. Zimmer.** 1992. Ribosomal RNA as a phylogenetic tool in plant systematics, p. 50-91. *In* P.S. Soltis, D.E. Soltis and J.J. Doyle (Eds.), Molecular Systematics of Plants. Chapman and Hall, New York.

13.**Hansen, H., H. Lemke and U. Bodner.** 1993. Rapid and simple purification of PCR products by direct band elution during agarose gel electrophoresis. BioTechniques *14*:28-29.

14.**Kretz, K.A., G.S. Carson and J.S. O'Brien.** 1989. Direct sequencing from low-melt agarose with Sequenase®. Nucleic Acids Res. *17*:5864.

15.**Krowczynska, A.M. and M.B. Henderson.** 1992. Efficient purification of PCR products using ultrafiltration. BioTechniques *13*:286-289.

16.**Kusukawa, N., T. Uemori, K. Asada and I. Kato.** 1990. Rapid and reliable protocol for direct sequencing of material amplified by the polymerase chain reaction. BioTechniques *9*:66-72.

17.**Lindahl, T.** 1993. Instability and decay of the primary structure of DNA. Nature *362*:709-715.

18.**Liston, A., L.H. Riesberg, R.P. Adams and Z. Ge-lin.** 1990. A method for collecting dried plant specimens for DNA and isozyme analyses, and the results of a field test in Xinjiang, China. Ann. Missouri Bot. Gard. *77*:859-863.

19.**Lizardi, P.M., R. Binder and S.A. Shart.** 1984. Preparative isolation of DNA and biologically active mRNA from diethylaminoethyl membrane. Gene Anal. Tech. *1*:33-39.

20.**Mullis, K., F. Faloona, S. Scharf, R. Saiki, G. Horn and H. Erlich.** 1986. Specific enzymatic amplification of DNA in vitro: the polymerase chain reaction. Cold Spring Harb. Symp. Quant. Biol. *51*:263-273.

21.**Nickrent D.L. and C.R. Franchina.** 1990. Phylogenetic relationships of the Santalales and relatives. J. Mol. Evol. *31*:294-301.

22.**Olmstead R.G., H.J. Michaels, K.M. Scott and J.D. Palmer.** 1992. Monophyly of the Asteridae and identification of its major lineages inferred from *rbc*L sequences. Ann. Missouri Bot. Gard. *79*:249-265.

23.**Pikaart, M.J. and B. Villeponteau.** 1993. Suppression of PCR amplification by high levels of RNA. BioTechniques *14*:24-25.

24.**Pyle, M.M. and R.P. Adams.** 1989. *In situ* preservation of DNA in plant specimens. Taxon *38*:576-581.

25.**Reichardt, J.K.V. and S.L.C. Woo.** 1990. Direct double-stranded sequencing of PCR products. US Biochem. Comments *17*:21-22.

26.**Rogers, S.O. and A.J. Bendich.** 1985. Extraction of DNA from milligram amounts of fresh, herbarium, and mummified plant tissues. Plant Mol. Biol. *5*:69-76.

27.**Rogstad, S.H.** 1992. Saturated NaCl-CTAB solution as a means of field preservation of leaves for DNA analysis. Taxon *41*:701-708.

28.**Rowland, L.J. and B. Nguyen.** 1993. Use of polyethylene glycol for purification of DNA from leaf tissue of woody plants. BioTechniques *14*:735-736.

29.**Sambrook, J., E.F. Fritsch and T. Maniatis.** 1989. Molecular Cloning: A Laboratory Manual. Cold Spring Harbor Laboratory Press, Cold Spring Harbor, NY.

30.**Zhang, H., R. Scholl, J. Browse and C. Somerville.** 1988. Double stranded DNA sequencing as a choice for DNA sequencing. Nucleic Acids Res. *16*:1220.

Update to:
From Field to Film: Rapid Sequencing Methods for Field-Collected Plant Species

Daniel L. Nickrent
Southern Illinois University, Carbondale, Il, USA

The methods outlined in the original paper (8) are still being used in our laboratory. This update includes references to improvements made to specific protocols and describes additional methods that we have incorporated since 1994.

PLANT DNA PRESERVATION AND EXTRACTION

Desiccation of plant tissue on silica gel (2) remains the most commonly used method for field-preservation of DNA. Liquid preservation methods (such as in alcohol) are less frequently used since they often result in low yields of high molecular weight DNA. A recent study showed, however, that this stems not from the inability of alcohol to preserve DNA but subsequent failure to extract the DNA from the preserved sample. The addition of proteinase (Pronase E) resulted in high DNA yields from alcohol-preserved spinach, broccoli and juniper samples (3). Since PCR requires very little template DNA, interest has increased in obtaining DNA from very small (milligram) samples (5) or from relatively crude preparations that allow rapid processing of numerous samples (11). Obtaining adequate amounts and quality of DNA from cultivated species is often routine, but such is not the case for many wild species, especially woody tropical plants. Removal of polysaccharides and secondary metabolites is often necessary to obtain usable DNA for PCR and restriction digestions. The use of polyethylene glycol (PEG) to precipitate the nucleic acid has proven effective for particularly difficult species (1,9). Another approach is to first precipitate the polysaccharides using 0.35 volumes of ethanol, thus allowing the DNA and RNA to remain in solution (7). Alternately, the membrane-bound organelles (containing secondary compounds) can be isolated prior to conducting a modified CTAB extraction (10). Our lab has encountered difficulties with mistletoes (Loranthaceae) in that the typical 2× CTAB method (8) results in unusually large, often crystalline precipitates. We found that dialysis (membrane molecular weight cut-off 12 000–14 000) removes these salts, leaving the DNA inside the dialysis membrane that can be ethanol precipitated.

If DNA concentration in the sample is low or if only a small of amount of DNA is required (e.g., for PCR), gel purification is a very effective method to remove contaminants. We use a modification of the method described by He et al. (4) where the DNA is electrophoresed in a 1% Tris-acetate agarose gel containing ethidium bromide; the band is excised and then placed in an agarose spin column (e.g., GenElute™; Supelco Park, Bellefonte PA, USA). The tube is frozen briefly, thawed and then centrifuged at 6000 rpm for 1 min. The ethidium bromide is removed from the resulting DNA solution with water-saturated butanol. This DNA is then suitable for PCR amplifications or other uses.

PCR METHODS

The routine symmetrical PCR protocol described in the original paper is still being used with essentially no modification. We have found that optimization of $MgCl_2$ concentration and annealing temperature is best accomplished using a thermocycler with a temperature gradient block (e.g., the Robocycler Gradient 40®; Statagene, La Jolla, CA, USA). This machine allows a number of variables to be changed during the same run.

SEQUENCING

For purifying PCR products for sequencing, we continue to use diethylaminoethyl (DEAE) membranes in the gel isolation protocol described in the original paper (8). The major advantage of this technique is the ability to select a specific band among others present on the gel. A heat denaturation/snap-cooling method is still exclusively used for sequencing double-stranded PCR products. Bands on agarose gels that appear single may upon isolation and sequencing give double ladders, thus indicating the presence of PCR products that differ in sequence and/or length. In these instances, it is necessary to clone the PCR products prior to sequencing. We have had good success with TA cloning as described by Mead et al. (6) using the TA Cloning Kit® (Invitrogen, San Diego, CA USA). Our experience with cloning PCR products eluted from DEAE membranes indicates that ligation efficiency is lowered, possibly because of the loss of A overhangs on the fragments following electrophoresis and precipitation. A simple alternative to this is to remove the band from the gel and isolate the DNA using a spin column as described above.

DNA templates that contain secondary structure (e.g., ribosomal RNA genes) often result in especially strong compressions at particular positions. We have overcome this problem using cycle sequencing as implemented with the ThermoSequenase™ cycle sequencing kit (Amersham Life Science, Cleveland, OH USA). Compressions are reduced owing to the use of a thermostable polymerase and the presence of base analogs such as 7-deaza-dGTP. One limitation of this method is that the template must have a span of at least 8 nucleotides downstream from the primer that lacks one of the four nucleotides. This allows sufficient incorporation of radioactive label during the primer extension step.

REFERENCES

1. **Agudo, L.C., I. Gavidia, P. Pérez-Bermúdez and J. Segura.** 1995. PEG precipitation, a required step for PCR amplification of DNA from wild plants of *Digitalis obscura* L. BioTechniques *18*:766-768.
2. **Chase, M.W. and H.H. Hills.** 1991. Silica gel: an ideal material for field preservation of leaf samples for DNA studies. Taxon *40*:215-220.
3. **Flournoy, L.E., R.P. Adams and R.N. Pandy.** 1996. Interim and archival preservation of plant specimens in alcohols for DNA studies. BioTechniques *20*:657-660.
4. **He, M., H. Liu, Y. Wang and B. Austen.** 1992. Optimized centrifugation for rapid elution of DNA from agarose gels. Genet. Anal. Tech. Appl. *9*:31-33.
5. **Knapp, J.E. and J.M. Chandlee.** 1996. RNA/DNA mini-prep from a single sample of orchid tissue. BioTechniques *21*:54-56.
6. **Mead, D.A., N.K. Pey, C. Herrnstadt, R.A. Marcil and L.M. Smith.** 1991. A universal method for the direct cloning of PCR amplified nucleic acid. Biotechnology (NY) *9*:657-663.
7. **Michaels, S.D., M.C. John and R.M. Amasino.** 1994. Removal of polysaccharides from plant DNA by ethanol precipitation. BioTechniques *17*:274-276.
8. **Nickrent, D.L.** 1994. From field to film: rapid sequencing methods for field collected plant species. BioTechniques *16*:470-475.
9. **Schierenbeck, K.A.** 1994. Modified polyethylene glycol DNA extraction procedure for silica gel-dried tropical woody plants. BioTechniques *16*:392-394.
10. **Scott, K.D. and J. Playford.** 1996. DNA extraction technique for PCR in rain forest plant species. BioTechniques *20*:974-979.
11. **Thompson, D. and R. Henry.** 1995. Single-step protocol for preparation of plant tissue for analysis by PCR. BioTechniques *19*:394-400.

An Improved Method for Detection of 5-Methylcytosine by PCR-Based Genomic Sequencing

Mati Reeben[1,2] and Hans Prydz[1]

[1]University of Oslo, Oslo, Norway, [2]Institute of Chemical Physics and Biophysics, Tallinn, Estonia

Mammalian DNA is often methylated at position 5 of the cytosine moiety in the dinucleotide CpG. Several lines of evidence suggest that methylation patterns, in many cases, correlate with the expression of eukaryotic genes and that DNA methylation plays an important role in regulating the state of differentiation of mammalian cells (1).

A powerful procedure for analysis of 5-methylcytosine in genomic DNA has been described recently (2). The method is based on bisulfite modification of genomic DNA, under conditions whereby cytosine is converted to uracil but 5-methylcytosine remains unaltered. The sequence under investigation is then amplified by the polymerase chain reaction (PCR) to yield a fragment on which all uracil residues have been amplified as thymine and only 5-methylcytosine residues have been amplified as cytosine. Direct sequencing of PCR products provides a strand-specific average sequence for the population of molecules, whereas cloning and sequencing provide methylation maps of single DNA molecules. Here we report that the use of a glass milk procedure simplifies and speeds up the protocol significantly.

The bisulfite reaction was carried out on 2 µg of genomic DNA plus 8 µg of carrier plasmid (usually pUC19) or on 6–10 µg of genomic DNA without carrier. DNA was alkali-denatured, neutralized and precipitated. Denatured DNA was incubated in a total volume of 1.2 mL with freshly prepared 3.1 M sodium bisulfite (Sigma Chemical, St. Louis, MO, USA), 0.5 mM hydroquinone, pH 5.0, at 50°C for 24 h under mineral oil, followed by cooling to room temperature and incubation for 15 min with 10 µL glass milk (Geneclean®, Bio 101, La Jolla, CA, USA). After 3 washes, DNA was eluted with 100 µL of 10 mM Tris-HCl, 0.1 mM EDTA, pH 7.5 (TE buffer) at 50°C. NaOH was added to a final concentration of 0.3 M at room temperature for 10 min (alkali treatment completes the bisulfite reaction), followed by the

(Reprinted from BioTechniques 16:416-417, 1994)

addition of ammonium acetate (pH 7.0) to a final concentration of 3 M and precipitated with 4 volumes of ethanol for 30 min at -20°C, resuspended in 100 μL of TE buffer (see above), and stored at -20°C. Amplification reactions were carried out using 5 μL of DNA.

Under the bisulfite reaction conditions, DNA effectively binds to glass milk, the recoveries are similar to those when dialysis is used (original method in Reference 2) and the procedure does not interfere with the chemical reaction. Only 1 out of about 400 cytosine nucleotides can remain unreacted and all 8 methylcytosines in *dcm* (DNA cytosine methylation) methylase recognition sites (CCWGG) in a control plasmid remained unreacted (data not shown). Instead of time-consuming dialysis with several buffer changes and vacuum drying, the glass milk procedure can be completed in about 30 min.

In applying this method to several genes (rat light neurofilament, human lecithin cholesterol acyltransferase, human apolipoprotein E, to be published elsewhere), we found there were some specific points to be taken into consideration in the design of primers for bisulfite-reacted DNA. Bisulfite converts C to U, which is amplified as T. Instead of a four-letter code, mainly three nucleotides appear—A, G and T (all nonmethylated Cs appear as T). This "three-letter code" significantly reduces primer specificity. With this

```
GTTAAAAGCTTATGTGTTAGTAGGGATTT
---------------------------->
GTTAAAAGTTGATGTGTCAGTAGGGACCTATTATTTGGAAATTTTAAGGGTCCTAGTGAG -281
||||||||||||||||| ||||||||  |||||||||||||||||||||||  |||||||
GTTAAAAGTTGATGTGTTAGTAGGGATTTATTATTTGGAAATTTTAAGGGTTTTAGTGAG

                                            *       *      *
----TATGGGGGTTTCAGAGAAAGTAGGGGGCCTCAGAGAAAAGCGGCCAGGCGGGAGCC -161
    ||||||||||| |||||||||||||||  | ||||||||||||  ||| ||||| |
----TATGGGGGTTTTAGAGAAAGTAGGGGGTTTTAGAGAAAAGCGGTTAGGTGGGAGTC

      *             *   *     *             *
GCAGTCCGCTAGCCTCCTGTCGCTCGCCCCCGCCTTTGCTCTTGCGCAGAATCCTCCCCT -101
| ||| || |||  |  ||||| | |    ||  |||| | ||||| |||||  |    |
GTAGTTCGTTAGTTTTTTGTCGTTTGTTTTCGTTTTTGTTTTTGCGTAGAATTTTTTTTT

              *                  *    * *   *            *
TGGCTGCAGCAGCACGCTGCCCCCACTGGCCTGCGCAACGCGATCGATCACAGTCTGCGT  -41
||| || || || ||| ||     | |||  || | || | ||| ||| | ||| || ||
TGGTTGTAGTAGTACGTTGTTTTTATTGGTTTGTGTAATGTGATTGATTATAGTTTGTGT

         *    *   *             *         *    *
----AGGGGCGCACCGACCGCCACCATGAGTTCGTTCAGCTACGAGCCGTACTTT     +135
    ||||| | |  ||  |  |  ||||||| ||| || || |||  ||| |||
----AGGGGTGTATTGATTGTTATTATGAGTTTGTTTAGTTATGAGTTGTATTTT
                         <------------------------------
                      3'-TACTCAAACAAATCAATACTTAAGATAAAA-5'
```

Figure 1. Cluster of methylated CpGs in the rat light neurofilament gene promoter (EMBL Accession No. X53981) from the nonexpressing glioma C6 cell line. Sequence comparison of the promoter region before (upper sequence) and after (lower sequence) bisulfite treatment. 5-Methylcytosine is represented as C in the lower sequence (underlined), all other Cs are amplified as T. Arrows show replacement of PCR primers for the upper strand of genomic DNA. *Hin*dIII and *Eco*RI sites have been included in primers to facilitate cloning. Asterisks indicate CpG dinucleotides.

predominantly three-letter code, the undesirable homology of primers with repeated elements occurs significantly more often than in conventional PCR. Even less desirable is the homology of primers to plasmids used as carrier DNA. Because this carrier plasmid is in large excess compared with a single copy gene, similarity of primers to the carrier plasmid sequence can produce extra PCR bands. We strongly recommend confirming the absence of homology between the primers and the carrier plasmid sequences where C nucleotide has been replaced by T, or to avoid carrier DNA completely. In PCR of bisulfite-reacted DNA, hot start improved the results with most primer pairs tested. Without hot start, mispriming may easily take place, especially if carrier plasmid is used.

Under the bisulfite reaction conditions described, some breakage of DNA occurs (2). We were able to amplify 500-bp fragments but not a 1-kb fragment from modified DNA.

ACKNOWLEDGMENTS

We thank Drs. Marianne Frommer and Susan Clark for advice in genomic sequencing and making available their method before publication. We thank Juhani Jänne and Alan Schulman for critical comments.

REFERENCES

1. **Bird, A.** 1992. The essentials of DNA methylation. Cell *70*:5-8.
2. **Frommer, M., L.E. McDonald, D.S. Millar, C.M. Collis, F. Watt, G.W. Grigg, P.L. Molloy and C.L. Paul.** 1992. A genomic sequencing protocol that yields a positive display of 5-methylcytosine residues in individual DNA strands. Proc. Natl. Acad. Sci. USA *89*:1827-1839.

Update to:
An Improved Method for Detection of 5-Methylcytosine by PCR-Based Genomic Sequencing

Mati Reeben

University of Helsinki, Helsinki, Finland

The original method published by Frommer et al. (Reference 2 in original article) has been developed further to maximize the efficiency of denaturation, bisulfite conversion and amplification of DNA, to permit methylation mapping of single genes from small amounts of genomic DNA that are readily available from germ cells and embryonic organs (1). A modified protocol that uses glass beads for the subsequent purification of the DNA in a similar fashion to our work has been published (2).

In a recent work, Raizis et al. (3) showed that the bisulfite reaction results in a significant level of template degradation due to DNA depurination. They suggested that the DNA fragmentation which occurs limits the sensitivity of the method and describe a simple solution to limit degradation of the DNA sample. They recommend the following bisulfite reaction conditions: 5 M sodium bisulfite (pH 5.0) at 50°C for 4 h, which is superior in their hands compared to 3.1 M sodium bisulfite at 55°C for 16 h as described in Reference 1 or 3.5 M sodium bisulfite at 0°C for 24 h as described in Reference 2.

Certain methyl groups can influence the bisulfite reactivity of neighboring cytosines within a DNA sequence that has the potential of forming secondary structures (4).

REFERENCES

1. **Clark, S.J., J. Harrison, C.L. Paul and M. Frommer.** 1994. High sensitivity mapping of methylated cytosines. Nucleic Acids Res. *22*:2990-2997.
2. **Feil, R., J. Charlton, A.P. Bird, J. Walter and W. Reik.** 1994. Methylation analysis on individual chromosomes: improved protocol for bisulphite genomic sequencing. Nucleic Acids Res. *22*:695-696.
3. **Raizis, A.M., F. Schmitt and J.-P. Jost.** 1995. A bisulfite method of 5-methylcytosine mapping that minimizes template degradation. Anal. Biochem. *226*:161-166.
4. **Rother, K.I., J. Silke, O. Georgiev, W. Schaffner and K. Matsuo.** 1995. Influence of DNA sequence and methylation status on bisulfite conversion of cytosine residues. Anal. Biochem. *231*:263-265.

General Method for PCR Amplification and Direct Sequencing of mRNA Differential Display Products

Steven A. Reeves, Mari-Paz Rubio and David N. Louis

Massachusetts General Hospital and Harvard Medical School, Charlestown, MA, USA

The recently described technique of mRNA differential display (1–3) has introduced a powerful means of identifying differentially expressed genes. However, the characterization of potentially interesting polymerase chain reaction (PCR)-amplified differential display cDNAs is time-consuming and sometimes problematic. We describe a simple, rapid and novel technique, ligation linked PCR (LLR), to characterize PCR products generated when using short oligonucleotide primers, such as those used in mRNA differential display. LLR optimizes PCR amplification of cDNAs that have been generated using mRNA differential display. Also, since LLR incorporates a T7 promoter primer at the upstream end of the amplified cDNAs, direct DNA sequencing can be performed on multiple cDNAs using a common and highly specific T7 sequencing primer, facilitating characterization of a large number of differentially expressed cDNAs.

In order to clone and characterize partial cDNAs generated using the mRNA differential display technique, bands are first excised from the sequencing gel, purified and then re-amplified with PCR. In our experience, the currently used method to re-amplify and characterize these cDNAs (1–3) results in poor yields, in addition to generating nonspecific PCR products due to the use of an arbitrary hexamer primer and high PCR-cycle numbers (80 cycles). With LLR, cDNAs extracted from a differential mRNA display gel are first transiently ligated into the PCR fragment cloning vector pCR™II (Invitrogen, San Diego, CA, USA) and then directly PCR-amplified using a poly(dT) primer and a T7 promoter primer that anneals immediately adjacent to the cloning site in the pCRII vector. Transient ligation to pCRII and the use of the specific PCR primers results in increased yield and specificity. LLR facilitates sequencing and characterization of PCR products since sequence obtained with a T7 promoter primer will produce the sequence farthest 5′ or

(Reprinted from BioTechniques 18:18-20, 1995)

upstream of the cDNA. This is an important feature of LLR since readable DNA sequence is often not obtainable when using the small arbitrary hexamer primers [incorporated into the cDNA during the initial reverse transcription (RT)-PCR]. In addition, the majority of the sequence obtained from the poly(dT) primer may be of limited utility when screening for homology with known genes since it corresponds primarily to the 3′-untranslated region of the cDNA.

LLR is also applicable to the analysis of PCR products obtained from other techniques such as 3′ RACE (rapid amplification of complementary ends). In this technique, potentially degenerate oligonucleotide primers corresponding to conserved motifs are used in conjunction with a poly(dT) primer in order to PCR-amplify, at low annealing temperatures, related members of a family of proteins. Direct sequencing of the products of 3′ RACE reactions is generally difficult because the degenerate oligonucleotide may not have enough sequence homology with the RACE product for use in sequencing reactions. Similarly, the poly(dT) primer is not generally appropriate for use in sequencing reactions. LLR can be applied to 3′ RACE products by transient ligation into pCRII followed by PCR amplification using T7 and poly(dT) primers. The products can then be sequenced directly with a T7 primer to obtain 5′ sequence data.

mRNA differential display (RT-PCR). For mRNA differential display, we used the RNAmap™ protocol and kit from GenHunter (Brookline, MA, USA). Briefly, 200 ng of total RNA (after DNase treatment with the MessageClean™ Kit; GenHunter) from the human glioma cell lines SNB19 and U343 were reverse-transcribed in a total reaction volume of 20 μL containing 1× RT buffer (125 mM Tris-HCl pH 8.3, 188 mM KCl, 7.5 mM $MgCl_2$, 25 mM dithiothreitol), 100 units of the Moloney murine leukemia virus reverse transcriptase (Life Technologies, Gaithersburg, MD, USA), 20 μM deoxyribonucleoside triphosphates (dNTP) and 20 pmol oligo(dT) primer (T12MN). Programmable Thermal Controller (MJ Research, Watertown, MA, USA) conditions were 65°C for 5 min, 37°C for 1 h and then 95°C for 5 min; reverse transcriptase was added after 10 min at 37°C.

One tenth of the cDNA was then amplified in a total reaction volume of 20 μL containing 1× PCR buffer (10 mM Tris-HCl, 1.5 mM $MgCl_2$, pH 9.0), 20 pmol of the oligo(dT) primer (T12MN), 4 pmol of an arbitrary hexamer primer, 2 μM dNTP, 1 μL ^{35}S-dATP (1200 Ci/mmol; Du Pont NEN, Boston, MA, USA) and 1 U *Taq* DNA polymerase (AmpliTaq®; Perkin-Elmer, Norwalk, CT, USA). Programmable Thermal Controller conditions were 95°C for 30 s, 40°C for 2 min and 72°C for 30 s with a total of 40 cycles. Two microliters of loading dye were added to 3.5 μL of the amplification products, heated at 80°C for 2 min and then fractionated on a 6% DNA sequencing gel. The gel was run at 60 W constant power for 3–4 h, dried down and then exposed to x-ray film (XAR-5; Eastman Kodak, Rochester, NY, USA).

Bands representing differentially expressed cDNAs were excised from the gel, soaked in 100 µL of distilled H_2O (dH_2O) for 10 min, boiled for 15 min and centrifuged for 2 min to clarify particulate material. DNA was precipitated from the supernatant with 10 µL of 3 M sodium acetate, 5 µL of glycogen (10 mg/mL; Sigma Chemical, St. Louis, MO, USA) and 450 µL of 100% ethanol for 30 min at -80°C. After centrifugation for 10 min at 4°C, the pellet was rinsed with 200 µL of ice-cold 85% ethanol and dissolved in 10 µL of dH_2O.

PCR re-amplification without prior ligation to pCRII. For each band, 4 µL of extracted DNA was re-amplified in a total reaction volume of 40 µL using the same primer set and PCR conditions used in the initial RT-PCR. As can been seen in Figure 1A (lanes 1–4), only one of the extracted DNAs was amplified after 40 cycles of PCR amplification (lane 2). A second round of PCR amplification (40 cycles) did not significantly improve the yield. Moreover, nonspecific amplification products were routinely observed with this extended number of PCR cycles (data not shown).

PCR re-amplification after transient ligation to pCRII. For each band, 3 µL of extracted DNA were ligated into pCRII for 30 min at room temperature in a total reaction volume of 10 µL containing 1× ligation buffer, 1 µL 10 mM ATP, 2 µL pCRII (20 ng) and 4 U of T4 DNA ligase (Stratagene, La Jolla, CA, USA), followed by heating at 65°C for 5 min. The ligation reaction was then used for standard PCR in a total reaction volume of 40 µL containing 20 µM dNTP, 40 pmol of oligo(dT) primer

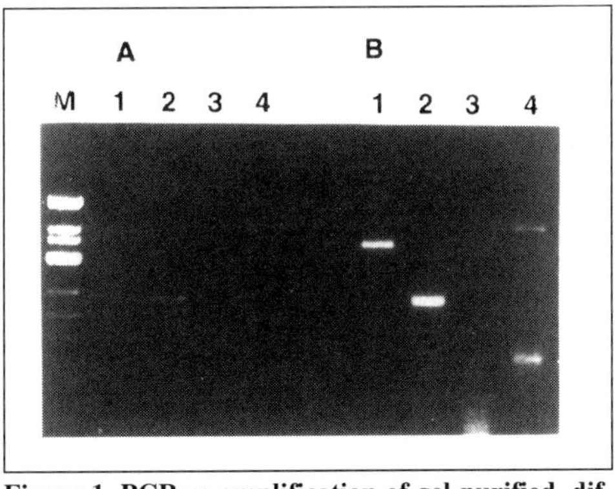

Figure 1. PCR re-amplification of gel-purified, differentially expressed cDNAs without prior ligation to pCRII and using oligo(dT) and arbitrary hexamer primers (A) or with prior ligation to pCRII (LLR) using oligo(dT) and T7 promoter primers (B). All PCRs were done in parallel. cDNAs isolated from the gel were as follows: lane 1, S1; lane 2, S2; lane 3, U1; lane 4, U4.

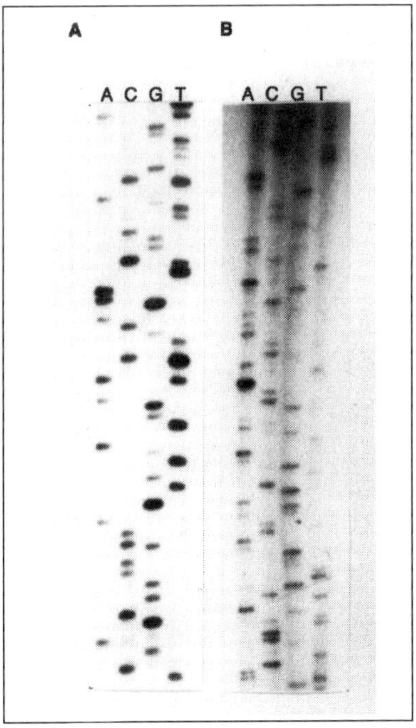

Figure 2. Direct sequencing of purified LLR re-amplified S1 and S2 cDNAs. DNA was purified from the gel shown in Figure 1B (lanes 1 and 2) and directly sequenced with either a T7 promoter primer (A) or a pCII polylinker primer (B).

(T12MN), 20 pmol of T7 primer and 2 U *Taq* DNA polymerase, with thermocycler conditions identical to the initial RT-PCR. Figure 1B (lanes 1–4) shows that specific PCR products were generated for all four of the differentially expressed cDNAs. Although the band in lane 3 is light, sequence data could easily be obtained with direct PCR sequencing of this band. Note also the significant increase in yield in LLR-derived PCR products when compared with yield obtained using the arbitrary hexamer primer (Figure 1A, lanes 1–4).

DNA sequencing. LLR re-amplified S1, S2, U1 and U4 cDNAs were prepared for direct sequencing as follows: 10 µL of PCR-LLR product, 19 µL of H_2O, 11 µL of 7.5 M ammonium acetate, 1 µL of 20 mg/mL glycogen and 40 µL of isopropanol were mixed and incubated at room temperature for 10 min. DNA was precipitated by spinning at 12 000 rpm for 10 min, washed twice with 70% ethanol, dried and resuspended in 30 µL of H_2O. Three microliters of the resuspended cDNAs were sequenced directly using a standard cycle-sequencing protocol, $Vent_R$® (exo^-) DNA polymerase (New England Biolabs, Beverly, MA, USA), [γ-^{32}P]ATP (Du Pont NEN) and the CircumVent™ Thermal Cycle DNA Sequencing Kit (New England Biolabs). Figure 2 shows typical sequence data obtained when using either a [γ-^{32}P] ATP end-labeled T7 promoter primer (Figure 2A) or a pCRII polylinker primer (5′-CGATTGGGCCCTCTAGA) (Figure 2B), which anneals between the T7 promoter and the cloning site. Although not shown, arbitrary hexamer primer sequence was present 3′ to T7 or pCRII polylinker sequences in all PCR products examined. In multiple LLR amplification and direct sequencing of differential display products S1, S2, U1 and U4, we consistently obtained the same sequence for each of the PCR products, demonstrating that potential re-amplification of contaminating bands is rare.

REFERENCES

1. **Liang, P. and A.B. Pardee.** 1992. Differential display of eukaryotic messenger RNA by means of the polymerase chain reaction. Science *257*:967-971.
2. **Liang, P. and A.B. Pardee.** 1992. Differential display and cloning of messenger RNAs from human breast cancer versus mammary epithelial cells. Cancer Res. *52*:6966-6968.
3. **Liang, P. and A.B. Pardee.** 1993. Distribution and cloning of eukaryotic mRNAs by means of differential display: refinements and optimization. Nucleic Acids Res. *21*:3269-3275.

A Method for PCR Sequencing of the p53 Gene from a Single 10-μm Frozen or Paraffin-Embedded Tissue Section

Fazlul H. Sarkar, Yi-Wei Li and John D. Crissman

Department of Pathology, Wayne State University School of Medicine, Detroit, MI, USA

A recent explosion of literature has demonstrated the importance of the mutation of the p53 gene that has been found in a wide variety of tumors and appears to represent a common event in the development of cancer (2,4,5). The detection of the p53 gene mutation has been greatly simplified by the introduction of the polymerase chain reaction (PCR) and direct sequencing, techniques that have provided a sensitive and reliable method. Unfortunately, the technique requires considerable time and relatively large amounts of tumor tissue for DNA extraction, PCR amplification and gene sequencing. These lengthy procedures preclude the analysis of the p53 sequencing in large numbers of tumors. An approach has recently been reported in *BioTechniques* for direct sequencing of the p53 gene from frozen tumor tissue embedded in OCT® (optimal cutting temperature) compound (polyvinyl alcohol/benzalkonium chloride/polyethylene glycol; Miles, Elkhart, IN, USA) using PCR (6). However, the DNA extraction and purification method is still lengthy and time-consuming for the analysis of a large number of specimens. We have, therefore, developed a rapid and simple method for direct sequencing of PCR-amplified p53 gene from 10-μm frozen or paraffin-embedded tissue sections. The method combines modified DNA extraction, PCR and direct sequencing and generates reliable and reproducible results. It allows the analysis of the sequence of multiple exons of the p53 gene from small tissue samples. In addition, the procedure can be adapted for use in large numbers of tumor specimens.

We have used a single 10-μm section (ca. 5 × 8 mm) of breast carcinoma tissue embedded in OCT compound or tissue from formalin-fixed paraffin-embedded tumor, sectioned on a cryostat or microtome for the application of our simplified method. The section was incubated in 200 μL of lysis buffer [10 mM Tris-HCl (pH 8.3) 50 mM KCl, 2.5 mM $MgCl_2$, 0.45% Tween® 20, 0.1 mg proteinase K] with constant shaking at 55°C for 2 h. After the incuba-

(Reprinted from BioTechniques 15:36-38, 1993)

tion the sample was boiled for 15 min. An aliquot of 10 µL of the solution containing genomic DNA was subjected to PCR-SSCP (single-strand conformation polymorphism) in a total volume of 50 µL of reaction mixture containing 10 mM Tris-HCl (pH 8.3), 50 mM KCl, 1.5 mM $MgCl_2$, 0.01% gelatin, 1.0 µM of each primer, 200 µM of deoxyribonucleoside triphosphates (dNTPs), 0.5 µL of [α-^{32}P]-dATP and 1.25 units of *Taq* DNA polymerase (Perkin-Elmer, Norwalk, CT, USA). Primer 1 (TCTGTTCACTTGTGCCCTGA) and primer 2 (CAACCAGCCCTGTCGTCTCT) flank exon 5 of the p53 gene. Primer 3 (TGATTCCTCACTGATTGCTC) and primer 4 (GACAACCACCCTTAACCCCT) flank exon 6. Primer 5 (CCTCATCTTGGGCCTGTGTT) and primer 6 (CCAGGGGTCAGCGGCAAGCA) flank exon 7. Primer 7 (ACTGCCTCTTGCTTCTCTTT) and primer 8 (AAGTGAATCTGAGGCATAAC) flank exon 8. PCR was carried out in a DNA Thermal Cycler (Perkin-Elmer) for 35 cycles. Each cycle consisted of denaturation at 94°C for 1 min, annealing at 55°C for 1 min and extension at 72°C for 1 min. After the last cycle of amplification, the extension was continued for an additional 7 min at 72°C. Two microliters of the PCR product were added to 9 µL of loading buffer (95% formamide, 10 mM NaOH, 0.05% bromophenol blue and 0.05% xylene cyanol), heated at 94°C for 2 min and chilled on ice. Five microliters of the sample were electrophoresed through 0.5× MDE™ gel (AT Biochem, Malvern, PA, USA) at 6 W for 14 h. The gel was autoradiographed for 16 h.

PCR sequencing was carried out using 20 µL of the genomic DNA that were subjected to PCR in a total volume of 100 µL of reaction mixture. The condition for PCR was the same as above except [α-^{32}P]-dATP was omitted. The PCR product was extracted with 100 µL of chloroform. An aliquot of 100 µL was electrophoresed on 2.5% SeaPlaque® Agarose (FMC BioProducts, Rockland, ME, USA) at 100 V for 1.5 h. The desired DNA band was visualized by staining the gel with ethidium bromide and excised using a blade. The DNA band was purified using Magic™ PCR Preps DNA Purification System (Promega, Madison, WI, USA). Each agarose slice was transferred to a microcentrifuge tube and incubated at 70°C for 10 min. One milliliter of DNA Purification System resin was added to the melted agarose slice and vortex mixed for 1 min. The mixture was pushed through a DNA Purification System minicolumn with a syringe followed by washing the column with 2 mL of DNA Purification System wash solution. Fifty microliters of dH_2O were added to the column to elute the bound DNA fragment. Approximately 45 µL of purified DNA were recovered from the column after centrifuging for 20 s in an Eppendorf centrifuge (Westbury, NY, USA).

The purified DNA fragment was subjected to direct sequencing using the *fmol*™ DNA Sequencing System (Promega), which takes advantage of the intrinsic properties of the *Taq* DNA polymerase and deaza nucleotide mixtures. For direct sequencing, 10 µL of purified DNA template, 1 µL (3 pmol) of primer (primer 1,3,5 or 7), 1 µL of [α-^{33}P]-dATP (1900 Ci/mmol), 4.25

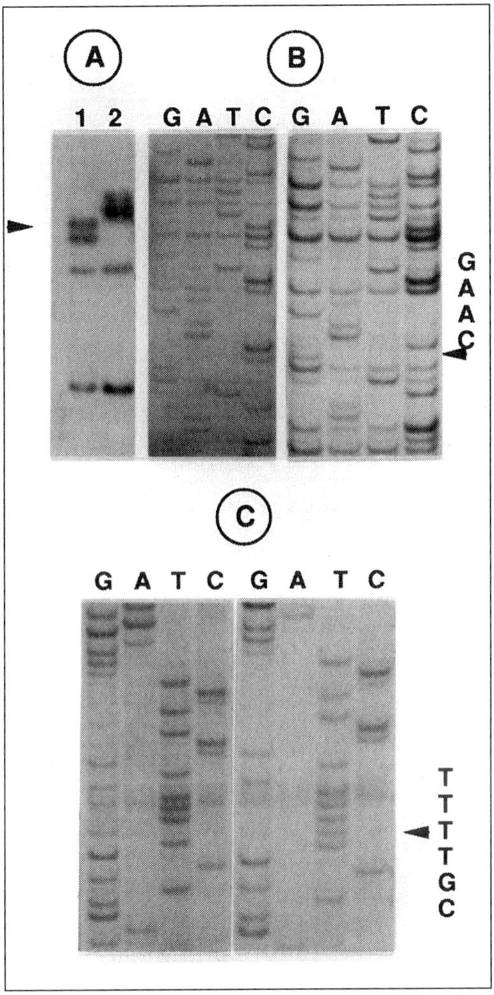

Figure 1. PCR SSCP (A) and PCR-sequencing (B,C) of the p53 gene. Panel A represents SSCP analysis of exon 5 where lane 1 represents PC-3 cells showing an abnormal banding pattern as compared with a breast carcinoma case (lane 2) with normal codon 138 (exon 5) of the p53 gene. Panel B represents DNA sequencing results of exon 5. Left side of the panel represents breast carcinoma case (identical to lane 2 of Panel A) with sequence of GC-CAAG as normal codon 138 and 139. Right side of the panel represents PC-3 cells DNA sequence of GC AAG in codon 138 and 139 showing a deletion as indicated by an arrowhead. Panel C represents DNA sequencing results of exon 8. Left side of the panel represents breast carcinoma case with a sequence of CGTGTT of codon 273 and 294 as normal p53 gene. Right side of the panel represents the exon 8 sequence of DU145 cells DNA showing the sequence of CGTTTT (mutation of G to T in codon 274) indicated by an arrowhead.

μL of 5× sequencing buffer and 1 μL (5 units) of sequencing grade *Taq* DNA polymerase were mixed. Four microliters of the mixture were separately mixed with 1 μL of 1:4 diluted each d/ddNTP(dideoxyribonucleoside triphosphate) and overlaid with 1 drop of mineral oil. The DNA sequencing reaction was carried out in a thermal cycler and initiated by 2 min of denaturation at 95°C and 30 cycles of reaction. Each cycle consisted of denaturation at 95°C for 30 s, annealing at 42°C for 30 s and extension at 70°C for 1 min, or the 42°C incubation step was omitted when the GC-content of the primer was greater than 50%. After the addition of 3 μL of stop solution, 5 μL of the reaction solution were electrophoresed on 6% polyacrylamide/urea gel for 2 h at 50 W. The gel was dried for 20 min in a gel dryer (Bio-Rad, Hercules, CA, USA) under vacuum, then exposed to Kodak XAR film overnight (Eastman Kodak, Rochester, NY, USA).

The p53 amplicons generated by PCR using primer 1/2 (flanking exon 5), primer 3/4 (flanking exon 6), primer 5/6 (flanking exon 7) and primer 7/8 (flanking exon 8) showed the expected size fragments of 247, 187, 209 and 236 bp, respectively, as visualized by ethidium bromide staining of an agarose gel (data not shown). These DNA fragments span the entire exons 5,6,7 and 8, including some intron in the two ends of the exons. These sizes are sufficient to determine sequence information of the entire exon 5-8 when DNA sequencing is carried out using these fragments as templates.

The results of PCR-SSCP and DNA sequencing are shown in Figure 1. The sequences found in a breast carcinoma are identical to the reported sequence of the wild-type p53 gene (1), while the sequences of the PC-3 and DU145 cell lines are mutated or deleted as reported previously (3).

The procedure described above for DNA extraction is quite simple and effective. The yield of DNA is enough for 10 PCRs so that multiple exons can be analyzed even though the tissue section is only 10 µm thick. The protocol excludes deparaffinization, phenol-chloroform extraction and ethanol precipitation that are usually included in the procedure for DNA extraction but are time-consuming. The paraffin in the tissue will float on surface of the lysis buffer during incubation and boiling. The amount of paraffin and OCT compound in the DNA extraction solution from a single 10-µm tissue section did not inhibit PCR in our experiments.

The introduction of *Taq* DNA polymerase in the sequencing reaction has simplified the sequencing process. The protocol takes advantage of the property of thermal stability of *Taq* DNA polymerase so that denaturation and polymerization can be repeated in each cycle in a programming thermal cycler and single-stranded DNA amplification can be omitted. Deaza nucleotide mixtures used in the sequencing reaction resolve band compressions, and ^{33}P-dATP incorporated in the amplified sequence improves autoradiographic resolution. From our experience, the sequence bands in the autoradiographic film are very easy to read.

This method is so simple that it only takes 2 1/2 days to perform the entire procedure from DNA extraction to reading sequence. It is also reliable and consistently yields reproducible results of DNA sequence. This modified technique will also allow investigators to analyze the sequences of multiple exons or multiple genes from a single 10-µm section of frozen or formalin-fixed, paraffin-embedded tissues allowing analysis of multiple tumors.

REFERENCES

1. **Buchman, V.L., P.M. Chumakov, N.N. Ninkina, O.P. Samarina and G.P. Georgiev.** 1988. A variation in the structure of the protein-coding region of the human p53 gene. Gene *70*:245-252.
2. **Hollstein, M., D. Sidransky, B. Vogelstein and C. Harris.** 1991. p53 Mutation in human cancers. Science *253*:49-53.
3. **Isaacs, W.B., B.S. Carter and C.M. Ewing.** 1991. Wild-type p53 suppresses growth of human prostate cancer cells containing mutant p53 alleles. Cancer Res. *51*:4716-4720.
4. **Miller, C., K. Simon, A. Aslo, K. Kok, J. Yokota, C. Buys, M. Terada and H.P. Koeffler.** 1992. p53 Mutation in human lung tumors. Cancer Res. *52*:1695-1698.
5. **Murakami, Y., K. Hayashi and T. Sekiya.** 1991. Detection of aberrations of the p53 alleles and the gene transcript in human tumor cell lines by single-strand conformation polymorphism analysis. Cancer Res. *51*:3356-3361.
6. **Sun, Y., G. Hegamyer and N.H. Colburn.** 1992. A simple method using PCR for direct sequencing of genomic DNA from frozen tumor tissue embedded in optimal cutting temperature compound. BioTechniques *12*:639-640.

Update to:

A Method for PCR Sequencing of the p53 Gene from a Single 10-μm Frozen or Paraffin-Embedded Tissue Section

Fazlul H. Sarkar, Yi-Wei Li and John D. Crissman

Department of Pathology, Wayne State University School of Medicine, Detroit, MI, USA

Recently we demonstrated the efficacy of the PCR-SSCP and sequencing method described in our original report by screening 61 cases of laryngeal squamous cell carcinoma for mutations in exon 5–8 of the p53 gene (1). We have also reduced the reagents needed for this method by using 5-μm paraffin-embedded tissue sections and by decreasing the amount of genomic DNA and reaction mixture used in PCR-SSCP.

Using the newest method, we are screening a large series of pancreatic adenocarcinoma cases for p53 (exon 2–11) and K-*ras* mutation using 5-μm paraffin-embedded tissue sections. Each section is incubated at 55°C in 120 μL of lysis buffer and placed on a shaker for 2 h. The extract is then boiled for 15 min to remove undesired proteases. For the PCR-SSCP step, 2.5 μL of the genomic DNA extract is used with 10 μL of reaction mixture containing 10 mM Tris-HCl, 50 mM KCl, 1.5 mM $MgCl_2$, 0.2 μm of each primer, 200 μM of each dNTP, 0.125 μL of α^{32}P-dATP, and 0.3 units of *Taq* DNA polymerase. Manual PCR-DNA sequencing was replaced by an automated method using the ABI PRISM™ 337 sequencer; the manufacturer's protocol is recommended. Automated PCR-DNA sequencing is rapid and efficient in sequencing gels and analyzing data.

As cancers are thought to develop from multiple genetic events that may involve activation of oncogenes and inactivation of tumor suppressor genes by point mutation and deletions, it has become necessary to screen for multiple genetic alterations within each tumor. The advantages of using the method described are apparent when sequencing multiple genes and exons from a large number of cases using minute amounts of tissue.

REFERENCE

1. **Sarkar, F.H., W.A. Sakr, Y.-W. Li, J. Jacobs and J.D. Crissman.** 1996. Tumor suppressor p53 gene mutation in squamous cell carcinoma of the larynx. Diagn. Mol. Pathol. *5*:201-205.

A Rapid and Efficient Protocol of the Inverted PCR Using Two Primer Pairs

Shigeru Takagi, Minoru Kimura[1] and Motoya Katsuki[2]

NTT Basic Research Laboratories, Musashino-Shi, Tokyo; [1]Division of Molecular Life Science, School of Medicine, Tokai University, Bohseidai, Isehara; [2]The Institute of Medical Science, The University of Tokyo, Minato-ku, Tokyo, Japan

The inverted polymerase chain reaction (IPCR), which allows amplification of the upstream and/or downstream region of known sequences, has many potential applications in molecular genetics. The feasibility of the IPCR was demonstrated with *Escherichia coli* DNA (2) and with a short (<500 bp) DNA fragment flanking exogenous sequences (ecotropic proviruses) integrated in the mouse genome (3).

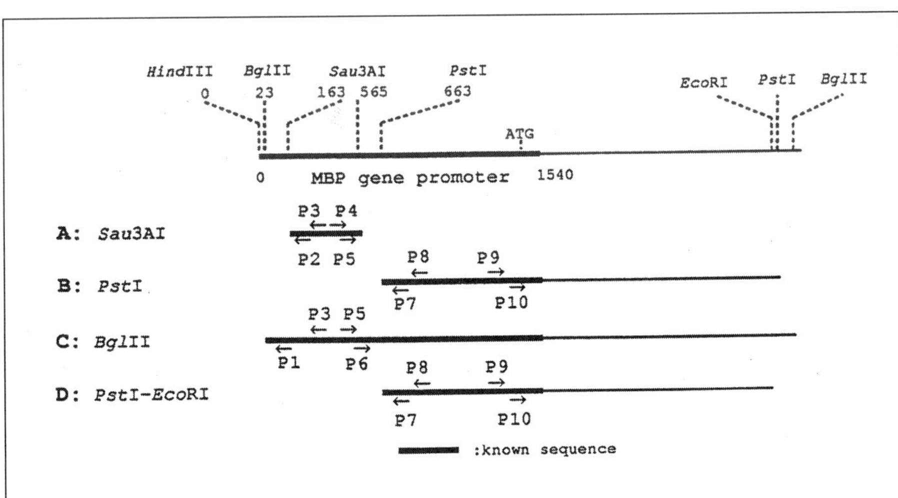

Figure 1. Restriction sites near the mouse MBP gene promoter region and primers of 20 bp (← and →) used for IPCR. Internal primers for samples A, B, C and D are (P3, P4),(P8, P9), (P3, P5) and (P8, P9), respectively. External primers for A, B, C and D are, respectively (P2, P5), (P7, P10), (P1, P6) and (P7, P10). 5′ positions of P1, P2, P3, P4, P5, P6, P7, P8, P9 and P10 primers are, respectively, nucleotides 140, 243, 280, 361, 451, 551, 830, 948, 1312 and 1413 (Reference 1).

(Reprinted from BioTechniques 13:176-178, 1992)

We have developed a high-sensitivity, low-background protocol for using IPCR when working with a more complex genome, having long fragments flanking both exogenous and endogenous sequences, and with a very small amount of specimen. This protocol, which uses two primer pairs (internal and external, see Figure 1), can amplify 0.2–2.3-kb fragments from <1 ng (0.05–1 ng) of mouse genomic DNA. The high sensitivity of the protocol facilitates the analysis of very small living things or mosaic insertional mutations. We also succeeded in using the IPCR with self-ligated DNA fragments whose different staggered ends, generated by two restriction endonucleases, were trimmed to blunt ends by using T4 DNA polymerase or Klenow fragments before ligation. The ability to use the IPCR on DNA with blunt ends greatly facilitates the selection of restriction sites and enzymes.

MATERIALS AND METHODS

DNA Samples

Restriction sites in the vicinity of the mouse myelin basic protein (MBP) gene promoter region are shown in Figure 1. Four samples were prepared from mouse genomic DNA (A: *Sau*3AI digested, B: *Pst*I digested, C: *Bgl*II digested and D: *Pst*I-*Eco*RI digested followed by forming blunt ends with T4 DNA polymerase). These samples were diluted to 0.05–1 ng/µL in TE buffer (10 mM Tris-HCl, pH 8.0, 1 mM EDTA).

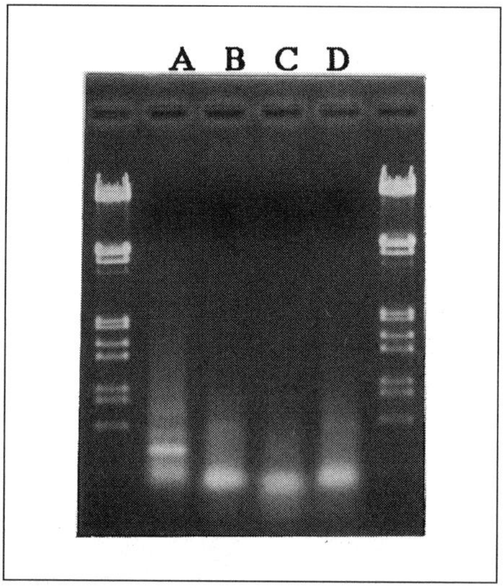

Figure 2. Products of conventional IPCR protocol using one primer pair and typical amplification conditions. 100 ng of each DNA sample were self-ligated in 50 µL of reaction kit agent (Toyobo) overnight, precipitated with ethanol, washed with 70% ethanol, dried under vacuum, suspended in 20 µL of TE buffer, heated at 95°C for 30 min, mixed with 80 µL of 1.25× PCR buffer containing 50 pM of internal primers ([P3, P4] for A, [P8, P9] for B, [P3, P5] for C and [P8, P9] for D) and amplified for 40 cycles, a 10-µL aliquot of PCR product was analyzed on 0.7% agarose gel. With samples B, C and D, DNA fragments of expected sizes (ca. 1.4 kb, ca. 2.3 kb and ca. 1.4 kb, respectively) were not seen with sample A (ca. 240 bp), several nonspecific bands were observed. A similar result was also obtained when external primers were used (data not shown).

Protocol

Ten primers were prepared for amplifying these DNA samples (Figure 1). One microliter of each sample DNA was self-ligated overnight in 10 µL of reaction kit agent (Toyobo, Chuo-ku, Tokyo, Japan) with 1 µL of T4 DNA ligase. The reaction mixtures were overlaid with mineral oil and heated at 95°C for 30 min to inactivate the enzyme and to introduce nicks on the circular DNA. The first

PCR amplification was performed after adding 90 µL of reaction mixture (1.1× PCR buffer [Stratagene, La Jolla, CA, USA], 100 nM of each deoxynucleoside triphosphate [dNTP], 25 pM of internal primers and 2.5 units of *Taq* DNA polymerase [Stratagene]). We used 30 cycles consisting of denaturation at 95°C for 2 min, primer annealing at 58°C for 2 min and extension by *Taq* DNA polymerase at 70°C for 4 min. For the second amplification, a 1–4 µL aliquot of the first PCR mixture was added to 96 µL of reaction mixture (1.1× PCR buffer, 200 nM of each dNTP, 50 pM of external primers and 2.5 units of *Taq* DNA polymerase). The samples were amplified for 30 cycles (30 s denaturation at 96°C, 30 s annealing at 57°C and x/1000-min extension at 72°C, where x = fragment length in bp.

RESULTS AND DISCUSSION

Figure 2 shows the results of the conventional IPCR using one primer pair and typical amplification conditions. With the longer samples (B, C and D), DNA fragments of the expected size were not observed. This was presumably because the PCR amplification efficiency for long fragments was low and nonspecific short background products competed more efficiently for reactants. With the shortest sample (A: ca. 240 bp), several bands were observed and a careful analysis like Southern blot hybridization would be needed to confirm the proper fragment.

Figure 3 illustrates the performance of our IPCR protocol. No band was observed in the first-round PCR products (lanes 1–4). DNA fragments of the

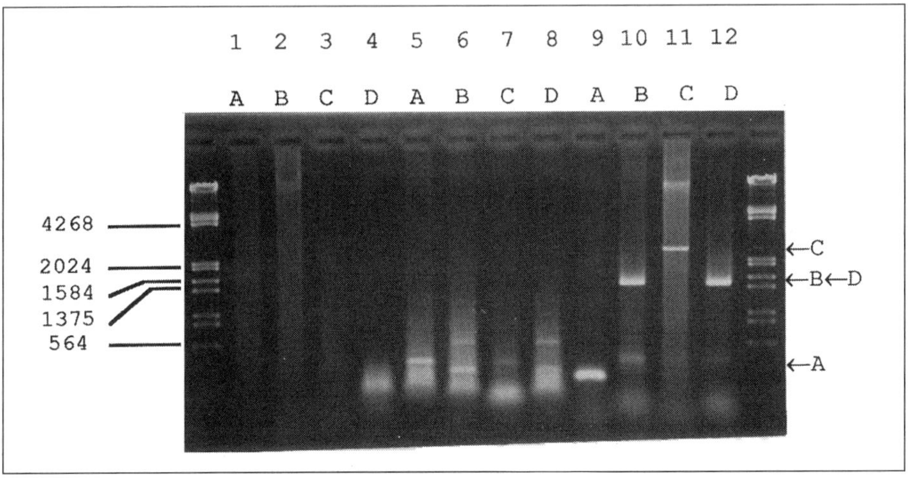

Figure 3. Performance of the proposed two primer pair IPCR protocol. Small amounts of the ligated DNA fragment (A: 0.3 ng, B: 0.6 ng, C: 1 ng, D: 1 ng) were used. Lanes 1–4: A 2-µL aliquot of first-round PCR product with the internal primers was analyzed on 0.7% agarose gel. Lanes 5–8: Products of the second-round PCR with the same set of internal primers. Lanes 9–12: Products of the second-round PCR with the external primers. DNA fragments of the expected size are indicated by arrows (←A, ←B, ←C, ←D).

expected size were not observed in the products of the second-round PCR with the same internal primers (lanes 5–8) but were observed in the products of the second-round PCR with the external primers (lanes 9–12). This suggests that the use of a second set of primers (external primers) is responsible for the higher sensitivity and consistency of the amplification.

Effects of various amounts of aliquot (1, 2, 4, 8 and 16 µL) of the first PCR product added to the second PCR buffer were tested for sample B. In all cases, fragments of the expected size were clearly identified, though the background increased slightly as the amount of aliquot increased (data not shown). The optimum amount of aliquot is between 1 and 4 µL. The sensitivity of the protocol was tested using sample B and was <0.05 ng (data not shown).

DNA fragments of lanes 9–12 in Figure 3 were recovered from the gel slices, reamplified with the external primers and subjected to direct sequencing. Primer proximal sequences were compared with known MBP DNA sequence (1). Sequencing patterns for fragments in lane 11 (sample C) and lane 12 (sample D) of Figure 3 are shown in Figure 4. Because of the 5′→3′ exonuclease activity of T4 DNA polymerase, some *Eco*RI-digested, 5′-protruding termini were clipped by one base resulting in the double-banding pattern of the autoradiogram (Figure 4, sample D).

In conclusion, the IPCR protocol using two primer pairs is highly sensitive and can amplify a wide range of DNA fragments from very small amounts of specimen that are difficult to amplify with the conventional IPCR protocol. Our protocol is quite simple and reliable because it is designed to exclude any additional steps (such as phenol extraction, ethanol precipitation or enrichment of the expected size of fragment) that are slow and that cause DNA loss.

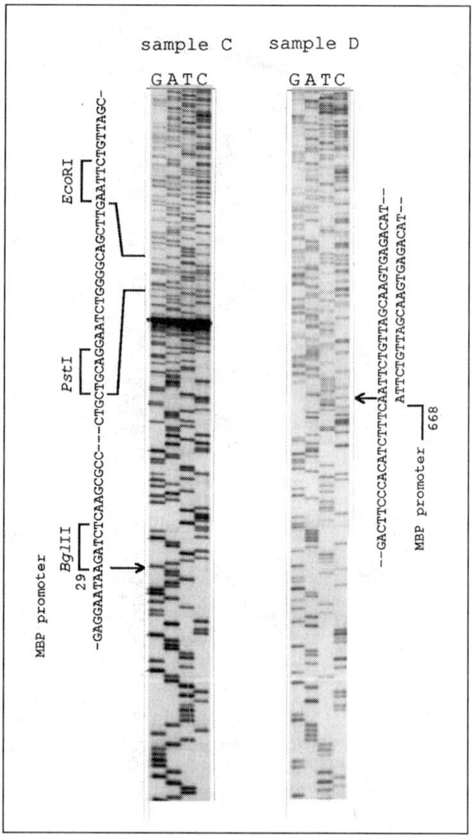

Figure 4. Each fragment of lanes 9–12 in Figure 3 was recovered from the gel slices, reamplified with the external primers and sequenced with P1 for C and P7 for D as a sequencing primer using Sequenase® (United States Biochemical, Cleveland, OH, USA). The self-ligated points are indicated by arrows. For the reason described in the text, the upper part of autoradiogram D shows doublet sequences (unclipped and 1 base clipped).

REFERENCES

1. **Miura, M., T. Tamura, A. Aoyama and K. Mikoshiba.** 1989. The promoter elements of the mouse myelin basic protein gene function efficiently in NG108-15 neuronal/glial cells. Gene *75*:31-38.
2. **Ochman, H., A.S. Gerber and D.L. Hartl.** 1988. Genetic applications of an inverse polymerase chain reaction. Genetics *120*:621-623.
3. **Silver, J. and V. Keerikatte.** 1989. Novel use of polymerase chain reaction to amplify cellular DNA adjacent to an integrated provirus. Virology *63*:1924-1928.

Update to:

A Rapid and Efficient Protocol of the Inverted PCR Using Two Primer Pairs

Minoru Kimura
Division of Molecular Life Science, School of Medicine, Tokai University, Bohseidai, Isehara, Japan

This method can be used for determining the junction sequences between the endogeneous genome and transgene introduced in transgenic animals. For example, using this protocol, we have compared the structure of the post- and premutation locus caused by transgene insertion in a transgenic line (1).

REFERENCES

1. **Naora, H., M. Kimura, H. Otani, M. Yokoyama, T. Koizumi, M. Katsuki and O. Tanaka.** 1994. Transgenic mouse model of hemifacial microsomia: cloning and characterization of insertional mutation region on chromosome 10. Genomics *23*:515-519.

CHAPTER 9

Direct Sequencing of DNA Isolated from mRNA Differential Display

Xinkang Wang and Giora Z. Feuerstein

SmithKline Beecham Pharmaceuticals, King of Prussia, PA, USA

ABSTRACT

A recently developed mRNA differential display technique has a potential power for identifying genes that are differentially expressed in a variety of in vitro and in vivo systems. One critical feature of this technique is to display most of the mRNA population on a sequencing gel after polymerase chain reaction using a 5′ decamer and a 3′ T_{12}MN anchored primer. However, these primers are too small to be successfully used as a sequencing primer using the classical sequencing protocol. In the present report we described the application of an extended primer set for the re-amplification after mRNA differential display, which renders the amplified DNA suitable for direct sequencing using either of these extended oligonucleotides as a primer without further subcloning. This improved technique will greatly save the time, cost and labor-intensive work for the discovery of genes using mRNA differential display.

INTRODUCTION

Polymerase chain reaction (PCR)-based mRNA differential display, since it was initially described by Liang and Pardee (6), has become an increasingly popular alternative technique for isolating genes of interest in a variety of in vitro (4) and in vivo (8,11) systems compared to the conventional techniques such as differential screening and subtractive hybridization. The advantages of this mRNA differential display technique include simplicity, sensitivity, versatility and low RNA requirements (1,5,6,11). However, some problems certainly exist in this technique such as a high incidence of false positives and the labor-intensive work involved in characterizing the genes after the differential display, especially when it is used for a relatively large-scale screening. Because a 10-mer upstream primer and a T_{12}MN (M = G, A or C; N = G, A, T or C) downstream anchored primer are used for the differential display PCR, and these primers are too small to be successfully used as a sequencing primer, the amplified-DNA fragments cannot be directly sequenced using the classical procedure. Therefore, to identify an amplified-

(Reprinted from BioTechniques 18:448-453, 1995)

DNA band by differential display, one has to subclone it into a vector prior to any sequencing analysis, which is time-consuming and labor-intensive. In the present report, we describe a simplified method using extended primers for the re-amplification of the isolated DNA bands from a differential displaying sequencing gel. These re-amplified DNA fragments and extended primers are now suitable for direct DNA sequencing analysis.

MATERIALS AND METHODS

Isolation of RNA

Aorta dissected from spontaneously hypertensive rats (SHR) and Wistar-Kyoto rats (WKY) were cultured in Dulbecco's modified Eagle's medium (DMEM) (Life Technologies, Gaithersburg, MD, USA) supplemented with 10% fetal bovine serum (FBS) in the presence of indomethacin (5 µM/L) and stimulated with 30 µg/mL lipopolysaccharide (LPS) for 4 h or without LPS as a sham-control. Total cellular RNA was isolated by homogenizing the aorta in an acid guanidinium thiocyanate solution and extracted with phenol and chloroform as described previously (3,10).

mRNA Differential Display

Differential display experiments were carried out using an RNAmap™ kit (GenHunter, Brookline, MA, USA) according to the manufacturer's specification. Briefly, total cellular RNA isolated from LPS-stimulated or unstimulated aorta was reverse-transcribed with T_{12}MA primers (M = G, A or C) in the presence of Maloney murine leukemia virus (MMLV) reverse transcriptase and then amplified with a 5′ decamer arbitrary primer (5′-GACCGCTT-GT-3′) and a 3′ T_{12}MA primer in the presence of [^{35}S]dATP (1200 Ci/mmol; Amersham, Arlington Heights, IL, USA) and 0.2 µL AmpliTaq® DNA polymerase (5 U/µL, Perkin-Elmer, Norwalk, CT, USA). PCR was carried out for 40 cycles as follows: 94°C for 30 s for denaturing; 40°C for 2 min for annealing; and 72°C for 30 s for extension, followed by 1 cycle for extension at 70°C for 10 min. The amplified products were resolved in an 8 M urea, 6% polyacrylamide DNA sequencing gel and analyzed by autoradiography. The band of interest was isolated, re-amplified and used for Northern analysis.

Re-Amplification and Isolation of the DNA Fragments

The re-amplification was carried out essentially as for the initial amplification except that the bands isolated from sequencing gel were used as a template for the re-amplification. For sequencing purposes, a pair of modified primers were synthesized and used for re-amplification. The modified 5′ primer contains the original 10 bases at its 3′ end and extended 9 bases at its 5′ end (5′-GCGAATTCCGACCGCTTGT-3′). Similarly, the modified 3′

anchored primer includes an extra 8 bases at its 5′ end (5′-CGGATCC-CTTTTTTTTTTTTMA-3′; M = G, A or C). Theoretically, these extended bases can be any sequence, but for subcloning convenience, if needed, we added an *Eco*RI site at the 5′ primer and a *Bam*HI site at the 3′ primer. After re-amplification, the DNA bands were resolved on a 6% polyacrylamide gel and isolated as described previously (9). Briefly, after staining the gel with ethidium bromide, the bands were cut out and minced into small pieces. The gel slices were then incubated in about 3 volume of 500 mM NH_4Ac and 1 mM EDTA for overnight at 37°C. Samples were microcentrifuged for 3 min (with several repeats if necessary) at room temperature, and the supernatants were collected, phenol-extracted and precipitated with ethanol. After resuspension in TE (10 mM Tris-HCl, pH 7.5, 1 mM EDTA, pH 8.0), the DNA was ready for either sequencing or probe-labeling.

DNA Sequencing Analysis

DNA sequencing was carried out using a double-stranded (ds)DNA cycle sequencing system (Life Technologies) according to the manufacturer's specification. Specifically, 0.1 μg DNA was used as a template in the presence of 10 ng of ^{32}P-end-labeled, modified 19-mer 5′ primers or 30 ng of the end-labeled, 3′-modified primers, which contain a mixture of 3 primers different for the penultimate base at the 3′ end. Comparatively, ^{32}P-end-labeled, 10-mer 5′ primers or 3′ T_{12}MN primers were used to prime the sequencing reaction. The cycle-sequencing was carried out as follows: initial cycle, 94°C for 4 min, 42°C for 30 s and 70°C for 1 min; followed by 29 cycles of 94°C for 30 s, 42°C for 30 s and 70°C for 1 min. The samples were resolved in an 8 M urea, 8% polyacrylamide sequencing gel, dried and subjected to autoradiography.

Northern Blot Analysis

RNA samples (10 μg/lane) extracted from LPS-stimulated or unstimulated aorta were resolved by electrophoresis, transferred onto a GeneScreen™ Plus membrane (Du Pont NEN, Boston, MA, USA) and subjected to Northern hybridization as described in detail previously (9,10). Ribosomal protein L32 (rpL32) gene (7), which exhibits a relatively constant expression in this experimental condition, was used for normalizing the samples loaded in each lane (10).

RESULTS AND DISCUSSION

Differential display was carried out using RNA samples isolated from cultured aorta of SHR and WKY stimulated with LPS or sham samples. One of the differentially expressed bands, designated as LPS-7, was demonstrated to be up-regulated in response to LPS (Figure 1A). This DNA fragment was isolated from a displaying sequencing gel, re-amplified and used as a probe for Northern analysis. Figure 1B shows the autoradiograph of Northern hybridization using LPS-7 cDNA as a probe, which confirms the up-regulation of this gene in response to LPS stimulation in the cultured aorta.

As usual, the next step is to determine the identity of this cDNA by sequencing analysis. However, because a decamer (at upstream) and a $T_{12}MA$ primer (at downstream) were used for the differential display PCR, neither of these primers is long enough to specifically prime the sequencing reaction using normal sequencing protocol (data not shown). Therefore, routinely, the isolated DNA fragment from differential display has to be subcloned into a vector, such as subcloning into a TA-cloning system (6,11). This procedure is not only time-consuming but also labor-intensive.

In the present report, we described utilization of a set of modified primers for the re-amplification after mRNA differential display (Figure 2). These modified primers contain identical sequences at the 3′ ends of the oligonucleotides but extended several bases at their 5′ ends (A′ and B′ in Figure 2A). These extended primers can efficiently (in fact even better than the 10-mer primers) amplify the DNA bands isolated from the differential display sequencing gel. Figure 2B shows the electrophoretic separation of the DNA bands amplified with the initial primers (i.e., A and B, used for the differential display experiment) (Figure 2B, lane 1) compared to the modified primers (i.e., A′ and B′) (Figure 2B, lane 2). A relatively stronger intensity for the band

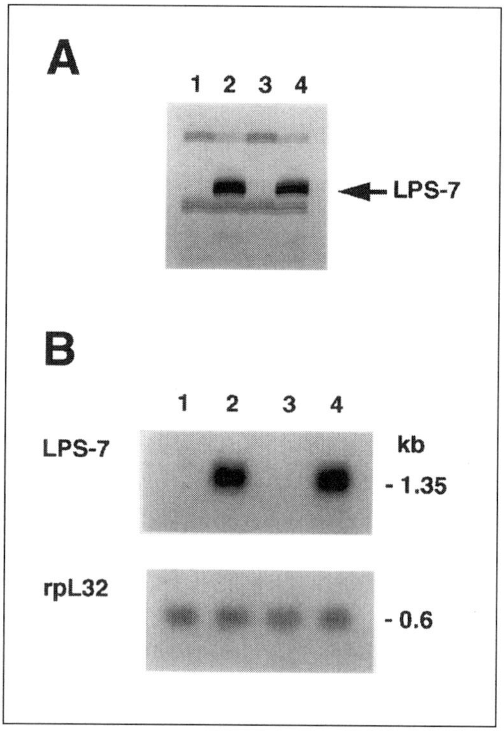

Figure 1. Identification of up-regulated gene expression in cultured rat aorta stimulated with LPS by mRNA differential display. A) The differential display was carried out as detailed in Materials and Methods. PCR products were resolved in a sequencing gel in the following order: unstimulated (lane 1) and stimulated aorta (lane 2) from SHR, unstimulated (lane 3) and stimulated aorta (lane 4) from WKY. Note the band(s) indicated with an arrow (designated as LPS-7) showing a marked induction in response to LPS stimulation on the differential displaying gel. B) Northern analysis of LPS-7 mRNA expression in cultured aorta stimulated with LPS. The RNA samples (10 μg/lane) were loaded in the same order as shown in A. Northern hybridization was carried out sequentially to the cDNA probes of LPS-7 and rpL32. The mRNA size was determined by comparing with the migration of an RNA ladder (Life Technologies) and the number of kilo bases (kb) marked on the right.

amplified with the modified primers was observed than when using the original primers, which is likely to be caused by a better annealing condition for the extended primers during the re-amplification. Moreover, standard conditions can be used for a further round of large-scale amplification to generate more DNA for making a probe to screen a library.

It should also be pointed out that the extended sequences of the modified oligonucleotide can be designed in many

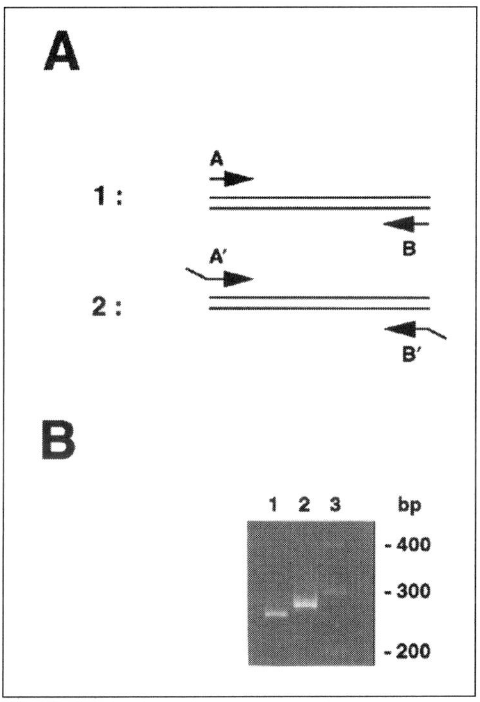

Figure 2. Re-amplification of the LPS-7 band with differential displaying primers and extended modifying primers. A) Schematic graph showing the re-amplification of the differential display bands with the initial primers (A and B) (top) or with the modified primers (A′ and B′) (low). B) Electrophoresis of the reamplified DNA fragments using primers A/B and A′/B′ as illustrated in A. The PCR samples were resolved in a 6% polyacrylamide gel and stained with ethidium bromide. Lane 1, using primers A/B; lane 2, using primers A′/B′; lane 3, DNA markers (United States Biochemical, Cleveland, OH, USA), their sizes are indicated on the right. Note, the DNA band generated by primers A′/B′ (lane 2) is relatively larger and more intensive than that of primers A/B (lane 1).

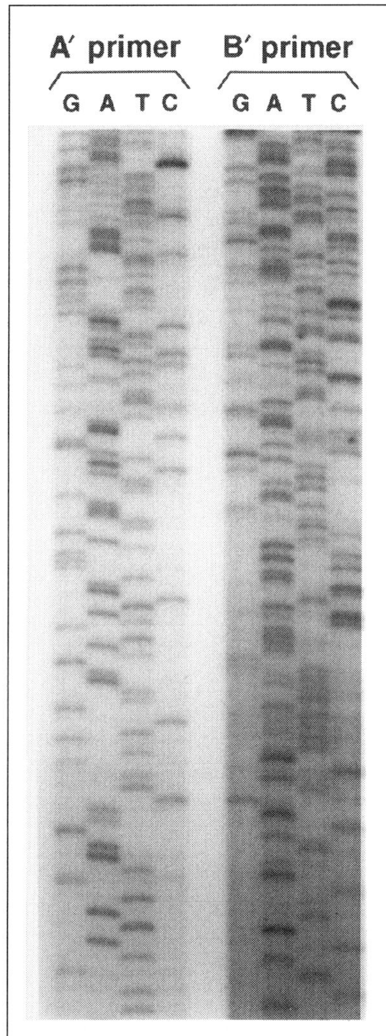

Figure 3. Direct sequencing of LPS-7 DNA amplified with the modified primers. LPS-7 DNA fragment was re-amplified with the modified primers (A′/B′), isolated from 6% polyacrylamide gel and used as a template for DNA sequencing in the presence of ^{32}P-end-labeled primer A′ or B′. Sequencing reaction was carried out using a dsDNA cycle sequencing system (Life Technologies) as detailed in Materials and Methods.

different ways to meet specific applications. In the present report, we added an *Eco*RI site to the upstream primer and a *Bam*HI site to the downstream primer because of its convenience for further subcloning, in case some differentially displayed band contains more than two different genes. The length of the modified primers and GC vs. AT ratios of the sequence should be considered to optimize the modified primers for sequencing.

The dsDNA was isolated from polyacrylamide gel after re-amplification and used as a template for direct sequencing in the presence of the modified primers (either the upstream or the downstream primers). Part of the sequencing data is shown in Figure 3 for the band LPS-7. As noted, sequencing primed with the downstream oligonucleotides usually has a relatively high background, which is likely caused by the presence of other primers (see Materials and Methods). The present data also suggest that the extended primers can be used as universal sequencing primers of which the DNA fragments were amplified. It is especially a case for the downstream primers since four 3′ anchored primers can prime all the mRNA population for differential display (6).

It should also be pointed out that sometimes a single band on displaying sequencing gel may contain more than one DNA species (References 1 and 2 and our observations), which is usually reflected by the appearance of multiple RNA species on Northern hybridization. In that particular case, a subcloning step must be carried out in order to make the differentially displayed band suitable for either sequencing or library screening.

In summary, the present report demonstrated the application of the extended primer sets for the re-amplification following mRNA differential display, which renders possible direct DNA sequencing analysis without subcloning. This modification should facilitate more efficient utilization of the mRNA differential display technique to identify genes of interest in a wide range of applications.

REFERENCES

1. **Bauer, D., H. Muller, J. Reich, V. Ahrenkiel, P. Warthoe and M. Strauss.** 1993. Identification of differentially expressed mRNA species by an improved display technique (DDRT-PCR). Nucleic Acids Res. *21*:4272-4280.
2. **Callard, D., B. Lescure and L. Mazzolini.** 1994. A method for the elimination of false positives generated by the mRNA differential display technique. BioTechniques *16*:1096-1103.
3. **Chomczynski, P. and N. Sacchi.** 1987. Single-step method of RNA isolation by acid guanidinium thiocyanate-phenol-chloroform extraction. Anal. Biochem. *162*:156-159.
4. **Liang, P., L. Averboukh, K. Keyomarsi, R. Sager and A.B. Pardee.** 1992. Differential display and cloning of messenger RNAs from human breast cancer versus mammary epithelial cells. Cancer Res. *52*:6966-6968.
5. **Liang, P., L. Averboukh and A.B. Pardee.** 1993. Distribution and cloning of eukaryotic mRNAs by means of differential display: refinements and optimization. Nucleic Acids Res. *21*:3269-3275.
6. **Liang, P. and A.B. Pardee.** 1992. Differential display of eukaryotic messenger RNA by means of the polymerase chain reaction. Science *257*:967-971.
7. **Meyuhas, O. and R.P. Perry.** 1980. Construction and identification of cDNA clones for mouse ribosomal proteins: applications for the study of 4 protein gene expression. Gene *10*:113-129.

8. **Utans, U., P. Liang, L.R. Wyner, M.J. Karnovsky and M.E. Russel.** Chronic cardiac rejection: identification of five up-regulated genes in transplanted hearts by differential mRNA display. Proc. Natl. Acad. Sci. USA *91*:6463-6467.
9. **Wang, X.K., G. Lee, S.A. Liebhaber and N.E. Cooke.** 1992. Human cysteine-rich protein: a member of the LIM/double-finger family displaying coordinate serum induction with c-myc. J. Biol. Chem. *267*:9176-9184.
10. **Wang, X.K., A.L. Siren, Y. Liu, T.L. Yue, F.C. Barone and G.Z. Feuerstein.** 1994. Upregulation of intercellular adhesion molecule 1 (ICAM-1) on brain microvascular endothelial cells in rat ischemic cortex. Mol. Brain Res. *26*:61-68.
11. **Zimmermann, J.W. and R.M. Schultz.** 1994. Analysis of gene expression in the preimplantation mouse embryo: use of mRNA differential display. Proc. Natl. Acad. Sci. USA *91*:5456-5460.

Update to:

Direct Sequencing of DNA Isolated from mRNA Differential Display

Xinkang Wang
SmithKline Beecham Pharmaceuticals, King of Prussia, PA

Since the initial introduction of mRNA differential display, it has been extensively refined to reduce the incidence of false positives and increase the efficiency of the techniques (for review, see Reference 2). The use of alternative, longer primers, e.g., 18–20 mers, in some laboratories have been applied directly for both RT and PCR of the differential display (1,5,6). The use of elongated primers usually generates less displayed bands but may increase the reproducibility of differential display and make it possible for direct sequencing of differential display products after reamplification. Nevertheless, the protocol outlined in this article (4) should be feasible for the analysis of differential display PCR products using both short (i.e., 10 mers) and elongated primers, as well as the PCR products obtained by other techniques such as RNA fingerprinting by arbitrary primed PCR (3).

REFERENCES

1. **Diachenko, L.B., J. Ledesma, A.A. Chenchik and P.D. Siebert.** 1996. Combining the technique of RNA fingerprinting and differential display to obtain differentially expressed mRNA. Biochem. Biophys. Res. Comm. *219*:824-828.
2. **Liang, P. and A.B. Pardee.** 1995. Recent advances in differential display. Curr. Opin. Immunol. *7*:274-280.
3. **McClelland, M., F. Mathieu-Daude and J. Welsh.** 1995. RNA fingerprinting and differential display using arbitrarily primed PCR. Trends Genet. *11*:242-246.
4. **Wang, X.K. and G.Z. Feuerstein.** 1995. Direct sequencing of DNA isolated from mRNA differential display. BioTechniques *18*:448-452.
5. **Yoshikawa, T., G.Q. Xing and S.D. Detera-Wadleigh.** 1995. Detection, simultaneous display and direct sequencing of multiple nuclear hormone receptor genes using bilaterally targeted RNA fingerprinting. Biochim. Biophys. Acta *1264*:63-71.
6. **Zhao, S., S.L. Ooi and A.B. Pardee.** 1995. New primer strategy improves precision of differential display. BioTechniques *18*:842-850.

CHAPTER 10

Extraction, PCR Amplification and Sequencing of Mitochondrial DNA from Human Hair Shafts

M.R. Wilson, D. Polanskey, J. Butler[1], J.A. DiZinno, J. Replogle and B. Budowle

FBI Academy, Quantico, VA and [1]University of Virginia, Charlottesville, VA, USA

ABSTRACT

Techniques have been developed for extracting, amplifying and directly sequencing mitochondrial DNA (mtDNA) from human hair shafts. The hair shaft is ground in a glass micro-tissue grinder, and the DNA is extracted with organic solvent and purified by filtration. The filtrate subsequently provides the mtDNA template for the PCR. The two hypervariable segments of the mtDNA control region are amplified in four separate reactions. After a purification step to remove unincorporated PCR primers, amplified products are quantitated by capillary electrophoresis and subjected to cycle sequencing. The products are separated and analyzed on an automated DNA sequencer. The mtDNA sequences from the hair shaft match the mtDNA sequences from blood samples taken from the same donor.

INTRODUCTION

Hairs are essentially extracellular projections of cellular entities (follicles) and are comprised chiefly of keratin (protein), trace metals and other metabolic by-products, sacks of air (medullae) and granules of pigment (eumelanin and phaeomelanin) (13,15, 18). Hairs associated with a crime or crime scene can be important forensic evidence (10). The principal method employed in forensic hair analysis is microscopic comparison of unknown hairs to known hair exemplars. With the advent of the polymerase chain reaction (PCR), the possibility of analyzing DNA in hair is possible (6,20). However, the pigment melanin is a strong inhibitor of the PCR process (22,25). Therefore, the challenge when attempting to type DNA from naturally shed hairs or hair shafts is extracting enough DNA from a source with very little DNA and overcoming inherent PCR inhibitors.

(Reprinted from BioTechniques 18:662-669, 1995)

Table 1. Primers and Nomenclature Used in This Study

HV I	
L15997	5′- CAC CAT TAG CAC CCA AAG CT-3′
L16159	5′- TAC TTG ACC ACC TGT AGT AC-3′
H16236	5′- CTT TGG AGT TGC AGT TGA TG-3′
H16401	5′- TGA TTT CAC GGA GGA TGG TG-3′
HV II	
L048	5′- CTC ACG GGA GCT CTC CAT GC-3′
L172	5′- ATT ATT TAT CGC ACC TAC GT-3′
H285	5′- GGG GTT TGG TGG AAA TTT TTT TG-3′
H408	5′- CTG TTA AAA GTG CAT ACC GCC A-3′
The following primer pairs are employed:	
HV1-A	L1599/H16236
HV1-B	L16159/H16401
HV2-A	L048/H285
HV2-B	L172/H408

Many, if not most, human hairs recovered from crime scenes are telogen phase (naturally shed) hairs, which contain little if any nuclear DNA (6). Therefore, extraction and amplification of nuclear DNA from these types of hairs have been problematic. By examining mitochondrial DNA (mtDNA), however, it is possible that a greater percentage of evidentiary hairs will be amenable to DNA typing. There are few reports of successful DNA typing of hairs, but these hairs usually contain sheath material (17,21–23). To date there has been very little information published on extracting, amplifying and typing mtDNA from hair shafts (6,8,22). However, most of these methods utilize multiple hairs with roots, and hence the possibility exists that some of the hairs could contain follicular tissue. These methods rely on chemical and enzymatic digestion of the hair prior to DNA extraction.

We have observed that certain chemically treated hairs, such as those which are chemically "straightened" with commercially available straighteners, are resistant to chemical and enzymatic digestion methods that utilize dithiothreitol, proteinase K, detergent, and heat to dissolve the hair. The method described herein yields a higher PCR amplification success rate on hair shafts than strictly chemical/enzymatic methods, and relies on physical crushing to obviate the potential lack of digestibility of some treated hairs. We are currently exploring a method that uses both physical crushing followed by chemical/enzymatic digestion.

The forensic potential of the control region (D-loop) of human mtDNA utilizing PCR has been described previously (2,6–8,15,19,23). The chief advantage of mtDNA typing over nuclear DNA typing is the high copy number of mtDNA per cell. This high copy number results in an increased sensitivity of detection, making mtDNA typing the method of choice when the amount of extracted DNA from a sample is extremely low or when the extracted DNA is significantly degraded.

MATERIALS AND METHODS

Extraction

The hair is handled with clean forceps and examined under a dissecting microscope for the presence of sheath material, debris or extracellular contaminants. Two centimeters of the root portion of the hair shaft are used as a DNA source for mtDNA analysis (in some cases less than two centimeters of starting material have successfully been analyzed). When the hair roots microscopically appear to possess follicular tissue, the root can be excised and extracted separately for nuclear DNA analysis if desired.

The 2-cm hair shaft is cut into two pieces approximately 1 cm in length. The pieces are washed in 100% ethanol in a small sterile test tube. They are then briefly rinsed in sterile, deionized and distilled water (ddH_2O). The fragments are then observed again under the stereomicroscope for the presence of external contaminants. When the hair is free of external contamination, the fragments are then placed into a sterilized 0.5-mL micro-tissue grinder (Kontes Glass, Vineland, NJ, USA) containing 100 µL of sterilized TE buffer (10 mM Tris-HCl, 0.1 mM EDTA, pH 8.0).

The homogenate is then transferred into a sterile, 0.5-mL tube for organic solvent extraction. One hundred microliters of an organic mixture of phenol:chloroform:isoamyl alcohol (PCIA) in the volumetric ratios of 25:24:1 are added to the homogenate, and the suspension is vortex mixed for 30 s. The aqueous and organic phases are separated by microcentrifugation at 10 000× *g* for 2 min. The aqueous phase (approximately 90 µL) is removed and transferred to a Microcon™-100 device (Amicon, Beverly, MA, USA) that contains 200 µL of sterile TE buffer. An additional 100 µL of TE buffer are added to the PCIA and the organic solvent extraction procedure is repeated. The second aqueous phase is also added to the same Microcon-100 device. Reagent blanks, consisting of ground TE buffer with no added hair, are used to control for possible contamination of the glass grinders. The ground TE buffer is placed in the same grinder used to crush the hair, then passed through the extraction method and amplified exactly as are the hair samples.

Grinders are cleaned and sterilized by soaking overnight in 100% ethanol. They are then rinsed in ddH_2O, washed three times in hot detergent (Alconox; Alconox, New York, NY, USA) and the inside surface scrubbed with flexible pipe cleaners. Approximately 200 µL of 4 *N* H_2SO_4 are then added, and after at least 20 min, the pestles are gently ground in their matching tubes. Another ddH_2O rinse follows. Prior to use, the grinders are placed in a microwave oven and exposed for 5 min on a high setting.

The DNA extract is dialyzed and concentrated by centrifuging the Microcon-100 at 2000× *g* for 5 min. The wash solution is discarded, and 400 µL of sterile TE buffer are added to the retentate in the Microcon-100 device. Again, the Microcon-100 is subjected to centrifugation at 2000× *g* for 5 min.

Finally, 60 mL of hot (80°–90°C) sterile TE buffer are added to the Microcon-100 filter. The Microcon-100 device is then subjected to vortex mixing for 30 s and inverted into the provided collection tube. The device is centrifuged for 3 min at 10 000× *g*. The provided cap is then placed over the retentate tube for storage of the sample until needed for PCR. Amplifications are usually performed in the same day. However, if necessary, extracts can be frozen at -4°C.

Amplification

The mtDNA control region subjected to analysis is 16 024–16 365 (HV I) and 73–340 (HV II) (1). This numbering system is derived from Reference 1, which is commonly referred to as the "Anderson sequence". In HV I, two sets of PCR primers are used to amplify overlapping segments. Similarly, in HV II, two sets of primers are used. mtDNA primers are named from the particular strand of mtDNA from which they are derived, in numerical form from the terminal 3′ base of the primer. The convention is L for light chain and H for heavy chain, followed by the number of that base as originally published (1). Primer sequences used for amplifying the regions of the mtDNA control region for forensic purposes have been previously reported (7,19,20). The specific primers used in this study, along with their nomenclature (24), are shown in Table 1.

Amplification is performed in 25-µL reactions using a GeneAmp® PCR System 9600 DNA Thermal Cycler (Perkin-Elmer, Norwalk, CT, USA). Each reaction contains 50 mM KCl, 10 mM Tris-HCl pH 8.3, 1.5 mM $MgCl_2$, 0.001% gelatin, 5.0 pmol of each primer (0.2 µM), 4 µg bovine serum albumin (BSA) (Sigma Chemical, St. Louis, MO, USA), 200 µM each deoxyribonucleoside triphosphate (dNTP) and 5 U *Taq* DNA polymerase (Perkin-Elmer). Ten microliters of the hair extract are used in four separate PCRs encompassing HV1 and HV2. Thermal cycling conditions are 1 min at 95°C, followed by 36 cycles of 95°C for 10 s, 60°C for 30 s, and 72°C for 30 s. Following the thermal cycling, the samples are maintained at 15°C for 10 min.

Post-Amplification Processing

The quality and quantity of PCR-generated amplification products are assessed by capillary electrophoresis (CE). Prior to analysis by CE, the amplified product is purified and separated from unreacted primers by filtration in a Microcon-100 device as described above, except that two 500-µL washes are employed and the final retentate volume is 40 μL.

A Beckman P/ACE™ System 2050 with a Laser Module 488 argon ion laser (Beckman Instruments, Palo Alto, CA, USA) is used with a DB-17-coated capillary (J&W Scientific, Folsom, CA, USA). The electrophoretic buffer consisted of 100 mM Tris-borate, 2 mM EDTA and contained additives of 1% (wt/vol) hydroxyethyl cellulose (Aldrich, Milwaukee, WI, USA) and 50 ng/mL of the intercalating dye YO-PRO-1 (Molecular Probes, Eugene, OR, USA) (11,12). Detection occurs by means of laser-induced fluorescence of the DNA-intercalating dye complex. Under the pressure injection scheme utilized, the detection limit is 700 pg with a linear range up to 250 ng (3).

One microliter of the PCR sample is mixed with 24 μL of deionized water containing 0.52 ng/μL of a 200-bp internal standard (M200; BioVentures, Murfreesboro, TN, USA). The ratio of the PCR product peak area to the internal standard peak area determines the quantity of the targeted PCR product (3). The size of the PCR product can be evaluated by comparing the migration time of the peak of interest to the migration time of a sizing standard of known length (5). The presence of PCR by-products and residual PCR primer contamination can also be determined by this method.

Sequencing

Sequencing is performed using the Perkin-Elmer/Applied Biosystems Division (PE/ABD) Taq DyeDeoxy™ Terminator Cycle Sequencing Kit (Foster City, CA, USA) according to the recommendations of the manufacturer, except that 100 mM ammonium chloride is substituted for 100 mM ammonium sulfate in the reaction buffer (personal communication, S. Spurgeon, PE/ABD). After quantitation by CE, 7 μL of a 3–5 ng/μL sample dilution (final amount 21–35 ng) are utilized as a DNA template in the cycle sequencing reaction.

An automated DNA sequencer (Model 373A; PE/ABD) with fluorescently labeled dideoxy nucleotide terminators was used to conduct sequencing. Cycle sequencing reactions are performed according to the instructions provided by the manufacturer. Both strands of the native double-stranded template are sequenced in separate sequencing reactions utilizing the same PCR product as a DNA template. Sequencing primers are identical to those used in the amplification. Sequence comparisons are made by using the software package SeqEd™ (Version 1.0.3; PE/ABD).

RESULTS AND DISCUSSION

Sequencing

Figure 1 shows a portion of a sequencing chromatogram from the amplified product of mtDNA extracted from a human pubic hair shaft. The sequencing primers used in both the forward and reverse directions (3.5 pmol per reaction) are the same as those used in the amplification reactions (L16159 and H16401). With respect to the Anderson sequence (1), three polymorphisms were observed in the sequence from this donor, at positions 16224 (T to C), 16292 (C to T) and 16311 (T to C). In order to determine the sequence, information from both strands, sequenced separately using the same purified PCR product as DNA template, are aligned with the program SeqEd. We have confirmed the findings of others that information from both strands greatly enhances the ability to unambiguously sequence a PCR-derived template (9). Multiple chromatograms can thereby be viewed and editing of bases conducted until all ambiguities are resolved.

DNA was extracted from a blood sample from the same donor referred to above. In this case, amplification of the entire control region was performed using primers L15997 and H408. The amplified product, or amplicon, was sequenced with primers from both strands. The edited sequence, in the form of a SeqEd document, is shown in Figure 2. The deviations from the standard sequence (1) are boxed.

To edit a sequence using SeqEd, the individual sequencing runs are col-

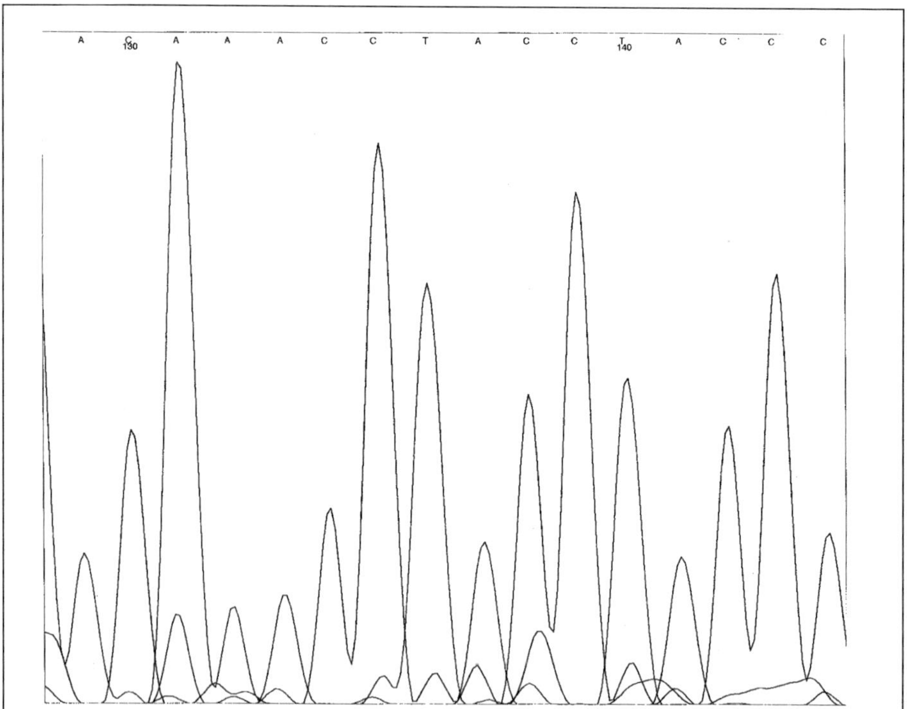

Figure 1. Chromatogram of a DNA sequencing run from a Model 373A Automated DNA Sequencer, with the base calls above the colored peaks. The sequence shown here was obtained from a PCR product utilizing two centimeters of pubic hair shaft as DNA template for PCR amplification. The region shown is in HV1, from position 16281–16296.

lected into a single SeqEd layout. After alignment and orientation, the sequence chromatograms from multiple runs are observed together, and base calls are made by the editor in regions where the analysis software is unable to make a nucleotide designation, or in areas where the nucleotide designations made by the analysis software differ from the complementary strands of the PCR product. This is a necessary step in the process, because currently base-calling software is not designed to make base calls from multiple chromatograms aligned from the same template DNA.

The polymorphisms with respect to the Anderson sequence from this individual are identical in the DNA extracted from both the blood and hair shaft. We have observed identical mtDNA sequence from DNA extracted from hair shaft and blood in approximately 30 individuals. Efforts to establish a database of human mtDNA control region sequences for forensic statistical purposes are continuing, and currently available data suggest that enough variation exists in this region to be of value in forensic identification (16,19).

```
                              30        40        50        60        70        80
1 FORWARD PRIMER 1       TTCTTTC ATGGGGAAGC AGATTTGGGT ACCACCCAAG TATTGACTCA CCCATCAACA A
2 FORWARD PRIMER 2
3 REVERSE PRIMER         TTCTTTC ATGGGGAAGC AGATTTGGGT ACCACCCAAG TATTGACTCA CCCATCAACA A
4 EDITED SEQUENCE        TTCTTTC ATGGGGAAGC AGATTTGGGT ACCACCCAAG TATTGACTCA CCCATCAACA A
5                        ------- ---------- ---------- ---------- ---------- ---------- -
6 ANDERSON 16024-16365   TTCTTTC ATGGGGAAGC AGATTTGGGT ACCACCCAAG TATTGACTCA CCCATCAACA A

                              90       100       110       120       130       140
1 FORWARD PRIMER 1       CCGCTATGT ATTTCGTACA TTACTGCCAG CCACCATGAA TATTGTACGG TACCATAAAT A
2 FORWARD PRIMER 2
3 REVERSE PRIMER         CCGCTATGT ATTTCGTACA TTACTGCCAG CCACCATGAA TATTGTACGG TACCATAAAT A
4 EDITED SEQUENCE        CCGCTATGT ATTTCGTACA TTACTGCCAG CCACCATGAA TATTGTACGG TACCATAAAT A
5                        --------- ---------- ---------- ---------- ---------- ---------- -
6 ANDERSON 16024-16365   CCGCTATGT ATTTCGTACA TTACTGCCAG CCACCATGAA TATTGTACGG TACCATAAAT A

                             150       160       170       180       190       200
1 FORWARD PRIMER 1       CTTGACCAC CTGTAGTACA TAAAAACCCA ATCCACATCA AAACCCCCTC CCCATGCTTA C
2 FORWARD PRIMER 2                                                AAACCCCCTC CCCATGCTTA C
3 REVERSE PRIMER         CTTGACCAC CTGTAGTACA TAAAAACCCA ATCCACATCA AAACCCCCTC CCCATGCTTA C
4 EDITED SEQUENCE        CTTGACCAC CTGTAGTACA TAAAAACCCA ATCCACATCA AAACCCCCTC CCCATGCTTA C
5                        --------- ---------- ---------- ---------- ---------- ---------- -
6 ANDERSON 16024-16365   CTTGACCAC CTGTAGTACA TAAAAACCCA ATCCACATCA AAACCCCCTC CCCATGCTTA C

                             210       220       230       240       250       260
1 FORWARD PRIMER 1       AAGCAAGTA CAGCAATCAA CCCCCAACTA TCACACATCA ACTGCAACTC CAAAGCCACC C
2 FORWARD PRIMER 2       AAGCAAGTA CAGCAATCAA CCCCCAACTA TCACACATCA ACTGCAACTC CAAAGCCACC C
3 REVERSE PRIMER         AAGCAAGTA CAGCAATCAA CCCCCAACTA TCA
4 EDITED SEQUENCE        AAGCAAGTA CAGCAATCAA CCCCCAACTA TCACACATCA ACTGCAACTC CAAAGCCACC C
5                        --------- ---------- ---*------ ---------- ---------- ---------- -
6 ANDERSON 16024-16365   AAGCAAGTA CAGCAATCAA CCCTCAACTA TCACACATCA ACTGCAACTC CAAAGCCACC C

                             270       280       290       300       310       320
1 FORWARD PRIMER 1       CTCACCCAC TAGGATACCA ACAAACCTAC CTACCCTTAA CAGTACATAG CACATAAAGC C
2 FORWARD PRIMER 2       CTCACCCAC TAGGATACCA ACAAACCTAC CTACCCTTAA CAGTACATAG CACATAAAGC C
3 REVERSE PRIMER         <==
4 EDITED SEQUENCE        CTCACCCAC TAGGATACCA ACAAACCTAC CTACCCTTAA CAGTACATAG CACATAAAGC C
5                        --------- ---------- ---------- -*-------- ---------- *--------- -
6 ANDERSON 16024-16365   CTCACCCAC TAGGATACCA ACAAACCTAC CCACCCTTAA CAGTACATAG TACATAAAGC C

                             330       340       350       360       370       380
1 FORWARD PRIMER 1       ATTTACCGT ACATAGCACA TTACAGTCAA ATCCCTTCTC GTCCC
2 FORWARD PRIMER 2       ATTTACCGT ACATAGCACA TTACAGTCAA ATCCCTTCTC GTCCC
3 REVERSE PRIMER         <==
4 EDITED SEQUENCE        ATTTACCGT ACATAGCACA TTACAGTCAA ATCCCTTCTC GTCCC
5                        --------- ---------- ---------- ---------- -----
6 ANDERSON 16024-16365   ATTTACCGT ACATAGCACA TTACAGTCAA ATCCCTTCTC GTCCC
```

Figure 2. Edited DNA sequencing layout from HV1. The final sequence was determined using both forward and reverse sequencing primers. The reverse sequencing reactions had to be reverse-complemented prior to alignment. The DNA template was a PCR product from DNA extracted from blood stains. Polymorphisms with respect to the Anderson sequence are shown in boxes.

Post-Amplification Purification

Many strategies have been used to sequence PCR products (4). The PCR should be sufficiently optimized to eliminate spurious amplification products so that a sequencing template of the requisite quality is produced. CE of the purified PCR product provides information regarding both the quantity and quality of the purified PCR product. CE analysis also enables a determination of whether or not PCR primers were removed prior to sequencing. If the PCR primers remain with the amplicon during the sequencing reaction, mispriming can result and thus yield uninterpretable results in the presence of dye-labeled terminators. Removal of primers is accomplished through microfiltration of the PCR product prior to the sequencing reaction. The quality of the

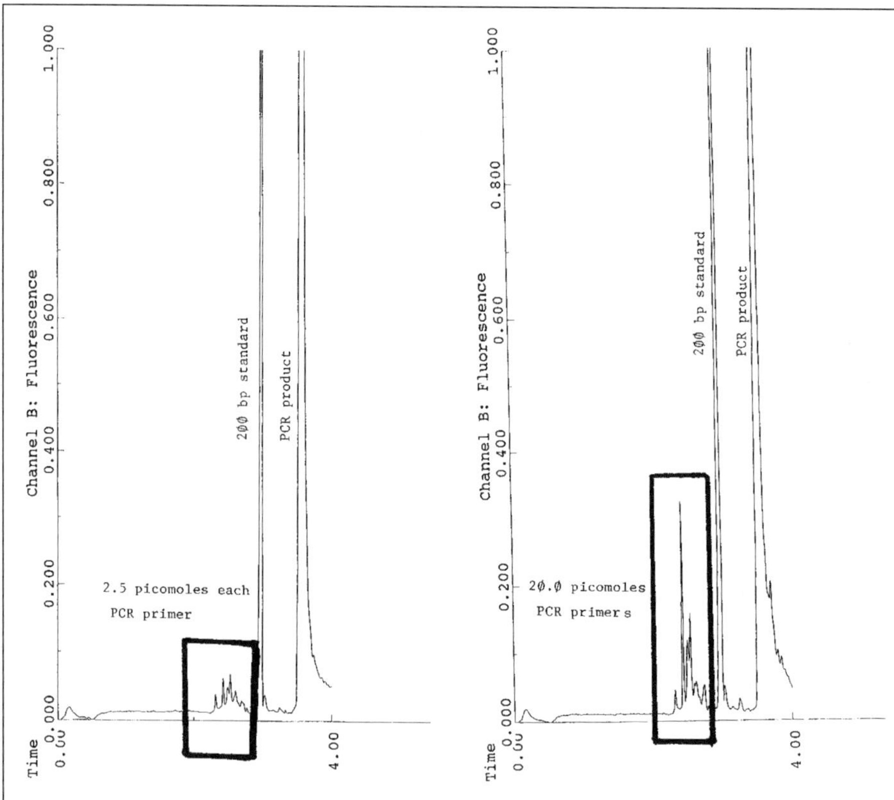

Figure 3. Capillary electropherogram showing Microcon-100-purified PCR products. The left panel shows purified PCR products from an amplification reaction containing 2.5 pmol of each PCR primer. The right panel shows an identical capillary electropherogram of purified PCR products amplified with 20.0 pmol of each primer. The boxes show PCR primer carryover after filtration. Excessive PCR primer present in sequencing reactions results in noisy, uninterpretable sequencing results. CE conditions are as follows: capillary: 0.05 mm × 27 cm DB-17 column; buffer: 1% HEC, 100 mM Tris-borate, 2 mM EDTA, pH8.1, 50 ng•mL YO-PRO-1. Pressure injection is at 0.5 psi for 45 s; separation voltage is 15 kV (22 μA), temperature 25°C, fluorescence detection is at 520 nm.

PCR amplicon can be assessed by evaluating the shape of the amplicon peak. Our observations suggest that the most important predictor of quality sequence results is the ratio of the PCR product peak relative to the primer carryover peak in the CE electropherogram. The presence of extraneous PCR peaks, such as "shouldering", does not seem to significantly alter the quality of the subsequent sequencing chromatograms.

Figure 3 shows post-amplification CE results from PCR products amplified with different initial concentrations of PCR primers. The CE results displayed in Figure 3 show that unreacted PCR primers can at times remain with the amplicon after purification. The template amplified with 20 pm of each primer, when sequenced, yielded a sequencing chromatogram of lesser quality than the same template amplified with 2.5 pm of each primer. Careful PCR optimization, together with thorough post-PCR Microcon-100 filtration, is necessary to ensure that excess unreacted PCR primers are not present following amplification and prior to cycle sequencing.

Contamination

Because of the sensitivity of this method, the issue of detection and quantitation of external contamination is important. We have observed low levels of amplified product in negative controls and reagent blanks when amplifying human mtDNA. Careful laboratory practices designed to completely overcome the problem of contamination are essential in a forensic laboratory and cannot be overstressed. However, in cases where low-level contamination is found, the experiments discussed herein suggest that with careful quantitation, correct typing results can be achieved when the ratio of sample to contaminant is greater than or equal to 10:1.

Previously sequenced amplicons were quantitated by CE and two samples with different mtDNA control region sequences were chosen for this study. Amplified DNA was mixed together from the two samples in ratios of 4:1, 8:1, 12:1 and 16:1. Sequencing reactions were performed on the mixtures as previously described. Editing of the analyzed data from the sequence of the mixed samples was conducted by two individuals.

Ambiguous bases (bases which could not be designated) and errors (bases which were called incorrectly) were noted. In the 4:1 mixture, both ambiguities and errors were observed. The 8:1 mixture yielded only two ambiguous calls from both editors at positions that differed in sequence between the mixed templates. No errors were noted at the 8:1 mixture. All base calls were completely consistent with the more abundant sequence at ratios at and above 8:1.

With automated fluorescent sequencing with *Taq* DNA polymerase and labeled terminators, the chromatograms consist of peaks of unequal amplitude (see Figure 1). The amplitude of any peak within the chromatogram is a function of three parameters: *(i)* the sequence of the template, *(ii)* the primer used for sequencing and *(iii)* the nature of the enzyme. Assuming that the enzyme and primer are constant, the amplitude will be dependent on the sequence of the template, which, in the case of mixtures, is mixed. Assuming that the mixture is of two different templates, one has a contribution to peak intensity at each position from two templates. Since the intensity is a function of the sequence, at points where the two templates differ in sequence, the chromatogram from the mixture will contain a sequence-dependent contribution from each of the two templates. However, because the amplitude is sequence-dependent, at a given position the amplitude will not necessarily directly reflect the relative amount of the components in the mixture. Therefore, at mixtures of 4:1, for example, it is conceivable that base-calling errors can be made, because the amplitude of the minor component at a particular position could be stronger than the major component. We have shown empirically that under the conditions described in this paper, base-calling errors do not occur at mixtures greater than 8:1, and hence we have adopted a 10:1 working ratio.

In order to confirm that correct typing results can be achieved using this ratio, additional tests were conducted. Amplified DNA samples from a variety of donors were mixed in 10:1 ratios with other amplified DNAs. The dominant and minor samples were blindly reversed to the sequence editors. At the 10:1 ratio in a total of five such tests, all the base calls from both editors were consistent with the known sequence of the more abundant sample with no errors or ambiguous calls. The ability to independently quantitate any amplified contaminant DNA from the sample-derived DNA is important in forensics. Should contamination be found, the results reported herein show that correct DNA typing results can be obtained if the ratio of the amplified contaminant to sample, as assessed by quantitating the blank and sample amplicons by CE, remains at or below 1:10. It should be stressed, however, that with evidentiary material such as hair, contamination by adherence should also be thoroughly addressed with properly designed validation studies. Hairs contaminated with semen, blood, etc., should be cleaned and typed in order to confirm that the cleaning procedure effectively removes adhering contaminants prior to DNA extraction.

ACKNOWLEDGMENTS

The authors wish to express their appreciation to Dr. F. Samuel Baechtel, Dr. Catherine T. Comey, Dr. Bruce R. McCord and Dr. Jenifer Lindsey for careful review of the manuscript and helpful comments and discussions. This work was supported in part by the National Institute of Justice (NIJ) Grant 93-IJ-CS-0030 provided to Mr. Butler. The names of commercial manufacturers are provided for identification only, and inclusion does not imply endorsement by the Federal Bureau of Investigation.

REFERENCES

1. **Anderson, S., A.T. Bankier, B.G. Barrell, M.H.L. de Bruijn, A.R. Coulson, I.C. Drouin, I.C. Eperon, D.P. Nierlick, B.A. Roe, F. Sanger, P.H. Schreier, A.J.H. Smith, R. Staden and I.G. Young.** 1981. Sequence and organization of the mitochondrial genome. Nature *290*:457-465.
2. **Budowle, B., D.E. Adams, C.T. Comey and C.R. Merril.** 1990. Mitochondrial DNA: a possible genetic material suitable for forensic analysis, p. 76-97. *In* H.C. Lee and R.E. Gaensslen (Eds.), Advances in Forensic Sciences. Year Book Medical Publishers, Chicago.
3. **Butler, J.M., R.O. Allen, B. McCord, J.M. Jung, M.R. Wilson and B. Budowle.** 1994. Quantitation of PCR products by capillary electrophoresis using laser fluorescence. J. Chromatogr. B:*658*:271-280.
4. **Ellingboe, J. and U.B. Gyllensten (Eds.).** 1992. The PCR Technique: DNA Sequencing. Eaton Publishing, Natick, MA.
5. **Goodall, D.M., S.J. Williams and D.K. Lloyd.** 1991. Quantitative aspects of capillary electrophoresis. Trends Anal. Chem. *10*:272-279.
6. **Higuchi, R.G., C.H. vonBeroldigen, G.F. Sensabaugh and H.A. Ehrlich.** 1988. DNA typing from single hairs. Nature *332*:543-546.
7. **Holland, M.M., D.L. Fisher, L.G. Mitchell, W.C. Rodriguez, J.J. Canik, C.R. Merril and V.W. Weedn.** 1993. Mitochondrial DNA sequence analysis of human skeletal remains: identification of remains from the Vietnam War. J. Forensic Sci. *38*:542-553.
8. **Hopgood, R., K.M. Sullivan and P. Gill.** 1992. Strategies for automated sequencing of human mitochondrial DNA directly from PCR products. BioTechniques *13*:82-92.
9. **Hyder, S.M., C. Hu, Y. Needleman, X.-Y. Sonoda and V.V. Baker.** 1994. Improved accuracy in direct automated DNA sequencing of small PCR products by optimizing the template concentration. BioTechniques *17*:478-482.
10. **Kirk, P.L.** 1941. Human hair studies I, general considerations of hair individualization and its forensic importance. J. Criminal Law, Criminol. Police Sci. *31*:486-496.
11. **McCord, B.R., J.M. Jung and E.A. Holleran.** 1993. High resolution capillary electrophoresis of forensic DNA using a non-gel sieving buffer. J. Liq. Chromatogr. *16*:1963-1981.
12. **McCord, B.R., D.M. McClure and J.M. Jung.** 1993. Capillary electrophoresis of PCR-amplified DNA using fluorescence detection with an intercalating dye. J. Chromatogr. *652*:75-82.
13. **Montagna, W. and E.J. Van Scott.** 1958. The anatomy of the hair follicle, p. 41. *In* W. Montagna and Richard A. Ellis (Eds.), The Biology of Hair Growth. Academic Press, New York.
14. **Orrego, C. and M.C. King.** 1990. Determination of familial relationships, p. 416-426. *In* M.A. Innis, D.H. Gelfand, J.J. Sninsky and T.J. White (Eds.), PCR Protocols: A Guide to Methods and Applications. Academic Press, London.
15. **Perkons, A.K. and R.E. Jervis.** 1966. Trace elements in human head hair. J. Forensic Sci. *2*:50-63.
16. **Piercy, R., K.M. Sullivan, N. Benson and P. Gill.** 1993. The application of mitochondrial DNA typing to the study of white Caucasian genetic identification. Int. J. Legal Med. *106*:85-90.
17. **Pilkington, S., C. Summers, S.L. Thein, N.T. O'Connor and J.S. Wainscoat.** 1987. Hair root DNA: a source of constitutional DNA in leukemia. Lancet *8524*:112.
18. **Pillsbury, S.** 1956. Hair, p. 40-56. *In* D.M. Pillsbury, W.B. Shelly and A.M. Kligman (Eds.), Dermatology. W.A. Saunders Publishing, Philadelphia, PA.
19. **Stoneking, M., D. Hedgecock, R.G. Higuchi, L. Vigilant and H.A. Ehrlich.** 1991. Population variation of human mtDNA control region sequences detected by enzymatic amplification and sequence specific oligonucleotide probes. Am. J. Hum. Genet. *48*:370-382.
20. **Sullivan, K.M., R. Hopgood and P. Gill.** 1992. Identification of human remains by amplification and automated sequencing of mitochondrial DNA. Int. J. Leg. Med. *105*:83-86.
21. **Thomson, D.M., N.N. Brown and A.E. Clague.** 1992. Routine use of hair root or buccal swab specimens for PCR analysis: advantages over using blood. Clin. Chem. Acta *207*:169-174.
22. **Uchici, R., K. Tamaki, T. Kojima, T. Yamamoto and Y. Katsumata.** 1992. Deoxyribonucleic acid (DNA) typing of human leukocyte antigen (HLA)-DQA1 from single hairs in Japanese. J. Forensic Sci. *37*:853-859.
23. **Vigilant, L.M., R. Pennington, H. Harpending, T.D. Kocher and A.C. Wilson.** 1989. Mitochondrial DNA sequences in single hairs from a southern African population. Proc. Natl. Acad. Sci. USA *86*:9350-9354.
24. **Wilson, M.R., M. Stoneking, M.M. Holland, J.A. DiZinno and B. Budowle.** 1993. Guidelines for the use of mitochondrial DNA sequencing in forensic science. Crime Lab. Digest *20*:68-77.
25. **Wilson, M.R. and B. Budowle.** 1993. Inhibition of *Taq* Polymerase by Melanin: Possible Solutions for Forensic Applications. 45th Annual Meeting of the American Academy of Forensic Sciences, Boston, MA. Williams Printing, Colorado Springs.

Update to:

Extraction, PCR Amplification and Sequencing of Mitochondrial DNA from Human Hair Shafts

M.R. Wilson
Hairs and Fibers Unit, FBI Headquarters, Washington, DC

A number of improvements have been made to the mtDNA typing protocol since the publication of this article. Some of the important changes are described below, along with a list of supplies and reagents.

DNA EXTRACTION: HAIRS

1. Prepare a wash solution of 5% (wt/vol) Terg-a-zyme™ detergent (Alconox, New York, NY).
2. Place an approximately 2-cm-long segment of a single hair into a 1.5-mL plastic tube filled with the Terg-a-zyme solution and place in a rack in an ultrasonic water bath. Agitate for approximately 20 min. Examine the hair under the stereomicroscope for the presence of surface debris. If necessary, continue to wash the hair in the Terg-a-zyme solution until free of surface debris.
3. Briefly rinse the hair in 100% ethanol, followed by ddH_20.
4. Prepare the reagent blanks by placing 200 µL of stain extraction buffer into each micro-tissue grinder to be used. Briefly simulate grinding. Remove the pestle and proceed to step 8.
5. Place the hair fragment into a micro-tissue grinder (Fisher Scientific; K885470-0000) containing 200 µL of stain extraction buffer. (Note: the grinders and pestles are matched sets).
6. Move the grinder up and down to force the hair into the bottom of the grinder. Grind the hair until completely dissolved.
7. Remove the pestle from the grinder. If liquid is adhering to the pestle head, gently pass it along the inner lip of the grinder cylinder until liquid flows down into the grinder.
8. Transfer the homogenate liquid or the reagent blank liquid into a sterile 1.5-mL plastic flip-top tube and add 5 µL of proteinase K (10 mg/mL).
9. Vortex mix on low speed and briefly centrifuge. Place the tube in a water bath and incubate at 56°C for a minimum of 2 h and a maximum of 24 h.
10. Remove the tube from the water bath and briefly spin the tube in a microcentrifuge to force condensate into the bottom of the tube.

11. Add 200 µL of phenol:chloroform:isoamyl alcohol (PCIA) (25:24:1) to the extract.
12. Vortex mix for 30 s to attain a milky emulsion and then spin the tube in a microcentrifuge for 3 min at 10 000× *g*.
13. To a Microcon-100 concentrator (Amicon; 42413) add 100 µL of ddH_2O. Carefully remove the aqueous phase (supernatant of approximately 200 µL) of the PCIA and place into the Microcon-100 concentrator. (Avoid drawing any of the proteinaceous interface into the pipet tip.)
14. Add 200 µL ddH_2O to the plastic flip-top tube containing the PCIA. Vortex mix 30 s to attain a milky emulsion. Spin the tube in a microcentrifuge for 3 min at 10 000× *g*. Carefully remove the aqueous phase (supernatant of approximately 200 µL) of the PCIA and place into the Microcon-100 concentrator. (Avoid drawing any of the proteinaceous interface into the pipet tip.)
15. Spin the Microcon-100 for 5 min at 2000× *g*.
16. Discard the wash and add 400 µL of filtered ddH_2O to the top.
17. Spin again at 2000× *g* for 5 min and discard the filtrate cup.
18. Add 60 µL of hot (80°–90°C), filtered ddH_2O to the filter side of the Microcon-100 concentrator and place the retentate cup on the concentrator.
19. Vortex mix the Microcon device with the retentate cup pointing upward (do not invert) on medium speed for approximately 30 s.
20. Invert the concentrator with the retentate cup and spin in a microcentrifuge at 10 000× *g* for 3 min.
21. Discard the concentrator. Cap the retentate cup.
22. The sample can then be amplified. It is recommended that extracted DNA be amplified immediately or as soon as practical. If necessary, the extract may be stored frozen until amplified.

It should be noted that the tissue grinders (Fisher Scientific; K885470-0000) consist of matched sets of grinders and pestles and should be used as such. To facilitate working with the grinders, the grinders may be placed in a small plastic tube, such as a 1.5-mL Sarstedt tube (Newton, NC, USA). Prior to beginning hair extraction procedures, the grinders should be cleaned using the following protocol.

1. When not in use, soak the grinders containing their matched pestles in approximately 100 µL of 100% ethanol.
2. Carefully rinse the grinders with ddH_20 and scrub the pestles and the insides of the grinders with cotton tip applicators and warm Alconox detergent.
3. Rinse the grinders with ddH_20 and add approximately 200 µL of 4 *N* H_2SO_4. (Note: when working with sulfuric acid, proper safety precautions should be observed. Gloves, protective eye wear and a lab coat should be worn and all work should should be performed under a fume

hood). Place the pestles in the matching grinders and briefly grind the pestles in grinders. Allow the grinders and pestles to soak in H_2SO_4 for a minimum of 20 min.

4. Remove the pestles and rinse the grinders with ddH_20. Place the grinders in microcentrifuge and pulse spin at high speed to collect the remaining water. Remove the grinders from microcentrifuge and remove the remaining water with a pipet.
5. Place the grinders with pestles in a nonmetallic rack and microwave on high for 2 min until dry.

AMPLIFICATION

PCR setup is to be done in the area designated for this activity using dedicated pipets under a dedicated laminar flow or dead space hood. All possible reagents, plastic tubes, PCR tubes/caps and PCR racks should be briefly exposed to UV light.

The primers and primer sets used for mtDNA amplification are listed below, followed by the amplification and post-amplification purification procedures.

Primers

Hypervariable Region 1

A1: (L 15997) 5′-CAC CAT TAG CAC CCA AAG CT-3′

B2: (H 16255) 5′-CTT TGG AGT TGC AGT TGA TG-3′

A2: (L 16159) 5′-TAC TTG ACC ACC TGT AGT AC-3′

B1: (H 16391) 5′-GAG GAT GGT GGT CAA GGG AC-3′

A4: (L 16209) 5′-CCC CAT GCT TAC AAG CAA GT-3′

B4: (H 16164) 5′-TTT GAT GTG GAT TGG GTT T-3′

Hypervariable Region 2

C1: (L 048) 5′-CTC ACG GGA GCT CTC CAT GC-3′

D2: (H 285) 5′-GGG GTT TGG TGG AAA TTT TTT TG-3′

C2: (L 172) 5′-ATT ATT TAT CGC ACC TAC GT-3′

D1: (H 408) 5′-CTG TTA AAA GTG CAT ACC GCC A-3′

For routine reference samples consisting of blood or saliva, and for questioned samples such as fresh bone, teeth or forcibly removed hairs where a potentially abundant amount of mtDNA template is present in the sample, HV1 and HV2 can be amplified using the primer sets listed below. For questioned samples such as hairs without tissue or degraded bone and teeth where little mtDNA template is present, HV1A, HV1B, HV2A and HV2B are amplified using the primer sets listed below.

Region	Primer Set
HV1	A1, B1
HV2	C1, D1
HV1A	A1, B2
HV1B	A2, B1
HV2A	C1, D2
HV2B	C2, D1

The primers A4 and B4 are utilized on samples that contain a homopolymeric stretch from positions 16184-16194. They are designed to bind near the homopolymeric region and prime DNA synthesis on their respective strands.

1. A Perkin-Elmer GeneAmp® 9600 thermal cycler is used. A total of 36 cycles are employed for hairs, bone and teeth. If the amount of template DNA has been determined, dilutions of 100 pg equivalent nuclear DNA are amplified for 32 cycles. Bodily fluid stains such as blood and saliva are typically diluted to a total of 100 pg and amplified for 32 cycles. The thermal cycling conditions are as follows:
    1. 95°C for 1 min
    2. Cycles:
        - 95°C for 10 s
        - 60°C for 30 s
        - 72°C for 30 s
    3. 15°C hold for 10 min.
2. Determine the number of samples to be amplified, including controls.
3. If samples have been stored frozen, vortex mix sample tubes for 10 s and spin briefly in a microcentrifuge at maximum speed.
4. If the amount of extracted DNA is known, amplification should be performed on 10 µL of an approximately 10 pg/µL solution of the extract. Calculate the amount of reagents in the PCR Master Mix using the following formula:

Amplification	Master Mix for 25 µL Reactions
ddH_20	6.0 µL
10× PCR buffer	2.5 µL
Each dNTP (×4)	0.5 µL
Each primer (10 µM) (×2)	0.5 µL
BSA (1.6 µg/µL)	2.5 µL
Taq DNA polymerase (5 U/µL)	1.0 µL
DNA template (extract)	10.0 µL

5. After the Master Mix has been prepared, aliquot 15 µL into each PCR

tube and place the reaction tubes in a PCR rack that has been cleaned with a 10% bleach solution.

6. Add 10 µL of the extract to each tube. Cap each sample and change pipet tips before proceeding to the next tube.
7. Add 10 µL of positive control DNA (total 100 pg) to the positive controls.
8. Add 10 µL of ddH_20 to the negative controls.
9. Cap the PCR tubes and carry the tubes in the PCR rack to the thermal cycler.
10. Place the top tray of the PCR rack containing the unamplified samples into the thermal cycler. Start the thermal cycler amplification program. Maintain a record of the heat block position of each tube.
11. The PCR amplification program is completed in approximately 2 h. The samples can be stored frozen in a space dedicated to post-amplification processing.

Amplicon purification

12. Add 500 µL of ddH_2O to a Microcon-100 concentrator. Add the entire amplified product to the Microcon-100 and spin in a microcentrifuge for 5 min at 2000× *g*.
13. Discard the wash solution and add another 500 µL of filtered ddH_2O to the Microcon concentrator. Spin in a microcentrifuge again for 5 min at 2000× *g*.
14. Discard the filtrate cup. Add 40 µL of hot (80°–90°C) filtered ddH_2O to the filter side of the Microcon-100 concentrator.
15. Place a retentate cup on the concentrator and vortex mix on medium speed with the retentate cup pointing upward for 30 s. Invert and spin for 3 min at 10 000× *g*.
16. Ensure that all sets of reagent blanks and samples have equal retentate volumes. If they do not have equal volumes, equalize by adding ddH_2O to the lesser volume(s).
17. Cap the retentate cup containing the purified amplicon for quantification and sequencing. Samples may be stored refrigerated or frozen.

SUPPLIES AND REAGENTS

BSA

8.0 mg/mL Bovine Serum Albumin: add 80 mg bovine serum albumin (e.g., Sigma Chemical, St. Louis, MO, USA; A3550) to 9.0 mL of sterile, deionized H_2O. Gently stir until the BSA is dissolved. Bring to a final volume of 10 mL with sterile, deionized H_2O to achieve a concentration of 8.0 mg/mL. Aliquot 100-µL portions of the solution into sterile screw-capped microcentrifuge tubes and store at -20°C.

DTT

1 M Dithiothreitol, 10 mL: dissolve 1.54 g dithiothreitol ($C_4H_{10}O_2S_2$; e.g., Sigma Chemical; D-9779, Molecular Biology Grade) in 10 mL sterile, deionized H_2O in a sterile, disposable plastic 15-mL tube. Do not autoclave. Store 100-µL aliquots in sterile 0.5 mL microfuge tubes at -20°C. Discard any unused portion of a thawed tube.

EDTA

0.5 M, pH 8.0, 1 L: slowly add 186.1 g ethylenediaminetetraacetic disodium salt, dihydrate ($C_{10}H_{14}N_2O_8Na_2$ $2H_2O$, e.g., Sigma Chemical; E-5134, Molecular Biology Grade) to 800 mL deionized water. Stir vigorously on a magnetic stirrer. Adjust to pH 8.0 by adding 10 *N* NaOH. Note: EDTA will not go into solution until pH adjustment. Adjust final volume to 1 L with sterile, deionized water. Autoclave. Store at room temperature.

PCIA

Phenol/Chloroform/Isolamyl alcohol (25:24:1; vol/vol): GIBCO-BRL (Gaithersburg, MD, USA), 5593UB. Caution: PCIA is an irritant and is toxic. its use should be confined to a fume hood.

Primers

Single-Stranded DNA Primers: primers are custom synthesized by Operon Technologies, Alameda, CA, USA. Synthesized primers are quantified by UV absorption at 260 nm and the quantity stated in picomoles. Lyophilized primers should be stored at -20°C. Each primer is reconstituted from the lyophilized form using ddH_2O to achieve a final concentration of 100 µm. Storage stocks should consist of 100 µL aliquots stored at -20°C.

Proteinase K

10 mg/mL, 10 mL: dissolve 100 mg proteinase K (e.g., Sigma Chemical P-4914, Molecular Biology Grade) in 10-mL sterile, deionized water in a sterile, disposable plastic 15-mL tube. Store 100-µL aliquots in sterile 0.5 mL microcentrifuge tubes as -20°C. Thaw tubes as needed for appropriate number of extractions. Discard any unused portions of thawed tubes. Caution: Powdered proteinase K and solutions of proteinase K can be irritating to mucous membranes. Wear safety glasses and gloves when handling.

Reaction Tubes

MicroAmp™ Reaction Tubes: Perkin-Elmer, N801-0533.

SDS (20% wt/vol)

Sodium Dodecyl Sulfate, 1 L: slowly dissolve 200 g sodium dodecyl sulfate $CH_3(CH_2)_{11}OSO_3Na$, electrophoresis grade (ultrapure), e.g., Boehringer Mannheim, 100155, in 800 mL sterile, deionized water. Solution may be heated. Adjust volume to 1 L. Store at room temperature. Caution: Wear a mask when working with powdered SDS.

Stain Extraction Buffer

10 mM Tris - 100 mM NaCl - 39 mM DTT - 10 mM EDTA - 2% SDS, 1 L: dissolve 5.84 g NaCl in 500 mL sterile, deionized H_2O. Add 10 mL 1 M Tris, 20 mL 0.5 M EDTA, and 100 mL 20% SDS. Titrate to pH 8.0 with HCl. Bring final volume of 1 L with deionized H_2O. Store at room temperature. Supplement with DTT before use. To 100 mL of the above solution, add 601.4 mg DTT and stir until dissolved. Store at room temperature. The complete solution is good for no more than two weeks.

Tris-HCl

1 M, pH 8.0, 1 L: Dissolve 121.1 g Tris base (tris[hydroxymethyl]-aminomethane, $C_4H_{11}NO_3$, e.g., Sigma Chemical, Trizma base S-8524, Molecular Biology Grade) in 800 mL sterile, deionized water. Adjust to pH 8.0 with concentrated HCl. Adjust final volume to 1 L. Autoclave. Store at room temperature.

U

V

W

X

Y

Z